Spectral Theory of Families of Self-Adjoint Operators

Mathematics and Its Applications (*Soviet Series*)

Volume 57

Spectral Theory of Families of Self-Adjoint Operators

by

Y. S. Samoilenko
Institute of Mathematics,
Academy of Sciences of the Ukrainian SSR,
Kiev, U.S.S.R.

KLUWER ACADEMIC PUBLISHERS
DORDRECHT / BOSTON / LONDON

ISBN 0-7923-0703-8

Library of Congress Cataloging-in-Publication Data

Samoĭlenko, Ĭŭ. S. (Ĭŭriĭ Stefanovich)
[Spektral'naîa teorîîa naborov samosoprîazhennykh operatorov. English]
Spectral theory of families of self-adjoint operators / by Yu.S. Samoilenko.
p. cm. -- (Mathematics and its applications. Soviet series ; 57)
Translation of: Élementy matematicheskoĭ teorii mnogochastotnykh kolebaniĭ.
Includes bibliographical references and index.
ISBN 0-7923-0703-8 (alk. paper)
1. Selfadjoint operators. 2. Spectral theory. I. Title.
II. Series: Mathematics and its applications (Kluwer Academic Publishers). Soviet series ; 57.
QA329.2.S2613 1991
515'.7246--dc20 90-26483

Published by Kluwer Academic Publishers,
P.O. Box 17, 3300 AA Dordrecht, The Netherlands.

Kluwer Academic Publishers incorporates
the publishing programmes of
D. Reidel, Martinus Nijhoff, Dr W. Junk and MTP Press.

Sold and distributed in the U.S.A. and Canada
by Kluwer Academic Publishers,
101 Philip Drive, Norwell, MA 02061, U.S.A.

In all other countries, sold and distributed
by Kluwer Academic Publishers Group,
P.O. Box 322, 3300 AH Dordrecht, The Netherlands.

Printed on acid-free paper

Translated from the Russian by E. V. Tisjachnij

This is the translation of the original work
ЭЛЕМЕНТЫ МАТЕМАТИЧЕСКОЙ ТЕОРИИ МНОГОЧАСТОТНЫХ КОЛЕБАНИЙ. ИНВАРИАНТНЫЕ ТОРЫ
Published by Nauka Publishers, Moscow, © 1987.

Printed in the Netherlands

SERIES EDITOR'S PREFACE

'Et moi, ..., si j'avait su comment en revenir, je n'y serais point allé.'

Jules Verne

The series is divergent; therefore we may be able to do something with it.

O. Heaviside

One service mathematics has rendered the human race. It has put common sense back where it belongs, on the topmost shelf next to the dusty canister labelled 'discarded non-sense'.

Eric T. Bell

Mathematics is a tool for thought. A highly necessary tool in a world where both feedback and non-linearities abound. Similarly, all kinds of parts of mathematics serve as tools for other parts and for other sciences.

Applying a simple rewriting rule to the quote on the right above one finds such statements as: 'One service topology has rendered mathematical physics ...'; 'One service logic has rendered computer science ...'; 'One service category theory has rendered mathematics ...'. All arguably true. And all statements obtainable this way form part of the raison d'être of this series.

This series, *Mathematics and Its Applications*, started in 1977. Now that over one hundred volumes have appeared it seems opportune to reexamine its scope. At the time I wrote

> "Growing specialization and diversification have brought a host of monographs and textbooks on increasingly specialized topics. However, the 'tree' of knowledge of mathematics and related fields does not grow only by putting forth new branches. It also happens, quite often in fact, that branches which were thought to be completely disparate are suddenly seen to be related. Further, the kind and level of sophistication of mathematics applied in various sciences has changed drastically in recent years: measure theory is used (non-trivially) in regional and theoretical economics; algebraic geometry interacts with physics; the Minkowsky lemma, coding theory and the structure of water meet one another in packing and covering theory; quantum fields, crystal defects and mathematical programming profit from homotopy theory; Lie algebras are relevant to filtering; and prediction and electrical engineering can use Stein spaces. And in addition to this there are such new emerging subdisciplines as 'experimental mathematics', 'CFD', 'completely integrable systems', 'chaos, synergetics and large-scale order', which are almost impossible to fit into the existing classification schemes. They draw upon widely different sections of mathematics."

By and large, all this still applies today. It is still true that at first sight mathematics seems rather fragmented and that to find, see, and exploit the deeper underlying interrelations more effort is needed and so are books that can help mathematicians and scientists do so. Accordingly MIA will continue to try to make such books available.

If anything, the description I gave in 1977 is now an understatement. To the examples of interaction areas one should add string theory where Riemann surfaces, algebraic geometry, modular functions, knots, quantum field theory, Kac-Moody algebras, monstrous moonshine (and more) all come together. And to the examples of things which can be usefully applied let me add the topic 'finite geometry'; a combination of words which sounds like it might not even exist, let alone be applicable. And yet it is being applied: to statistics via designs, to radar/sonar detection arrays (via finite projective planes), and to bus connections of VLSI chips (via difference sets). There seems to be no part of (so-called pure) mathematics that is not in immediate danger of being applied. And, accordingly, the applied mathematician needs to be aware of much more. Besides analysis and numerics, the traditional workhorses, he may need all kinds of combinatorics, algebra, probability, and so on.

In addition, the applied scientist needs to cope increasingly with the nonlinear world and the extra mathematical sophistication that this requires. For that is where the rewards are. Linear models are honest and a bit sad and depressing: proportional efforts and results. It is in the non-linear world that infinitesimal inputs may result in macroscopic outputs (or vice versa). To appreciate what I am hinting at: if electronics were linear we would have no fun with transistors and computers; we would have no TV; in fact you would not be reading these lines.

There is also no safety in ignoring such outlandish things as nonstandard analysis, superspace and anticommuting integration, p-adic and ultrametric space. All three have applications in both electrical engineering and physics. Once, complex numbers were equally outlandish, but they frequently proved the shortest path between 'real' results. Similarly, the first two topics named have already provided a number of 'wormhole' paths. There is no telling where all this is leading - fortunately.

Thus the original scope of the series, which for various (sound) reasons now comprises five sub-series: white (Japan), yellow (China), red (USSR), blue (Eastern Europe), and green (everything else), still applies. It has been enlarged a bit to include books treating of the tools from one subdiscipline which are used in others. Thus the series still aims at books dealing with:

- a central concept which plays an important role in several different mathematical and/or scientific specialization areas;
- new applications of the results and ideas from one area of scientific endeavour into another;
- influences which the results, problems and concepts of one field of enquiry have, and have had, on the development of another.

The spectral theory of a single self-adjoint operator in an Hilbert space constitutes a large, vastly important, and rich chapter in mathematics. Whole books are devoted to it and a substantial number of volumes could be filled with applications. However in many fields one meets not with a single operator but with a family of them. For instance, operator valued functions, families of commuting operators, a collection of operators satisfying the canonical commutation or anticommutation relations of quantum mechanics, etc., etc.

Such families are, for instance, encountered in quantum theory, in statistical mechanics, in population biology, in soliton theory, ...

For completely general families there is but little that can be said. The situation changes drastically as soon as there are algebraic relations, especially various kinds of commutation relations. For instance, when a countable family forms the basis of an infinite-dimensional Lie algebra such as a Kac-Moody algebra. (These algebras and certain generalisations currently attract a great deal of attention in, for example, string theory, lattice statistical mechanics, conformal quantum field theory etc.). In this case there are, of course, narrow relations with representation theory.

This unique book by a foremost investigator in the field is devoted to the spectral theory of countable families of self-adjoint operators. I consider it a most valuable addition to the literature.

The shortest path between two truths in the real domain passes through the complex domain.

J. Hadamard

La physique ne nous donne pas seulement l'occasion de résoudre des problèmes ... elle nous fait pressentir la solution.

H. Poincaré

Never lend books, for no one ever returns them; the only books I have in my library are books that other folk have lent me.

Anatole France

The function of an expert is not to be more right than other people, but to be wrong for more sophisticated reasons.

David Butler

Bussum, December 1990

Michiel Hazewinkel

CONTENTS

PART III
COLLECTIONS OF UNBOUNDED SELF-ADJOINT OPERATORS SATISFYING GENERAL RELATIONS

PART IV
REPRESENTATIONS OF OPERATOR ALGEBRAS AND NON-COMMUTATIVE RANDOM SEQUENCES

PREFACE

In this book, mathematical models of a number of physical systems are constructed and studied in the language of the theory of self-adjoint operators in a Hilbert space and their spectral properties. In particular, the classical monographs of N.I. Akhiezer, I.M. Glazman [1], F. Riesz, B. Sz.-Nagy [1], are devoted to the exposition of the theory of linear operators and the spectral theory of operators. The expansion in generalized eigenfunctions of self-adjoint operators is studied in the monograph of Yu.M. Berezanskiĭ [5]; the spectral theory of finitely many commuting self-adjoint operators is considered in the book by M.Sh. Birman, M.Z. Solomyak [1], application of self-adjoint operators in modern mathematical physics is discussed in the works of M. Reed, B. Simon [1-5].

Mathematical study of physical systems with many degrees of freedom as well as systems of statistical physics and field theory requires the development of a spectral theory for families of operators or, in other words, operator-valued functions, operators depending on parameters, etc. (See, for example, the monographs of N.N. Bogolyubov, A.A. Logunov, I.T. Todorov [1] and G. Emch [1].

The present book is devoted to spectral questions of countable families (countable collections) of self-adjoint operators on a Hilbert space H, which are pairwise connected by commutation, anticommutation or other algebraic relations.

In the introduction, we present the necessary definitions, formulate the statements of the problems, and also try to explain the choice of the subject matter in terms of general spectral theory. Solving the basic spectral questions even for pairs of self-adjoint operators of a general form is a complicated problem and at present it has not been solved. Therefore, it is quite natural to study families of self-adjoint operators connected, in particular, through algebraic relations. Today, the spectral theory of such families of operators is intensively developed in connection with its applications in the mathematical models of quantum systems with infinitely many degrees of freedom, where "strange" representations of commutation and anticommutation relations are used, in the description of interacted quantum fields with the aid of gauge groups and their representations, in the theory of exactly solvable nonlinear equations and their quantum analogues, in problems of control theory, etc. In fact, such families of operators appear as representations of generators of the corresponding algebraic objects such as Lie groups and Lie algebras (in particular, infinite-dimensional ones, if a family is infinite), graded ("coloured") Lie groups and algebras, etc.

The first part deals with countable collections of commuting self-adjoint operators and their joint spectral properties. Many spectral questions (including the theory of expansion in joint generalized eigenfunctions) have been studied for arbitrary systems of commuting self-adjoint operators in the monograph of Yu.M. Berezanskiĭ [9]. However, countable collections of commuting self-adjoint operators have a lot of additional properties which yield a simplification of the formulations of the known theorems and a number of new assertions, which are similar to their finite-dimensional analogues. There exists a comparable situation in the study of random sequences: all events which we usually come across, have quite definite probabilities (that is by no means true when considering the random functions). The above mentioned correspondence between problems in the theory of random processes and problems which arise when studying infinite families of commuting operators, is not accidental. One can consider a commuting family of self-adjoint operators as a "non-realized" random field. More precisely, the study of spectral properties of a countable collection of commuting self-adjoint operators (with a joint simple spectrum) is actually the study of those properties, which the corresponding random sequences almost certainly possesses. Therefore, it is natural that infinite-dimensional analysis and measure theory for sequence spaces are essential tools in the investigation of such operator families. The necessary information from measure theory for infinite-dimensional spaces is briefly presented in Part I, Chapter 1, Section 1.1, 1.3 and 1.5, Chapter 2, Section 2.1 and Chapter 3, Section 3.1, 3.2.

The theory of countable families of commuting self-adjoint operators is closely connected with the theory of unitary representations of inductive limits of commutative locally compact simply connected Lie groups (Chapter 2): a family (a collection) of commuting self-adjoint infinitesimal operators corresponds to each unitary representation of such groups and, conversely, each countable collection of commuting self-adjoint operators can be connected with the corresponding unitary representation of $\mathbb{R}_0^\infty$. The questions scrutinized in Chapter 2 for unitary representations of groups are analogous to the questions, which have been considered for the kind of families of self-adjoint operators studied in Chapter 1.

The study of differential operators with constant coefficients on functions of countably many variables in Chapter 3 is used as an illustration for both Chapter 1 and Chapter 2. A passage from finitely many variables to countably many variables essentially complicates the situation: there appear different unitary non-equivalent classes of differential operators with constant coefficients. This phenomenon is related to the existence of non-equivalent quasi-invariant measures on infinite-dimensional spaces.

Starting from the second part the exposition is divided into two. Firstly we consider finite and "tame" countable collections of operators connected by algebraic relations; secondly we study "wild" countable collections. The considered "tame" collections of self-adjoint operators still allow the major spectral questions to be solved. For the "wild" countable collections of self-adjoint operators, which are examined in this book, we consider only separate classes of representations. Spectral problems in these classes lead to the corresponding problems of the non-commutative theory of random sequences. In our approach to the study of the "wild" countable collections of operators we follow L. Gårding, A. Wightman [1,2] and I.M. Gelfand, N.Ya. Vilenkin [1], namely, we build commutative models for the considered collections of operators using the decomposition of the space H with respect to the spectrum of a countable subcollection of commuting self-adjoint operators. The algebraic relations connecting all the operators of the collection, are taken further into account.

The spectral questions for the self-adjoint representations of the generators in a finite-dimensional complex Lie algebra are directly related to the unitary representations of the corresponding real Lie group. Part II is devoted to representation theory of countable collections of self-adjoint operators which establish bases in countable-dimensional Lie algebras. But since neither the structure theory nor the Lie theory, nor the representation theory has been developed for general countable-dimensional Lie algebras, we shall confine ourselves to countable-dimensional Lie algebras, which are in fact inductive limits of finite-dimensional Lie algebras. So far the structure theory for such Lie algebras has not been developed either. But inductive limits of the corresponding Lie groups are naturally associated with the above-mentioned algebras and some examples of unitary representation classes for such Lie algebras have already been studied. Among them there are the canonical commutation relations of systems with countably many degrees of freedom (L. Gårding, A. Wightman [2], H. Araki [2], I.M. Gel'fand, N.Ya. Vilenkin [1], V.Ya. Golodets [1] and others), the infinite symmetrical groups (E. Thoma [2], A.M. Vershik, S.V. Kerov [1-4] and others), inductive limits of groups of operators (S. Strătilă, D. Voiculescu [1], G.I. Olshanskiĭ [1-5], A.M. Vershik, S.V. Kerov [5], I.M. Gel'fand, M.I. Graĭev [2] and others), etc. Other countable-dimensional Lie algebras, can be found in the book of V. Kac [1] which is devoted to the Kac-Moody Lie algebras and their representations and in the bibliography given there.

However, even in the class of inductive limits of Lie groups, it is not yet clear, whether their representations possess properties similar to properties of representations of finitely-dimensional Lie groups. We mention the existence of an invariant dense set of analytic vectors, the existence of a quasi-invariant measure on a natural completion of the group, etc.

The groups G_0^∞ of finite currents on a countable set with values in a fixed Lie group G are the simplest inductive limits of Lie groups. In Chapter 4 and 5 we consider simple examples of classes of unitary representations of the group G_0^∞ (in particular, the representations of the canonical commutation relations of systems with countably many degrees of freedom). Representations constructed with the aid of probability measures and cocycles come out here.

Then, in Part II we deal with a class of inductive limits of groups $\lim\limits_{\rightarrow} G_n$, for which $G_{n+1} = G^{(n)} \wedge G_n(x)$ ($\wedge$ is a semi-direct product), $n = 1,2,\ldots$ For these groups, there is a quite naturally constructed projective limit of groups $\lim\limits_{\leftarrow} G_n \supset \lim\limits_{\rightarrow} G_n$, $\lim\limits_{\rightarrow} G_n$-quasi-invariant measures on $\lim\limits_{\leftarrow} G_n$. We describe the simplest representations of $\lim\limits_{\rightarrow} G_n$, extendable by continuity to representations of $\lim\limits_{\leftarrow} G_n$. The analysis of the group of finite upper triangular matrices with units on the main diagonal, $B_0(\mathbb{N}, \mathbb{R})$ in Chapter 6, precedes the general study of this class of groups in Chapter 7.

For the study of representations of inductive limits of classical operator groups see A.A. Kirillov [4,6], G.I. Ol'shanskiĭ, [1-7], N.I. Nessonov [1-3], S. Strătilă, D. Voĭculesku [1], A.M. Vershik, S.V. Kerov [2], I.M. Gel'fand, M.I. Graĭev [2], and others.

We observe that there is a lack of information on the representations of continuous, smooth and other groups of currents and fermion currents (bigger than there is for a countable family of operators, connected by algebraic relations) similar to the given ones in the commutative situation described in Part I. For these problems, we refer to the work on group representations for continuous and smooth currents by I.M. Gel'fand, M.I. Graĭev [1], H. Araki [1], A.M. Vershik, I.M. Gel'fand, M.I. Graĭev [1,2,3,5], R.S. Ismagilov [5,7], K. Parthasarathy, K. Schmidt [1,2], S. Albeverio, R. Höegh-Krohn [1,2], S. Albeverio, R. Höegh-Krohn, D. Testard [1] and others. On representations of groups of diffeomorphisms and other "large" (infinite-dimensional) groups and Lie algebras, respectively, we refer to the work of R.S. Ismagilov [1-4,6,8], A.A. Kirillov [4-6], A.M. Vershik, I.M. Gel'fand, M.I. Graĭev [4], G. Segal [1], I. Frenkel [1] and others. Note, however, when studying representations of "large" groups it might also be useful to consider these groups as completions of certain inductive limits of Lie groups.

In Chapter 8 we give the definition of two anticommuting unbounded self-adjoint operators, study their properties and give an analogue of the spectral theorem. The same questions are studied in Chapter 9 for finite collections of graded-commuting (i.e. pairwise commuting or anticommuting) self-adjoint operators. Passing in Chapter 9 to countable collections of graded-commuting operators, first we outline a class of "tame" (classifiable) collections and receive the spectral representations for these collections. Then we consider representations of generators of the algebra of local observables of the spin system with countably many degrees of freedom. At present, the problem of describing all representations for these generators to within unitary equivalence has not been solved yet. We shall present a commutative model (the Gårding-Wightman form) for them and then consider representations with a simple spectrum of the commutative subalgebra. It should be emphasized that the study of representations with a simple spectrum agrees with the study of properties which the corresponding non-commutative random sequence possesses almost certainly. The simplest non-commutative probability measure, which appears here, consists of the usual probability measure on $\{-1,1\}^\infty$ combined with $\alpha_t(\cdot)$ $(t \in \{-1,1\}_0^\infty)$. The cocycle and the measure can be united into a single complex measure on $\{-1,1\}^\infty$ if the cocycle is trivial. The following constructions of cocycles are presented: product, Markov, locally dependent, etc.

An arbitrary countable collection of graded-commuting self-adjoint operators might be either "tame" (classifiable) or "wild" (the problem of constructing its spectral theory contains the very same problem as for an algebra of local observables of a spin system with countably many degrees of freedom) cf. Chapter 9.

The commutation or anticommutation relations is a particular case of the relations $AB = BF(A)$ $(F(\cdot): \mathbb{R}^1 \to \mathbb{R}^1)$ considered in Chapter 10. In this chapter we present and study the definition of such relations for unbounded self-adjoint operators A and B. We also prove a corresponding structure theorem.

In this chapter we also give a structure theorem for a countable collection of commuting self-adjoint operators $(A_k)_{k=1}^{\infty}$ and a finite collection of self-adjoint operators $(B_j)_{j=1}^{n}$ such that $A_k B_j = B_j F_{kj}(A_1, \ldots, A_n, \ldots)$ $(F_{kj}(\cdot): \mathbb{R}^{\infty} \to \mathbb{R}^1, k = 1,2,\ldots ; j = 1, \ldots, n)$. This theorem generalizes the structure theorem for finite collections of graded-commuting self-adjoint operators. Here, it is essential that the family $(B_j)_{j=1}^{n}$ is finite. Using the technique of expansion in joint generalized eigenvectors, in Chapter 10 we construct the commutative models for countable families, $(A_k)_{k=1}^{\infty}$ and $(B_j)_{j=1}^{\infty}$.

In future it would be natural to try to combine the exposition of the second part with the exposition of Part III on the basis of "coloured" infinite-dimensional Lie algebras or noncommutative dynamical systems and their representations. But this unification is possible even within the wider framework of the theory of $*$-algebras and the theory of their representations. Before the fourth part we do not use a technique of $*$-algebras and their representations. However, terms as non-commutative probability measure, non-commutative random sequence have the traditional meaning of states on the corresponding $*$-algebras (see numerous publications in journals "Communications in Mathematical Physics", "Journal of Functional Analysis", "Journal of Mathematical Physics", "Teoreticheskaya i matematicheskaya fizika", "Funktsionalni analiz i yego prilozheniya", etc.). Therefore, the fourth part is devoted to an introduction of non-commutative random sequences and constructive methods of their description (representations with a fixed cyclic vector, inductive limits of cyclic representations, moments, measures and cocycles).

In order to succesfully read this book, it is enough to be familiar with a basic university course including an exposition of the theory of unbounded self-adjoint operators. Nevertheless, some basic knowledge of measure theory on sequence spaces might be useful (we refer to the book of A.V. Skorohod [3]), when reading Part III it is necessary to know the fundamentals of the theory of unitary representations of Lie groups and Lie algebras (cf. the book by A.A. Kirillov [3] and D.P. Zhelobenko [1]); the fourth part requires knowledge of the fundamentals of the theory of $*$-algebras and their representations (cf. the book by J. Dixmier [1,3] and M.A. Naĭmark [4]).

The first part contains a fairly large number of examples that can serve as illustrations for mathematical notions, methods and assertions. Meanwhile, they are not often dealt with in detail which enables the aspiring reader to repeatedly check himself. In their turn, the phenomena studied in Part I-III, provide specific examples for the fourth part.

References to literature, often contained in the comments to the chapters, do not claim to be complete and, presumably, do not contain much bibliography on books and papers directly concerned with the questions touched upon in this book. Sometimes, the references to original sources are replaced by the references to available monographs or reviews containing additional bibliographical material.

The numbering of formulae, theorems, lemmas, definitions and assertions, examples and remarks is independent in every chapter. Referring to a theorem, a lemma, an assertion, etc. taken from another chapter, we also provide the number of the chapter. The sign $\square$ means the end of a proof or the end of a coherent part of the text.

The author is sincerely grateful to many mathematicians for their help in this work: to his teacher, Academician of the Academy of Sciences Ukr. SSR Yu.M. Berezanskiĭ, whose research by large defined the content of this book, for his kind attitude and patient attention. The author would also like to express his gratitude to Academician of the Academy of Sciences Ukr. SSR A.V. Skorokhod, prof. Yu.L. Daletskiĭ and prof. G.I. Kats and all the participants of the seminars on operators of mathematical physics, harmonic analysis, distributions in functional spaces, algebraic problems in mathematical physics, representation theory, who took part in discussions of many results in the book. The author is also thankful ot the reviewers of Russian edition of the book professors M.L. Gorbachuk and M.I. Yadrenko, to his colleagues Candidates of Physics and Mathematics L.I. Vainerman, Yu.A. Chapovsky, Yu.G. Kondrat'ev, V.I. Kolomytsev, S.A. Kruglyak, G.F. Us and his students Candidates of Physics and Mathematics A.Yu. Daletskiĭ, A.V. Kosyak, V.L. Ostrovskiĭ, B.L. Tsigan for a number of essential remarks.

INTRODUCTION

1. Let us look at the subject of the book from the point of view of the theory of a single in general unbounded self-adjoint operator. In the following we sum the problems of the spectral theory:

a) determination and study of the simplest operators: an operator of multiplication by a real number in $\mathbb{C}^1$ is the simplest self-adjoint operator on a separable Hilbert space H; the spectrum of any self-adjoint operator on H is a subset of the set $\mathbb{R}^1$ of all real numbers (i.e. a subset of the set of all simplest operators);

b) introduction of topological and measure structures on the spectrum;

c) proof of the spectral theorem on the decomposition of an operator into the simplest ones: the spectral theorem states that any self-adjoint operator on H can be "glued together" using the simplest operators. More precisely, any self-adjoint operator on H is unitarily equivalent to a direct sum (or an integral) of the simplest self-adjoint operators (which are not necessarily distinct). "The spectral theorem together with the multiplicity theory is one of the pearls of mathematics" (M. Reed, B. Simon [1 p. 259]);

d) description of unitary invariants of operators;

e) construction of functions of operators;

f) study of spectral properties of specific (generally differential) operators, etc.

2. Consider a family $(A_\alpha)_{\alpha\in\Lambda}$ of operators on H.

Everywhere in the book we study such sets up to unitary equivalence and take irreducible families to be the simplest ones (this is, of course, not the unique possible way to choose an equivalence relation between families of operators as well as to define the notion of "simplest family").

Definition 1. Two families of normal operators $(A_\alpha)_{\alpha\in\Lambda}$ on H and $(\tilde{A}_\alpha)_{\alpha\in\Lambda}$ on $\tilde{H}$ are called unitarily equivalent if there is a unitary operator $U : H \to \tilde{H}$ such that $A_\alpha = U^* \tilde{A}_\alpha U$ for all $\alpha \in \Lambda$.

Definition 2. A family $(A_\alpha)_{\alpha\in\Lambda}$ of normal operators on H is called irreducible if any bounded operator on H, which commutes with all of the operators in the set (i.e. with all their spectral projections) is a multiple of the unit operator.

A version for the spectral theory of a family of self-adjoint operators could contain:

a) to define and to study up to unitary equivalence the simplest families of self-adjoint operators;

b) to introduce topological and measure structures in simplest families of operators;

c) to prove the spectral theorem on a decomposition into the simplest families of any family of operators on H;

d) to describe unitary invariants of the family;

e) to develop functional calculus;

f) to study some particular families and so on.

3. It is a very difficult problem to classify up to a unitary equivalence all irreducible pairs (A_1, A_2) of bounded self-adjoint operators on H (or equivalently, to classify single operators $A_1 + i A_2$ on H).

The Cayley transformation $(A - iI)(A + iI)^{-1}$ (I denotes the identity) gives a one-to-one correspondence between the self-adjoint bounded operators and the unitary operators on H, whose spectrum does not contain 1. This correspondence preserves unitary equivalence of the families in the sense that unitarily equivalent families of self-adjoint operators correspond to unitarily equivalent families of unitary operators. Also irreducible families of self-adjoint operators correspond to irreducible families of unitary operators. Consequently, the classification problem for irreducible pairs of unitary operators contains up to unitary equivalence the classification problem for irreducible pairs of self-adjoint bounded operators.

The classification up to unitary equivalence of irreducible pairs of unitary operators on H is a known unsolved problem for a description of irreducible representations of a "wild" group, which is a free group with two generators.[1)] A classification subproblem for irreducible pairs of unitary operators with spectrum not containing 1, i.e. a classification problem for self-adjoint bounded operators, is also very complicated.

Let us show that a classification up to unitary equivalence of irreducible pairs of self-adjoint bounded operators would yield a classification up to unitary equivalence irreducible countable families $(A_k)_{k=1}^{\infty}$ of bounded operators with totally bounded norms on H.

To do this, we construct a pair of self-adjoint bounded operators $\mathbf{A}_{(A_k)}$, $\mathbf{B}_{(A_k)}$ on the space $\mathbf{H} = \oplus \sum_{i=1}^{\infty} H$ for any family of self-adjoint operators $(A_k)_{k=1}^{\infty}$ with totally bounded norms. We set

$$\mathbf{A}_{(A_k)} = \begin{bmatrix} I & & & & \\ & \frac{1}{2} I & & 0 & \\ & & \ddots & & \\ & 0 & & \frac{1}{n} I & \\ & & & & \ddots \end{bmatrix}$$

Solving some linear algebra problems, one takes another known unsolved problem of classification of a pair of matrices up to similarity to be a standard of complexity (as a standard "wild" problem). See I.M. Gel'fand, V.A. Ponamarev [1].

$$\mathbf{B}_{(A_k)} = \begin{bmatrix} 0 & I & A_1 & 0 & \cdots \\ I & 0 & I & A_2 & \cdots \\ A_1 & I & 0 & I & \cdots \\ 0 & A_2 & I & 0 & \cdots \\ \cdots & \cdots & \cdots & \cdots & \cdots \end{bmatrix}$$

(I is the identity on H)

The Hermitian operators $\mathbf{A}_{(A_k)}$ and $\mathbf{B}_{(A_k)}$ are bounded (but not compact) on $\mathbf{H}$ because the norms of self-adjoint operators $A_k (k=1,2,...)$ are totally bounded.

Proposition 1. The family $(A_k)_{k=1}^{\infty}$ of operators on H is irreducible if and only if the pair $\mathbf{A}_{(A_k)}$, $\mathbf{B}_{(A_k)}$ is irreducible on $\mathbf{H}$.

Proof. For any bounded operator $\mathbf{C}$ on $\mathbf{H}$ it follows from its commutativity with the operator $\mathbf{A}_{(A_k)}$ that its block structure in the direct sum $\oplus \sum_{1}^{\infty} H$ is given by

$$\mathbf{C} = \begin{bmatrix} C_1 & & & & \\ & C_2 & & 0 & \\ & & \ddots & & \\ & 0 & & C_n & \\ & & & & \ddots \end{bmatrix}$$

Commutativity of $\mathbf{C}$ with $\mathbf{B}_{(A_k)}$ on $\mathbf{H}$ means that $C_1 = C_2 = \cdots = C_n = \cdots = C$ and $C A_k = A_k C$ $(k=1,2,..)$.

Hence if any operator C on H, that commutes with the set $(A_k)_{k=1}^{\infty}$ is a multiple of the identity then any operator $\mathbf{C}$ on $\mathbf{H}$ that commutes with the pair $\mathbf{A}_{(A_k)}$, $\mathbf{B}_{(A_k)}$, is also a multiple of the identity and vice versa. □

For a family $(\tilde{A}_k)_{k=1}^{\infty}$ of self-adjoint operators on H with totally bounded norms, give the following pair of self-adjoint bounded operators $\mathbf{A}_{(\tilde{A}_k)}$, $\mathbf{B}_{(\tilde{A}_k)}$ on $\mathbf{H} = \oplus \sum_{1}^{\infty} H$:

$$\mathbf{A}_{(\tilde{A}_k)} = \begin{bmatrix} I & & & & \\ & \frac{1}{2}I & & 0 & \\ & & \ddots & & \\ & 0 & & \frac{1}{n}I & \\ & & & & \ddots \end{bmatrix} \qquad \mathbf{B}_{(\tilde{A}_k)} = \begin{bmatrix} 0 & I & \tilde{A}_1 & 0 & \\ I & 0 & I & \tilde{A}_2 & 0 \\ \tilde{A}_1 & I & 0 & I & \ddots \\ 0 & \tilde{A}_2 & I & 0 & \ddots \\ & 0 & \ddots & \ddots & \ddots \end{bmatrix}$$

Proposition 2. The families $(A_k)_{k=1}^{\infty}$ and $(\tilde{A}_k)_{k=1}^{\infty}$ of operators on H are unitarily equivalent if and only if the pairs $\mathbf{A}_{(A_k)}$, $\mathbf{B}_{(A_k)}$ and $\mathbf{A}_{(\tilde{A}_k)}$, $\mathbf{B}_{(\tilde{A}_k)}$ of bounded self-adjoint operators on $\mathbf{H}$ are unitarily equivalent.

Proof. We show that the unitary equivalence of the pairs $\mathbf{A}_{(A_k)}$, $\mathbf{B}_{(A_k)}$ and $\mathbf{A}_{(\tilde{A}_k)}$, $\mathbf{B}_{(\tilde{A}_k)}$ on H implies unitary equivalence of the families $(A_k)_{k=1}^{\infty}$ and $(\tilde{A}_k)_{k=1}^{\infty}$ on H. Let $\mathbf{U}$ be an operator that yields the unitary equivalence on $\mathbf{H}$ between $\mathbf{A}_{(A_k)}$ and $\mathbf{A}_{(\tilde{A}_k)}$. Because $\mathbf{A}_{(A_k)} = \mathbf{A}_{(\tilde{A}_k)}$, $\mathbf{U}$ commutes with $\mathbf{A}_{(A_k)}$, and so it has the following structure

$$\mathbf{U} = \begin{bmatrix} U_1 & & & & \\ & U_2 & & 0 & \\ & & \ddots & & \\ & 0 & & U_n & \\ & & & & \ddots \end{bmatrix}$$

The equality $\mathbf{U}\mathbf{B}_{(A_k)} = \mathbf{B}_{(\tilde{A}_k)}\mathbf{U}$ implies the equalities $U_1 = U_2 = \cdots = U_n = \cdots = U$ and $A_k = U^* \tilde{A}_k U$ $(k=1,2,\ldots)$. The converse is checked directly: if $A_k = U^* \tilde{A}_k U$ $(k=1,2,\ldots)$ then the operators

$$\mathbf{U} = \begin{bmatrix} U & & & \\ & U & 0 & \\ & & \ddots & \\ & 0 & U & \\ & & & \ddots \end{bmatrix}$$

is a unitary equivalence operator on $\mathbf{H}$ between the pairs $\mathbf{A}_{(A_k)}$, $\mathbf{B}_{(A_k)}$ and $\mathbf{A}_{(\tilde{A}_k)}$, $\mathbf{B}_{(\tilde{A}_k)}$. []

So we see that the problem of classification up to unitary equivalence for irreducible pairs of self-adjoint bounded operators on a separable Hilbert space H contains a very difficult problem of classification up to unitary equivalence for irreducible countable sets of operators. So far even the problem of classification up to unitary equivalence for irreducible families $(A_k)_{k=1}^{\infty}$ of anitcommutative self-adjoint operators on H which satisfy the additional condition $A_k^2 = I$ $(k=1,2,\ldots)$, has not been solved.

4. Therefore, it is natural to study the structural questions not for all the collections of self-adjoint operators but only for those which are interrelated, for example, which satisfy some algebraic relations. Essentially, in this book we study the structural questions for finite or countable collections of self-adjoint operators which commute or anticommute, or satisfy Lie or more general relations.

For such collections:

a) we give accurate definitions of the relations satisfied by the operators of the collection (because the operators are not assumed to be bounded);

b) if the collection is "tame", then we prove structural theorems which describe such irreducible and general collections as an integral of the simplest types.

c) if the collection is "wild", then we study the structure of only a subset of such collections (such subsets can be obtained by imposing the condition of simplicity for the joint spectrum of a fixed subcollection of commuting operators, or by requiring that the representation could be continuously extended);

d) we construct "commutative models".

COMMENTS TO THE INTRODUCTION

For the complexity of problems of unitary classification of general bounded operators (of pairs of self-adjoint operators) see, for example, A.M. Vershik [2], A.A. Kirillov [3], J. Ernest [4], and the bibliography given therein. The exposition given in Section 3 of the introduction follows the paper of S.A. Kruglyak, Yu.S. Samoĭlenko [1]. A proof of a problem complexity by means of isolation a standard difficult ("wild") subproblem is used, for instance, in P. Halmos and J. McLaughlin [1], where it is proved that the problem of unitary classification of partial isometies of a space H contains as a subproblem the difficult problem of unitary classifying pairs of self-adjoint operators.

Note that besides unitary representations of groups, $*$-representations of Lie algebras and graded (finite-dimensional and infinite-dimensional) Lie algebras, $*$-representations of $*$-algebras which are closely related to the subject of this book (see comments to the paragraphs), one also studies $*$-representations of algebraic objects of different origin. We mention $*$-representations of generalized shift operators, hypercomplex systems, and hypergroups as treated in J. Delsarte [1], B.M. Levitan [1-5], Yu.M. Berezanskiĭ [1-4], Yu.M. Berezanskiĭ and S.G. Kreĭn [1-3] and others. The bibliography related to this subject is given in the survey papers K. Ross [1], L.I. Vaĭnerman [2]. We mention representations of ring groups as considered in G.I. Kats [3,4], L.I. Vaĭnerman, G.I. Kats [1], E. Kirchberg [1], M. Takesaki [2], M. Enock, J.-M. Schwartz [1], and others. Also, $*$-representations of algebras with quadratic relations as studied in E.K. Sklyanin [1,2], A.M. Vershik [4], D.I. Gurevich [1], and others, $*$-representations of quantum groups as considered in V.G. Drinfel'd [2], S. Woronowicz [4], L.L. Vaksman, Ya.S. Soĭbelman [1] and others. One studies spectral theory for a family of operators which satisfy commutation relations related to a dynamic system in for example A.A. Lodkin [1]. Representations of general commutation relations, μ-structures, and their applications are studied in V.P. Maslov [1,3], M.V. Karasev, V.P. Maslov [1], V.P. Maslov and V.E. Nazaĭkinskiĭ [1], M.V. Karasev [1-4], and others. For $*$-representations of groupoids see J. Renault [1] and the bibliography given there. In A.V. Roĭter [1], S.A. Kruglyak [1,2,3], V.V. Sergeĭchuk [1], one also considers $*$-representations of involutive quivers, partially ordered sets, and boxes.

PART I
FAMILIES OF COMMUTING NORMAL OPERATORS

Chapter I
SPECTRAL ANALYSIS OF COUNTABLE FAMILIES OF COMMUTING SELF-ADJOINT OPERATORS (CSO)

In spectral theory of families of commuting self-adjoint operators (as in the theory of a single self-adjoint operator) one considers two kinds of problems: firstly, the study of their common spectral properties and secondly, the decomposition of H with respect to their common generalized eigenvectors. In what follows we study the classical spectral properties of countable families of commuting self-adjoint operators.

One of the ways to study families of CSO on H is to use the theorem which states that all of the operators in the family can be expressed as functions of a single bounded self-adjoint operator A. But simple examples show that the operator A and the functions which produce the operators of the family, could be very peculiar. Further, the spectral properties depend essentially on the form of A and of the functions.

In this book we use a joint resolution of the identity (a projection-valued measure on the measurable set of all real sequences) to study countable families of CSO in a similar way as a resolution of the identity (projection-valued measure on a real line) is used in the study of a self-adjoint operator. On one hand, such an approach gives a possibility to study many classical problems for families of CSO (spectral theorem, unitary invariants, construction of functions of a self-adjoint operator, and so on), and on the other hand it allows to use methods of measure theory on infinite dimensional spaces (random sequence measures) to study families of CSO.

Every section starts with a review of those spectral properties of a single operator that will be studied for a family of operators. Then we describe the case of a finite family of operators. After that we study the case of a countable family, give a summary of results for a general family or explain why some of the earlier obtained theorems do not hold.

1.1. The joint resolution of the identity for a countable family of CSO. Spectral theorem in terms of projection-valued measures.

For every, generally speaking unbounded, self-adjoint operator A_j on a separable complex Hilbert space H there is a resolution of the identity $E_j(\cdot)$ (see, for example, Akhiezer and Glazman [1]). It is a function on the Borel sets of $\mathbb{R}$ with the following properties:

1°) $E_j(\Delta)$ is a projection for all Borel sets $\Delta \subset \mathbb{R}$; $E_j(\varnothing)=0$, $E_j(\mathbb{R}^1)=I$ (I is the identity operator on H, 0 - the zero operator on H);

2°) E_j is σ-additive: for any sequence of Borel sets $(\Delta_k)_{k=1}^{\infty}$ such that $\Delta_i \cap \Delta_k = \varnothing$ $(i \neq k; i,k=1,2,...)$ we have

$$E_j(\bigcup_{i=1}^{\infty} \Delta_i) = \sum_{i=1}^{\infty} E_j(\Delta_i);$$

3°) E_j is orthogonal:

$$E_j(\Delta' \cap \Delta'') = E_j(\Delta')\, E_j(\Delta'')$$

where $\mathbf{B}(\mathbb{R}^1)$ denotes the Borel σ-algebra on the real line and $\Delta', \Delta'' \in \mathbf{B}(\mathbb{R}^1)$.

Conversely, if there is an operator-valued measure $E_j(\cdot)$ on the real line $\mathbb{R}^1$ with the properties 1°–3° then it is a resolution of the identity of the self-adjoint operator

$$A_j = \int_{\mathbb{R}^1} \lambda_j\, E_j(\lambda_j) \tag{1}$$

where $\lambda_j \in \mathbb{R}^1$.
Convergence of the integral in (1) is understood in the strong sense. The domain of the operator A_j is

$$\mathbf{D}(A_j) = \{f \in H \mid \int_{\mathbb{R}^1} \lambda_j^2\, d(E_j(\lambda_j)f,f) < \infty\}.$$

Let, further, $E_1(\cdot), \dots, E_n(\cdot)$ be a resolution of the identity of self-adjoint, generally unbounded, operators $A_1, \dots, A_n$ on H. We say that the operators $A_1, \dots, A_n$ mutually commute if their resolution of the identity mutually commute, i.e. $E_k(\Delta')$ and $E_j(\Delta'')$ commute for any $\Delta', \Delta'' \in \mathbf{B}(\mathbb{R}^1)$. If the operators A_k and A_j are bounded, then it follows that they commute:

$$[A_k, A_j] = A_k A_j - A_j A_k = 0. \tag{2}$$

Conversely, for bounded self-adjoint operators A_k, A_j from (2) it follows that the corresponding resolutions of the identity also commute[1)].

[1)] Note that commutativity of unbounded self-adjoint operators A and B on a dense invariant set $\Phi \subset H$ for the operators A and B does not, in general, imply commutativity of their spectral projections even if Φ is a domain on which A and B are essentially self-adjoint (see E. Nelson

For a family of commuting self-adjoint operators $A_1, \ldots, A_n$ we can construct a joint n-dimensional resolution of the identity. It is an operator-valued measure $E(\cdot)$ defined on a σ-algebra of Borel sets of $\mathbb{R}^n((\lambda_1, \ldots, \lambda_n) \in \mathbb{R}^n)$. This measure has the following properties:

1°) $E(\Delta)$ is a projection for all $\Delta \in \mathbf{B}(\mathbb{R}^n)$, $E(\varnothing) = 0$, $E(\mathbb{R}^n) = I$;

2°) $E(\cdot)$ is σ-additive:

$$E(\bigcup_{i=1}^{\infty} \Delta_i) = \sum_{i=1}^{\infty} E(\Delta_i) \quad (\Delta_j \cap \Delta_k = \varnothing \text{ for all } j \neq k);$$

3°) $E(\cdot)$ is orthogonal:

$$E(\Delta' \cap \Delta'') = E(\Delta')\, E(\Delta'') \quad (\Delta', \Delta'' \in \mathbf{B}(\mathbb{R}^n)).$$

For sets of the form $\Delta = \times_{j=1}^{n} [a_j, b_j]$ this measure is defined as $E(\Delta) = \prod_{j=1}^{n} E_j([a_j, b_j])$. Then it is extended on $\mathbf{B}(\mathbb{R}^n)$.

Conversely, every measure on $(\mathbb{R}^n, \mathbf{B}(\mathbb{R}^n))$ that has the properties 1°–3° is a joint resolution of the identity for a family of commuting self-adjoint operators

$$A_j = \int_{\mathbb{R}^n} \lambda_j \, dE(\lambda_j, \ldots, \lambda_n) \tag{1'}$$

(convergence of the integral is understood in strong sense). The domain of the self-adjoint operator A_j is

$$\mathbf{D}(A_j) = \{f \in H \mid \int_{\mathbb{R}^n} \lambda_j^2 \, d(E(\lambda_1, \ldots, \lambda_n)f, f) =$$

$$= \int_{\mathbb{R}^1} \lambda_j^2 \, d(E_j(\lambda_j)f, f) < \infty\}).$$

Now consider a countable family $(A_j)_{j=1}^{\infty}$ of commuting self-adjoint operators on H.

In the linear space $\mathbb{R}^\infty = \mathbb{R}^1 \times \mathbb{R}^1 \times \cdots$ of all real sequences (a point $\lambda \in \mathbb{R}^\infty$ could be written as $\lambda = (\lambda_1, \lambda_2, \ldots))$ introduce the product topology. This topology is induced by the metric

$$d(\lambda, \tilde{\lambda}) = \sum_{k=1}^{\infty} \frac{1}{2^k} \frac{|\lambda_k - \tilde{\lambda}_k|}{1 + |\lambda_k - \tilde{\lambda}_k|}.$$

It is well known that with such a topology $\mathbb{R}^\infty$ becomes a regular topological space with a countable base. Let $\mathbf{B}(\mathbb{R}^\infty)$ be the collection of Borel sets in $\mathbb{R}^\infty$.

---

[1]; M. Reed and B. Simon [1]).

Further, let $(\mathbb{R}^{\infty}, C_{\sigma}(\mathbb{R}^{\infty}))$ be the measurable space, where $C_{\sigma}(\mathbb{R}^{\infty})$ is the smallest σ-algebra which contain all cylindrical half-intervals, i.e. sets consisting of all $\lambda \in \mathbb{R}^{\infty}$ satisfying

$$a_j < \lambda_j \leq b_j \quad (j=1, \ldots, n; n=1,2,\ldots).$$

PROPOSITION 1.

$$\mathbf{B}(\mathbb{R}^{\infty}) = C_{\sigma}(\mathbb{R}^{\infty}). \tag{3}$$

It is sufficient to prove that every closed set $F \subset \mathbb{R}^{\infty}$ is a countable intersection of closed cylindrical sets and so belongs to $C_{\sigma}(\mathbb{R}^{\infty})$. But the mapping $\mathbb{R}^{\infty} \ni \lambda \mapsto \pi_n(\lambda) = (\lambda_1, \ldots, \lambda_n) \in \mathbb{R}^n$ is continuous and $F = \bigcap_{n=1}^{\infty} \overline{\pi_n(F)}$ ("$\overline{\cdot}$" denotes closure of a set). □

In particular, the following subsets of $\mathbb{R}^{\infty}$

$$l_2\left[\left(\frac{1}{N_k}\right)_{k=1}^{\infty}\right] = \left\{\lambda \in \mathbb{R}^{\infty} \,\middle|\, \|\lambda\|^2_{l_2(\frac{1}{N_k})} = \sum_{k=1}^{\infty} \frac{\lambda_k^2}{N_k} < \infty\right\},$$

$$l_{\infty}\left[\left(\frac{1}{N_k}\right)_{k=1}^{\infty}\right] = \left\{\lambda \in \mathbb{R}^{\infty} \,\middle|\, \|\lambda\|_{l_{\infty}(\frac{1}{N_k})} = \sup_k |\lambda_k| / N_k < \infty\right\}$$

($(N_k)_{k=1}^{\infty}$ is an arbitrary sequence of positive numbers) are $C_{\sigma}(\mathbb{R}^{\infty}) = \mathbf{B}(\mathbb{R}^{\infty})$-measurable. The subsets $C = \{\lambda \in \mathbb{R}^{\infty} \mid \lim_{n \to \infty} \lambda_n \text{ exists and is finite}\}$, $C_N = \{\lambda \in \mathbb{R}^{\infty} \mid \lim_{n \to \infty} \lambda_n = N\}$, $C_{\infty, N} = \{\lambda \in \mathbb{R}^{\infty} \mid \sup_n \lambda_n \leq N\}$ are $C_{\sigma}(\mathbb{R}^{\infty}) = \mathbf{B}(\mathbb{R}^{\infty})$-measurable as well.

Also note that from (3) it follows that $\mathbf{B}(\mathbb{R}^{\infty})$ is a σ-algebra with a countable base, i.e. it can be obtained by taking countable unions and intersections of a countable system of generating sets.

According to the Kolmogorov theorem, the probability measures $\mu(\cdot)$ on $(\mathbb{R}^{\infty}, \mathbf{B}(\mathbb{R}^{\infty}))$ are uniquely defined by their mutually agreed projections $\mu_n(\cdot)$ defined on $(\mathbb{R}^n, \mathbf{B}(\mathbb{R}^n))$ by

$$\mu_n((a_1, b_1] \times \cdots \times (a_n, b_n]) = \mu((a_1, b_1] \times \cdots \times (a_n, b_n] \times \mathbb{R}^1 \times \mathbb{R}^1 \times \cdots).$$

In particular, the product-measure $\bigotimes_{k=1}^{\infty} \omega_k(\cdot)$ on $(\mathbb{R}^{\infty}, \mathbf{B}(\mathbb{R}^{\infty}))$ is defined by a sequence of probability measures $\omega_k(\cdot)$ $(k=1,2,\ldots)$ on $(\mathbb{R}^1, \mathbf{B}(\mathbb{R}^1))$ and a sequence of mutually agreed measures $\bigotimes_{k=1}^{n} \omega_k(\cdot)$ defined on $(\mathbb{R}^n, \mathbf{B}(\mathbb{R}^n))$:

$$(\bigotimes_{k=1}^{n} \omega_k)\,((a_1, b_1] \times \cdots \times (a_n, b_n]) = \omega_1((a_1, b_1]) \cdot \cdots \cdot \omega_n((a_n, b_n]).$$

The measure $\mu(\cdot)$ on $(\mathbb{R}^\infty, \mathrm{B}(\mathbb{R}^\infty))$ is also uniquely determined by its characteristic function which is the positive-definite function on $\mathbb{R}_0^\infty$ satisfying

$$k(t) = \int_{\mathbb{R}^\infty} e^{i(\lambda, t)}\, d\mu(\lambda)\,, \tag{4}$$

where $\mathbb{R}_0^\infty$ is the set of all finite real sequences, $(\lambda, t) = \sum_{k=1}^{\infty} t_k \lambda_k$.

In particular, on $(\mathbb{R}^\infty, \mathrm{B}(\mathbb{R}^\infty))$, the Gaussian measure $g_B(\cdot)$ which has mean zero correlation matrix B is determined by the characteristic function

$$k(t) = e^{-\frac{1}{2}(Bt, t)} \quad (t \in \mathbb{R}_0^\infty).$$

Here B is an infinite, real, symmetric, positive definite (i.e. all of the main minors are positive) matrix.

The measure $\mu(\cdot)$ on $(\mathbb{R}^\infty, \mathrm{B}(\mathbb{R}^\infty))$ is closely related to its moments (if they exist)

$$s_\alpha^{\mu(\cdot)} = \int_{\mathbb{R}^\infty} \lambda_1^{\alpha_1} \cdots \lambda_n^{\alpha_n} \cdots d\mu(\lambda) = \int_{\mathbb{R}^\infty} \lambda^\alpha\, d\mu(\lambda)\,,$$

where $(\alpha_1, \ldots, \alpha_n, \ldots) = \alpha \in \mathbb{N}_0^\infty$ is a finite multi-index.

In particular, we have

$$s_{(0,\ldots,0,1,0,\ldots,0,1,0,\ldots)}^{g_B(\cdot)} = \int_{\mathbb{R}^\infty} \lambda_j \lambda_k\, dg_B(\lambda) = B_{jk}. \tag{5}$$

But it is not always possible to reconstruct the measure $\mu(\cdot)$ from its moments. A simple sufficient condition for the unique solvability of the countably dimensional moment problem is its unique solvability for the measures $\mu_n(\cdot)$ on $(\mathbb{R}^n, \mathrm{B}(\mathbb{R}^n))$ for all $n = 1,2,\ldots$. The latter corresponds to an n-dimensional moment problem. Sufficient (close to necessary) conditions are given, for example, in the monograph Yu.M. Berezanskiĭ [5].

In particular, sufficient conditions for the unique solvability of a countably dimensional moment problem are given by the following inequalities

$$\sum_{\substack{\alpha_1, \ldots, \alpha_k \\ \beta_1, \ldots, \beta_k}} s_{(\alpha_1+\beta_1, \ldots, \alpha_k+\beta_k, 0, \ldots)}^{\mu(\cdot)}\, c_{\alpha_1, \ldots, \alpha_k}\, \overline{c_{\beta_1, \ldots, \beta_k}} \geq 0 \tag{6}$$

($c_{\alpha_1, \ldots, \alpha_k}$ - an arbitrary complex number; $\alpha_1 = 0, 1, \ldots, n_1$; $\cdots$; $\alpha_k = 0, 1, \ldots, n_k$) and

$$|\, s_{(\alpha_1, \ldots, \alpha_k, 0, \ldots)}^{\mu(\cdot)} \,| \leq C_k M_k^{\alpha_1+\alpha_2+\cdots}\, \alpha_1^{\alpha_1} \cdots \alpha_k^{\alpha_k} \tag{7}$$

$$(\alpha = (\alpha_1, \ldots, \alpha_k, 0, \ldots),\ K = 1,2,\ldots;\ C_k, M_k > 0).$$

Then on the intervals

$$\Delta = \{\lambda \in \mathbb{R}^\infty \mid a_1 < \lambda_1 \leq B_1, \ldots, a_n < \lambda_n \leq B_n\}$$

define an operator-valued measure $E(\cdot)$ as

$$E(\Delta) = E_1((a_1, b_1]) \cdot \cdots \cdot E_n((a_n, B_n])$$

(here $E_j(\cdot)$ is a resolution of the identity for the operator A_j, $j = 1,2,\ldots$).

The function $E(\cdot)$ which is defined on intervals can be extended to an operator-valued measure on the measurable space $(\mathbb{R}^\infty, \mathbf{B}(\mathbb{R}^\infty))$. Indeed, according to the Kolmogorov theorem, the scalar measure $\rho_{f,f}(\cdot) = (E(\cdot)f, f)$, defined on the algebra of intervals, can be extended to a measure on the space $(\mathbb{R}^\infty, \mathbf{B}(\mathbb{R}^\infty))$. For a fixed $\Delta \in \mathbf{B}(\mathbb{R}^\infty)$ the quadratic form $\rho_{f,f}(\Delta)$ is nonnegative. The bilinear form

$$\rho_{f,g}(\Delta) = \frac{1}{4}[\rho_{f+g,f+g}(\Delta) - \rho_{f-g,f-g}(\Delta) + i\,\rho_{f+ig,f+ig}(\Delta) -$$

$$- i\,\rho_{f-ig,f-ig}(\Delta)]$$

is Hermitian, depends bilinearly on $f,g \in H$ and satisfies $|\rho_{f,g}(\Delta)|^2 \leq \|f\|^2\,\|g\|^2$. Consequently, there exists an operator-valued function $E(\cdot)$, defined on $\mathbf{B}(\mathbb{R}^\infty)$, such that $(E(\Delta)f,g) = \rho_{f,g}(\Delta)$. The function $E(\cdot)$ has the following properties:

1°. $E(\cdot)$ is a projection-valued function. Indeed, if for a certain $\Delta \in \mathbf{B}(\mathbb{R}^\infty)$ we can choose a sequence of cylindrical sets $\Delta^{(j)}$ $(j = 1,2,..)$ with $\Delta^{(1)} \supset \Delta^{(2)} \supset \cdots$ such that $\Delta = \bigcap_{j=1}^{\infty} \Delta^{(j)}$ then $E(\Delta^{(n)}) = \prod_{j=1}^{n} E(\Delta^{(j)})$ is a monotone sequence of projections that converges in weak (and consequently in strong) sense to a projection $E(\Delta)$. If for a certain $\Delta \in \mathbf{B}(\mathbb{R}^\infty)$ we can choose $\tilde{\Delta}^{(j)}$ with $\tilde{\Delta}^{(1)} \subset \tilde{\Delta}^{(2)} \subset \cdots$ such that $E(\tilde{\Delta}^{(j)})$ are projections $(j = 1,2,\ldots)$, then $E(\bigcup_{j=1}^{\infty} \tilde{\Delta}^{(j)})$ is the limit in weak (and strong) sense of a monotone sequence of projections $E(\bigcup_{j=1}^{n} \tilde{\Delta}^{(j)}) = E(\Delta^{(n)})$ and so it is a projection itself. For an arbitrary $\Delta \in \mathbf{B}(\mathbb{R}^\infty)$ we can consider a succession of similar limiting passages. We also have:

$$E(\varnothing) = 0, \quad E(\mathbb{R}^\infty) = E_1(\mathbb{R}^1)\,E_2(\mathbb{R}^1) \cdots = I.$$

2°. The projection-valued function $E(\cdot)$ is σ-additive on $\mathbf{B}(\mathbb{R}^\infty)$ because all the $\rho_{f,g}(f,g \in H)$ are σ-additive.

3°. The constructed projection-valued measure $E(\cdot)$ is orthogonal, that is we have

$$E(\Delta' \cap \Delta'') = E(\Delta')\,E(\Delta'')$$

for all $\Delta', \Delta'' \in \mathbf{B}(\mathbb{R}^\infty)$. Indeed, if $\Delta^{(n)}, \delta^{(n)}$ $(n = 1,2,\ldots)$ are monotone sequences of cylindrical sets and $\Delta = \bigcap_{j=1}^{\infty} \Delta^{(j)}$, $\delta = \bigcap_{j=1}^{\infty} \delta^{(j)}$ then $\Delta^{(n)} \cap \delta^{(n)}$ $(n = 1,2,\ldots)$ is a monotone creasing sequence of sets and $\Delta \cap \delta = \lim_{n \to \infty} (\Delta^{(n)} \cap \delta^{(n)})$. Thus

$E(\Delta \cap \delta) = \lim_{n\to\infty} E((\Delta^{(n)}) \cap \delta^{(n)}) = \lim_{n\to\infty} E(\Delta^{(n)})\, E(\delta^{(n)}) = E(\Delta)\, E(\delta)$ (we used the fact that $E(\Delta^{(n)}) \to E(\Delta)$ and $E(\delta^{(n)}) \to E(\delta)$ in the strong topology). This argument is then repeated in the next limiting passage and so on.

Consider the following definition.

DEFINITION 1. An operator-valued measure $E(\cdot)$ defined on $(\mathbb{R}^\infty, \mathbf{B}(\mathbb{R}^\infty))$ is called a resolution of the identity if:

1°) $E(\Delta)$ is a projection on H for all $\Delta \in \mathbf{B}(\mathbb{R}^\infty)$, $E(\emptyset) = 0$, $E(\mathbb{R}^\infty) = I$;

2°) the absolute additivity property holds, i.e. if $\Delta^{(j)} \in \mathbf{B}(\mathbb{R}^\infty)$ $(j = 1,2,\ldots)$ are mutually disjoint, then $E(\bigcup_{i=1}^{\infty} \Delta^{(j)}) = \sum_{j=1}^{\infty} E(\Delta^{(j)})$, where the series converges in strong sense;

3°) the orthogonality property holds, i.e. $E(\Delta' \cap \Delta'') = E(\Delta')\, E(\Delta'')$ $(\Delta', \Delta'' \in \mathbf{B}(\mathbb{R}^\infty))$.

We summarize everything said above in a theorem.

THEOREM 1. (spectral theorem in terms of projection-valued measures for a countable family of CSO). For every countable family $(A_j)_{j=1}^{\infty}$ of CSO there is a unique resolution of the identity $E(\cdot)$ defined on $(\mathbb{R}^\infty, \mathbf{B}(\mathbb{R}^\infty))$. Conversely, every such operator-valued measure is generated by a family of CSO in the above described way with

$$A_j = \int_{\mathbb{R}^\infty} \lambda_j \, dE(\lambda_1, \lambda_2, \ldots) \quad (j = 1,2,\ldots). \tag{8}$$

To prove the last statement of the theorem we use the given resolution of the identity $E(\cdot)$ on $(\mathbb{R}^\infty, \mathbf{B}(\mathbb{R}^\infty))$ to define one-dimensional resolutions of the identity $E_k(\cdot)$ by setting $E_k((a,b]) = E(\mathbb{R}^1 \times \cdots \times \mathbb{R}^1 \times (a,b] \times \mathbb{R}^1 \times \cdots)$ $(k = 1,2,\ldots)$. So we have the self-adjoint operators

$$A_k = \int_{\mathbb{R}^1} \lambda_k \, dE_k(\lambda_k) = \int_{\mathbb{R}^\infty} \lambda_k \, dE(\lambda_1, \lambda_2, \ldots).$$

Because the measure $E(\cdot)$ is orthogonal, the measures $dE(\lambda_1, \lambda_2, \ldots)$ and $\bigotimes_{k=1}^{\infty} dE_k(\lambda_k)$ coincide on the cylindrical sets, and consequently they coincide on the entire σ-algebra $\mathbf{B}(\mathbb{R}^\infty)$. □

Remark 1. Theorem 1 also holds for a general family $(A_\alpha)_{\alpha\in\Lambda}$ of CSO, i.e. there exists a spectral theorem for a joint resolution of the identity $E(\cdot)$ on the measurable space $(\mathbb{R}^\Lambda, C_\sigma(\mathbb{R}^\Lambda))$, where $C_\sigma(\mathbb{R}^\Lambda)$ is the σ-algebra generated by the cylindrical sets in $\mathbb{R}^\Lambda$ (see Yu.M. Berezanskiĭ [9]). Note, that in general, the σ-algebras $C_\sigma(\mathbb{R}^\Lambda)$ and $\mathbf{B}(\mathbb{R}^\Lambda)$ are not the same. For a family of CSO to determine the joint spectrum and to study its properties, the choice of a σ-algebra in $\mathbb{R}^\Lambda$ plays the most important role, Thus, for example, if $\Sigma_1(\Lambda_1)$, $\Sigma_2(\Lambda_2)$ are arbitrary σ-algebras of subsets in Λ_1 and Λ_2 and $\Sigma_1 \times \Sigma_2(\Lambda_1 \times \Lambda_2)$ is a σ-algebra in $\Lambda_1 \times \Lambda_2$, namely the product of $\Sigma_1(\Lambda_1)$ and $\Sigma_2(\Lambda_2)$, then the theorem stating that the product of the spectral measures $E(\cdot)$ defined on $(\Lambda_1, \Sigma_1(\Lambda_1))$, and $F(\cdot)$ defined on $(\Lambda_2, \Sigma_2(\Lambda_2))$, is an operator-valued measure defined on the measurable space $(\Lambda_1 \times \Lambda_2, \Sigma_1 \times \Sigma_2(\Lambda_1 \times \Lambda_2))$ does not hold any more. A counter-example is given in the paper M.Sh. Birman, A.M. Vershik, M.Z. Solomyak [1].

1.2. The joint spectrum of a family of CSO.

The spectrum $\sigma(A_j)$ of a self-adjoint operator A_j is a closed subset of $\mathbb{R}^1$. It is the complement of the open set of regular points of the operator. We note that a point $\lambda_j \in \mathbb{R}^1$ is regular if the operator $(A_j - \lambda_j I)^{-1}$ is an everywhere defined, bounded operator on H.

In other words, if $\lambda_j \in \mathbb{R}^1$ is a regular point of the operator A_j, then $\exists\, \varepsilon_{\lambda_j} > 0$ such that for any $f \in \mathbf{D}(A_j)$

$$\|(A_j - \lambda_j I) f\| \geq \varepsilon_{\lambda_j} \|f\|.$$

If $\lambda_j \in \sigma(A_j)$, then one can find a sequence of $f_n \in \mathbf{D}(A_j)$ such that $\|f_n\| = 1$ and $\|(A_j - \lambda_j I) f_n\| \to 0$.

Further, if λ_j is a regular point of the operator A_j, then there exists $\varepsilon_{\lambda_j} > 0$ such that $E_j((\lambda_j - \varepsilon_{\lambda_j}, \lambda_j + \varepsilon_{\lambda_j})) = 0$ (see N.I. Akhiezer and I.M. Glazman [1]). Because $\mathbb{R}^1$ is a topological space with a countable base, we can cover the set of regular points with a countable family of sets with spectral measure zero to get $E_j(\sigma(A_j)) = I$. This allows to define the spectrum of a self-adjoint operator A_j differently, namely, to be the intersection of all closed subsets of $\mathbb{R}^1$ with full $E_j(\cdot)$-measure

$$\sigma(A_j) = \bigcap_{\Delta_\alpha \subset \mathbb{R}^1} \Delta_\alpha (\Delta_\alpha \text{ is closed}, E_j(\Delta_\alpha) = I).$$

If $(A_k)_{k=1}^n$ is a finite family of CSO, then we define the joint spectrum $\sigma(A_1, \ldots, A_n)$ of the family to be the smallest closed subset of $\mathbb{R}^n$ with full operator-valued $\bigotimes_{k=1}^{n} dE_k(\lambda_k)$-measure.

A point $(\lambda_1, \ldots, \lambda_n)$ does not belong to $\sigma(A_1, \ldots, A_n)$ (regular point) if λ_k is a regular point of each operator A_k for $k = 1, \ldots, n$. Indeed, because $\sigma(A_1) \times \cdots \times \sigma(A_n)$ is a closed subset of $\mathbb{R}^n$ with full $\bigotimes_{k=1}^{n} dE_k(\lambda_k)$-measure, we have

$$\sigma(A_1, \ldots, A_n) \subset \sigma(A_1) \times \cdots \times \sigma(A_n). \tag{9}$$

However, the set of regular points of a family is not reduced to sets of this form in general. As a simple example take the family of equal operators $A_1, \ldots, A_1$ with joint spectrum on the diagonal $(\lambda_1, \ldots, \lambda_n) \in \mathbb{R}^n, \lambda_j \in \mathbb{R}^1$. In this case (9) is a genuine inclusion.

PROPOSITION 2. For $(\lambda_1, \ldots, \lambda_n) \in \sigma(A_1, \ldots, A_n)$ it is necessary and sufficient that there exists a sequence of unit vectors $f_m \in \mathbf{D}(A_k)$ $(m = 1, 2, \ldots)$ such that

$$\|(A_k - \lambda_k) f_m\| \overset{m\to\infty}{\longrightarrow} 0 \quad (k=1,\ldots,n). \tag{10}$$

We prove the assertion for the point $(0,\ldots,0) \in \mathbb{R}^n$. Assume that any neighborhood of zero has positive operator measure. Now choose spheres $S_{1/m}$ with center in $(0,\ldots,0)$ and radius $1/m$ and choose unit vectors f_m from H such that $E(S_{1/m})f_m = f_m$ (this is possible since $E(S_{1/m}) \neq 0$ for all $m=1,2,\ldots$). From the spectral representation (1'), we get

$$\|A_k f_m\|^2 = \int_{\mathbb{R}^n} \lambda_k^2 \, d(E(\lambda_1,\ldots,\lambda_n) f_m, f_m) =$$

$$= \int_{S_{1/m}} \lambda_k^2 \, d(E(\lambda_1,\ldots,\lambda_n) f_m, f_m) \le \frac{1}{m^2} \overset{m\to\infty}{\longrightarrow} 0 \quad (k=1,\ldots,n).$$

Conversely, assuming that there exists a number N such that $E(S_{1/N}) = 0$, we have for every unit vector $f \in \mathbf{D}(A_k)$,

$$\|A_k f\|^2 = \int_{\mathbb{R}^n / S_{1/N}} \lambda_k^2 \, d(E(\lambda_1,\ldots,\lambda_n) f, f) \ge \frac{1}{N^2} \|f\|^2 = \frac{1}{N^2}$$

$$(k=1,\ldots,n)$$

and consequently there is not a sequence $f_m \in \mathbf{D}(A_k)$ $(m=1,2,\ldots)$ such that $\|A_k f_m\| \overset{m\to\infty}{\longrightarrow} 0$ $(k=1,\ldots,n)$. The obtained contradiction proves the proposition. $\square$

The joint spectrum of a family $(A_k)_{k=1}^n$ of CSO has the full $dE(\lambda_1,\ldots,\lambda_n) = \bigotimes_{k=1}^{n} dE_k(\lambda_k)$-measure. The equality

$$E(\sigma(A_1,\ldots,A_n)) = I$$

follows from the countability of the base in the σ-algebra $\mathbf{B}(\mathbb{R}^n)$.

Now let $(A_k)_{k=1}^{\infty}$ be a countable family of CSO. We make the following definition.

DEFINITION 2. The joint spectrum $\sigma(A_1,\ldots,A_n,\ldots)$ of a countable family $(A_k)_{k=1}^{\infty}$, is defined to be the intersection of all closed subsets of $\mathbb{R}^\infty$ with full operator-valued $\bigotimes_{k=1}^{\infty} dE_k(\lambda_k)$-measure.

Because $\sigma(A_1)\times\cdots\times\sigma(A_n)\times\cdots$ is a closed subset of $\mathbb{R}^\infty$ with full measure, it follows that

$$\sigma(A_1,\ldots,A_n,\ldots) \subset \sigma(A_1)\times\cdots\times\sigma(A_n)\times\cdots \tag{11}$$

Besides that

$$\sigma(A_1,\ldots,A_n,\ldots) \subset \sigma(A_1,\ldots,A_{n_1})\times\sigma(A_{n_1+1},\ldots,A_{n_2})\times\cdots \tag{12}$$

for any sequence of natural numbers $n_1 < n_2 < \cdots$.

THEOREM 2. We have

$$\sigma(A_1,\ldots,A_n,\ldots)=\bigcap_{n=1}^{\infty}\sigma(A_1\times\cdots\times A_n)\times \mathbb{R}^1\times\cdots; \tag{13}$$

the measure $dE(\lambda)=dE((\lambda_1,\ldots,\lambda_n,\ldots))=\bigotimes_{k=1}^{\infty}dE_k(\lambda_k)$ of the joint spectrum of the CSO $(A_k)_{k=1}^{\infty}$ satisfies

$$E(\sigma(A_1,\ldots,A_n,\ldots))=I. \tag{14}$$

Proof. The inclusions

$$\sigma(A_1,\ldots,A_n,\ldots)\subset\sigma(A_1,\ldots,A_{n+1})\times\mathbb{R}^1\times\cdots\subset\sigma(A_1,\ldots,A_n)\times\mathbb{R}^1\times\cdots$$

follow from the definition.

Conversely, if $\lambda=(\lambda_1,\ldots,\lambda_n,\ldots)\notin\sigma(A_1,\ldots,A_n,\ldots)$, then there exists in $\mathbb{R}^\infty$ a subset which is open with respect to the product topology contains $\lambda=(\lambda_1,\ldots,\lambda_n,\ldots)$ and has zero $dE(\lambda)=\bigotimes_{k=1}^{\infty}dE_k(\lambda_k)$-measure. But then, there exists a number N such that the point $(\lambda_1,\ldots,\lambda_N)\in\mathbb{R}^N$ belongs to an open set in $\mathbb{R}^N$ with zero $\bigotimes_{k=1}^{N}dE_k(\lambda_k)$-measure. Consequently, $(\lambda_1,\ldots,\lambda_N)\times\mathbb{R}^1\times\cdots\notin\sigma(A_1,\ldots,A_N)\times\mathbb{R}^1\times\cdots$ and equality (13) is proved.

Property (14) follows from the countability of the base in the σ-algebra $\mathbf{B}(\mathbb{R}^\infty)$. Note that due to (14), the spectrum $\sigma(A_1,\ldots,A_n,\ldots)$ is not empty. □

It follows from Theorem 2 that the integrals with respect to the measure $E(\cdot)$ could be considered only on the joint spectrum of the operators of the family.

Example 1. Let $(P_k)_{k=1}^{\infty}$ be a family of commuting projections. Then the joint spectrum of the family is contained in the closed subset $\{0,1\}^\infty\subset\mathbb{R}^\infty$. □

Remark 2. For uncountable families $(A_\alpha)_{\alpha\in\Lambda}$ of CSO, it may happen that the intersection of the sets closed in the product topology with full spectral measure would not be measurable in the σ-algebra generated by cylindrical sets. The intersection may also be empty (see Yu.M. Berezanskiĭ [9]).

Example 2. (the joint resolution of the identity for CSO with empty support). For the CSO established by multiplication by x and by $\dfrac{1}{x-x_0}$ in $L_2([0,1],dx)$ ($x_0\in[0,1]$) the joint spectrum is empty. □

1.3. Countable families of CSO with a simple joint spectrum.

An operator A_j on H is an operator with a simple spectrum if there exists a vector $\Omega\in H$ such that the closed linear span (CLS) of $\{E(\Delta)\Omega\mid\Delta\in\mathbf{B}(\mathbb{R}^1)\}$ equals H. The structure of these operators is the most simple one. They are unitarily equivalent to the operators of multiplication by the independent variable λ_j in $L_2(\mathbb{R}^1, d\rho(\lambda_j))$ ($\rho(\cdot)$ is a finite measure on $(\mathbb{R}^1, \mathrm{B}(\mathbb{R}^1))$).

DEFINITION 3. We say that the joint spectrum of a family $(A_k)_{k=1}^{\infty}$ of CSO is simple if there exists a vector $\Omega \in H$ such that $\mathrm{CLS}\{E(\Delta)\Omega\}_{\Delta} = H$. The vector Ω is called a cyclic vector or a vacuum for the joint resolution of the identity $E(\cdot)$.

Example 3. As a family of CSO with a simple spectrum, consider the family $(\lambda_k)_{k=1}^{\infty}$ of CSO which are operators of multiplication by the independent variables in $H = L_2(\mathbb{R}^{\infty}, d\rho(\lambda))$. Here $\rho(\cdot)$ is a probability measure on $(\mathbb{R}^{\infty}, \mathrm{B}(\mathbb{R}^{\infty}))$. The vector $\Omega \equiv 1$ is a vacuum for the system $(\lambda_k)_{k=1}^{\infty}$ of CSO, because the characteristic functions of the cylindrical sets are total in H and because each projection $E(\Delta)$ is the operator of multiplication by the characteristic function $h(\Delta)$ of the set $\Delta \in \mathrm{B}(\mathbb{R}^{\infty})$.

DEFINITION 4. Two families of CSO, $(A_k)_{k=1}^{\infty}$ on H and $(\tilde{A}_k)_{k=1}^{\infty}$ on $\tilde{H}$ are called unitarily equivalent if there exists a unitary operator $U : H \to \tilde{H}$ such that for all $k = 1,2,\ldots$ $A_k = U^* \tilde{A}_k U$.

THEOREM 3. The systems of operators $(\lambda_k)_{k=1}^{\infty}$ on $L_2(\mathbb{R}^{\infty}, d\rho_1(\lambda))$, and $(\lambda_k)_{k=1}^{\infty}$ on $L_2(\mathbb{R}^{\infty}, d\rho_2(\lambda))$ are unitarily equivalent if and only if the measures $\rho_1(\cdot)$ and $\rho_2(\cdot)$ are equivalent (i.e. they have the same subsets of measure zero).

Proof. If the measures $\rho_1(\cdot)$ and $\rho_2(\cdot)$ on $\mathbb{R}^{\infty}$ are equivalent then there is an invertible operator $f(\lambda) \mapsto \sqrt{P(\lambda)}\, f(\lambda)$, where $P(\lambda) = \dfrac{d\rho_2}{d\rho_1}(\lambda)$. This operator establishes a unitary equivalence between the systems of operators $(\lambda_k)_{k=1}^{\infty}$ on $L_2(\mathbb{R}^{\infty}, d\rho_1(\lambda))$ and $(\lambda_k)_{k=1}^{\infty}$ on $L_2(\mathbb{R}^{\infty}, d\rho_2(\lambda))$. The proof of the converse statement is based on the following lemmas.

LEMMA 1. Resolutions of the identity of unitarily equivalent families of CSO are equivalent.

Proof. If $E(\cdot)$ is the resolution of the identity of the family $(A_k)_{k=1}^{\infty}$, then the resolution of the identity $\tilde{E}(\cdot)$ of the unitarily equivalent family $(\tilde{A}_k = U A_k U^*)_{k=1}^{\infty}$ equals $\tilde{E}(\cdot) = U E(\cdot) U^*$. So that if $E(\Delta)$ is equal to zero then the same is true for $\tilde{E}(\Delta)$ and vice versa. □

LEMMA 2. Let $(A_k)_{k=1}^{\infty}$ be a family of CSO with simple joint spectrum and with vacuum Ω. The projection-valued measure $E(\cdot)$ and the scalar measure $\rho(\cdot) = (E(\cdot)\Omega, \Omega)$ are equivalent.

Proof. It follows from the definition that the equality $E(\Delta_0) = 0$ implies $\rho(\Delta_0) = 0$. Conversely, if $\rho(\Delta_0) = 0$ then $E(\Delta_0)\Omega = 0$ and as Ω is a vacuum and $E(\Delta_0)H = E(\Delta_0)\,(\mathrm{CLS}\{E(\Delta)\Omega\}_{\Delta}) = \mathrm{CLS}\{E(\Delta \cap \Delta_0)\Omega\}_{\Delta} = 0$ we have that $E(\Delta_0) = 0$. □

In virtue of Lemma 1, unitary equivalence of the resolutions of the identity $E_1(\cdot)$ and $E_2(\cdot)$ implies their equivalence. Consequently, it follows from Lemma 2 that the scalar measures

$$\rho_1(\cdot) = (h(\cdot)\,1,\,1)_{L_2(\mathbb{R}^\infty, d\rho_1(\lambda))},$$

$$\rho_2(\cdot) = (h(\cdot)\,1,\,1)_{L_2(\mathbb{R}^\infty, d\rho_2(\lambda))}$$

are equivalent. □

DEFINITION 5. Any probability measure on $(\mathbb{R}^\infty, \mathbf{B}(\mathbb{R}^\infty))$, which is equivalent to a resolution of the identity $E(\cdot)$ of a family $(A_k)_{k=1}^\infty$ of CSO, is called a spectral measure of this family.

For a family of CSO with a simple joint spectrum and a vacuum Ω, the measure $\rho(\cdot) = (E(\cdot)\,\Omega, \Omega)$ is a spectral measure.

Regardless of its simplicity, Example 3 is general by its nature. In fact, we have the following theorem.

THEOREM 4. (the spectral theorem for a family of CSO of multiplications with simple joint spectrum). For any countable family $(A_k)_{k=1}^\infty$ of CSO with a simple joint spectrum on a separable H, there exists a finite measure $\rho(\cdot)$ on $(\mathbb{R}^\infty, d\rho(\lambda))$. The measure $\rho(\cdot)$ is determined up to equivalence.

Proof. Because $\mathrm{CLS}\{E(\Delta)\,\Omega\}_\Delta = H$, the scalar measure $\rho(\cdot) = (E(\cdot)\,\Omega, \Omega)$ and the operator measure $E(\cdot)$ are absolutely continuous with respect to each other. The isometric correspondence $E(\cdot)\,\Omega \mapsto h(\cdot)$ generates a unitary mapping from H into $L_2(\mathbb{R}^\infty, d\rho(\lambda))$. The image of the operator A_k is the operator λ_k of multiplication $(k = 1,2,\ldots)$ in $L_2(\mathbb{R}^\infty, d\rho(\lambda))$. Any other measure $\rho_1(\cdot)$ equivalent to the measure $E(\cdot)$, is equivalent to the measure $\rho(\cdot)$. □

Example 4. In Yu.M. Berezanskiĭ and G.F. Us [1,2] operators and operator functions are studied which act "on different variables". To be more precise, one considers the family $(A_k = I \otimes \cdots \otimes I \otimes \mathbf{A}_k \otimes I \otimes \cdots)_{k=1}^\infty$ of CSO on $H = \bigotimes_{k=1;e}^\infty \mathbf{H}_k$ with stabilization $e = (e_k \in \mathbf{H}_k)_{k=1}^\infty$. This family is generated by the operators $\mathbf{A}_k$ on $\mathbf{H}_k$. Let the operators $\mathbf{A}_k$ on $\mathbf{H}_k$ $(k = 1,2,\ldots)$ have a simple spectrum. In this case, the family $(A_k)_{k=1}^\infty$ of operators on $H = \bigotimes_{k=1;e}^\infty \mathbf{H}_k$ is unitarily equivalent to the family $(\lambda_k)_{k=1}^\infty$ of operators on $L_2(\mathbb{R}^\infty, d\rho(\lambda))$, where $\rho(\cdot)$ is the product measure on $\mathbb{R}^\infty$ calculated from the spectral measures of the operators $\mathbf{A}_k$ and the stabilization $e = (e_k)$ $(k = 1,2,\ldots)$. □

Example 5. Let $(P_k)_{k=1}^\infty$ be a system of commuting projections on H with simple joint spectrum. Then according to Theorem 4, the system is unitarily equivalent to the family $(\lambda_k)_{k=1}^\infty$, of operators on the space $L_2(\{0,1\}^\infty, d\rho(\lambda)$, where $\{0,1\}^\infty$ is the space of sequences of 0 and 1. The measures on $\{0,1\}^\infty$ arise naturally in the construction of models for statistical mechanics of classic lattice systems (see, for example, C. Preston [1] and the references given there). □

Example 6. Let H be a separable Hilbert space of functions with as natural involution, the complex conjugation for a function. Let $\otimes H^n$ be its tensor product of degree n (it is a space of functions of n variables), and $\circledS H^n$ its symmetric product of degree n, consisting of symmetric functions of n variables. In what follows, we write vectors $u \in \mathbf{F} = \oplus \sum_{n=0}^{\infty} \otimes H^n$ as sequences $(u_0, u_1, \ldots, u_n, \ldots)$ $(u_n \in \otimes H^n)$. In the space $\mathbf{F}$, we introduce in the usual way (see, for example, F.A. Berezin [1]) creation operators $A_+(u_1)$ and annihilation operators $A_-(u_1)$

$$A_+(u_1)u_n = \sqrt{n+1}\,(u_1 \otimes u_n) \in \otimes H^{n+1},$$

$$A_-(u_1)u_n = \sqrt{n}\,(u_1, u_n)_n \in \otimes H^{n-1}$$

where $u_1 \in H$, $u_n \in \otimes H^n$ and $(u_1, u_n)_n$ denotes a scalar product with respect to the n-th variable. The closure of the image of the operator $A_+(u_1) + A_-(u_1)$ acting on the set of finite sequences, will be called the Hermitian free scalar field denoted by $A(u_1)$.

Also consider a real basis $(e_k)_{k=1}^{\infty}$ in H and a family $(A_k = A(e_k))_{k=1}^{\infty}$ of self-adjoint operators on m. On the space $\mathbf{F}_B = \oplus \sum_{n=0}^{\infty} \circledS H^n$, one can define a countable family $(P_B A_k P_B)_{k=1}^{\infty}$ of CSO, where P_B is a projection of $\mathbf{F}$ into $\mathbf{F}_B$. The vector $\Omega = (1,0,0,\ldots)$ is cyclic for the family $(P_B A_k P_B)_{k=1}^{\infty}$ of CSO. For this family, the measure $g(\cdot) = (E(\cdot)\,\Omega, \Omega)$ is the Gaussian measure $dg(\lambda) = \otimes_{k=1}^{\infty} dg_k(\lambda_k) = \otimes_{k=1}^{\infty} \frac{1}{\sqrt{2\pi}} \rho^{\frac{-\lambda_k^2}{2}}\, d\lambda_k$ (see V.D. Koshmanenko and Yu.S. Samoĭlenko [1]). □

DEFINITION 6. A vector Ω_1 is called a joint cyclic vector of the family $(A_k)_{k=1}^{\infty}$ if all the operators $A_1^{\alpha_1} \cdots A_n^{\alpha_n}$ (with only a finite number of the α_k different from zero) contain Ω_1 in their domain and $\mathrm{CLS}\{A_1^{\alpha_1} \cdots A_n^{\alpha_n} \cdots \Omega_1\}_\alpha = H$ $(\alpha = (\alpha_1, \ldots, \alpha_n, \ldots) \in \mathbb{N}_0^{\infty})$.

If Ω_1 is a cyclic vector of the family, then $\mathrm{CLS}\{E(\Delta)\,\Omega\}_\Delta = H$, i.e. Ω_1 is also a vacuum. Indeed, it follows from the spectral representation

$$(A_1^{\alpha_1} \cdots A_n^{\alpha_n} \cdots \Omega_1, f) = \int_{\mathbb{R}^\infty} \lambda_1^{\alpha_1} \cdots \lambda_n^{\alpha_n} \cdots d(E(\lambda)\,\Omega_1, f)$$

that if f is orthogonal to $\mathrm{CLS}\{E(\Delta)\,\Omega\}_\Delta$, then it will be orthogonal to $\mathrm{CLS}\{A_1^{\alpha_1} \cdots A_n^{\alpha_n} \cdots \Omega_1\}$. So we would get a contradiction if a cyclic vector of the family would not be a vacuum.

However, if Ω is a cyclic vector for a joint decomposition of the identity, it does not necessarily imply that Ω is a joint cyclic vector for the family $(A_k)_{k=1}^{\infty}$. Indeed, for the family of commuting self-adjoint operators of multiplication by $\lambda_k (k = 1,2,,\ldots)$ on $L_2(\mathbb{R}^\infty, d\rho(\lambda))$ with $\rho(\cdot)$ an arbitrary probability measure on $(\mathbb{R}^\infty, \mathrm{B}(\mathbb{R}^\infty))$, the vector $\Omega \equiv 1$ will not be a joint cyclic vector of the family (polynomials may not belong to $L_2(\mathbb{R}^\infty, d\rho(\lambda))$, if there are no moments for the measure $\rho(\cdot)$), or may not be dense there. Nevertheless, we have the following theorem:

THEOREM 5. For a family $(A_k)_{k=1}^{\infty}$ with simple joint spectrum, there always exists a joint cyclic vector.

It is sufficient to prove Theorem 5 for the system $(\lambda_k)_{k=1}^{\infty}$ of operators on $L_2(\mathbb{R}^\infty, \mathbf{B}(\mathbb{R}^\infty))$, where $\rho(\cdot)$ is an arbitrary probability measure on $(\mathbb{R}^\infty, \mathbf{B}(\mathbb{R}^\infty))$. Let us show that for any measure $\rho(\cdot)$, there exists a sequence of positive integers $(N_k)_{k=1}^{\infty}$ such that the vector $e^{-\|\lambda\|^2_{\mathbf{H}_-}}$ is cyclic for the family $(\lambda_k)_{k=1}^{\infty}$ of operators on $L_2(\mathbb{R}^\infty, d\rho(\cdot))$. Here

$$\mathbf{H}_- = l_2\left[\left(\frac{1}{k^2 N_k}\right)_{k=1}^{\infty}\right] = \left\{\lambda \in \mathbb{R}^\infty \;\middle|\; \|\lambda\|^2_{\mathbf{H}_-} = \sum_{k=1}^{\infty} \frac{\lambda_k^2}{k^2 N_k} < \infty\right\}.$$

The proof is based on the following lemmas.

LEMMA 3. For any finite measure $\rho(\cdot)$ which is defined on the σ-algebra generated by the cylindrical subsets of $\mathbb{R}^\infty$, there exists a Banach sequence space

$$l_\infty\left[\left(\frac{1}{N_k}\right)_{k=1}^{\infty}\right] = \left\{\lambda \in \mathbb{R}^\infty \;\middle|\; \|\lambda\|_{l_\infty((\frac{1}{N_k}))} = \sup_k \frac{|\lambda_k|}{N_k} < \infty\right\}$$

with full $\rho(\cdot)$-measure.

Proof. To see that Lemma 3 holds we can use Theorem 1 of A.M. Vershik [1]. However, we give a direct proof, repeating in essence arguments of M. Reed [1].

Let $\rho(\cdot)$ be a finite measure on $(\mathbb{R}^\infty, \mathbf{B}(\mathbb{R}^\infty))$. We can assume that this measure is normed by one. Define a family $(B_l)_{l=1}^{\infty}$ of measurable sets in such a way that $B_l \subset B_{l+1}$ $(l=1,2,\ldots)$ and $\rho(B_l) \geq 1 - \frac{1}{2^l}$. To do that, we choose a sequence of numbers $(\tau(1,k))_{k=1}^{\infty}$ $(\tau(1,k) \leq \tau(1,k+1), k=1,2,\cdots)$ such that the complements of the sets $(F_{1,k})_{k=1}^{\infty} = (\{\lambda \in \mathbb{R}^\infty \mid |\lambda_k| \leq \tau(1,k)\})_{k=1}^{\infty}$ have $\rho(\cdot)$-measure not more than $\frac{1}{2} \cdot \frac{1}{2} k$.

It is clear that $\rho(B_1) = \rho(\{\lambda \in \mathbb{R}^\infty \mid |\lambda_k| \leq \tau(1,k), k=1,2,\ldots\}) = \rho(\bigcap_{k=1}^{\infty} F_{1,k}) \geq 1 - \frac{1}{2^2} - \frac{1}{2^3} - \cdots = \frac{1}{2}$. Now use induction to determine a sequence $(\tau(l,k))_{k=1}^{\infty}$ so that $\tau(l,k) \leq \tau(l+1,k)$, $\tau(l,k) \leq \tau(l,k+1)$ $(k,l=1,2,\cdots)$ and so that the measures of the complements of the sets $F_{l,k} = \{\lambda \in \mathbb{R}^\infty \mid |\lambda_k| \leq \tau(l,k)\}$ do not exceed $\frac{1}{2^l} \cdot \frac{1}{2^k}$. Consequently,

$$\rho(B_l = \bigcap_{k=1}^{\infty} F_{l,k}) \geq 1 - \frac{1}{2^l}\left(\frac{1}{2} + \frac{1}{2^2} + \cdots\right) = 1 - \frac{1}{2^l}.$$

So the sets B_l are such that $B_l \subset B_{l+1}$ and $\rho(\bigcup_{l=1}^{\infty} B_l) = 1$. It is clear that the Banach sequence space

$$l_\infty\left[\left(\frac{1}{\tau(k,k)}\right)_{k=1}^{\infty}\right] = \left\{\lambda \in \mathbb{R}^\infty \mid \|\lambda\|_{l_\infty((\frac{1}{\tau(k,k)}))} = \sup_k \frac{|\lambda_k|}{\tau(k,k)} < \infty\right\}$$

contains the sets B_l for all $l = 1,2,\ldots$. So this space has full $\rho(\cdot)$ measure. $\square$

Remark 3. For any finite measure $\rho(\cdot)$ on $\mathbb{R}^\infty$ the Hilbert space

$$\mathbf{H}_- = l_2\left(\left(\frac{1}{k^2 N_k}\right)_{k=1}^{\infty}\right) = \left\{\lambda \in \mathbb{R}^\infty \mid \sum_{k=1}^{\infty} \frac{\lambda_k^2}{k^2 N_k} < \infty\right\} \supset l_\infty\left(\left(\frac{1}{N_k}\right)_{k=1}^{\infty}\right)$$

has full $\rho(\cdot)$-measure.

Remark 4. In virtue of Lemma 3, any continuous positive definite function on $\mathbb{R}_0^\infty = \lim\limits_{\rightarrow} \mathbb{R}^n$ is continuous in $\mathbb{R}_0^\infty = \bigcap\limits_{(N_k)} l_2\left(\left(\frac{1}{N_k}\right)_{k=1}^{\infty}\right)$ with the topology induced by the projection limit of Hilbert spaces $l_2\left(\left(\frac{1}{N_k}\right)_{k=1}^{\infty}\right)$.

LEMMA 4. The functions $P(\lambda_1, \ldots, \lambda_m)\exp(-\|\lambda\|_{\mathbf{H}_-}^{\infty})$, where $\rho(\cdot)$ are all possible cylindrical polynomials are dense in $L_2(\mathbb{R}^\infty, d\rho(\lambda))$.

Proof. We can prove this lemma using a result from the book A.V. Skorokhod [3]. However, we give a direct proof. It is sufficient to show that the functions $P(\lambda_1, \ldots, \lambda_m)\, e^{-\|\lambda\|_{\mathbf{H}_-}^2}$ $(m = 1,2,\ldots)$ are dense in the set of cylindrical functions. Take an arbitrary nonzero cylindrical function $u_n(\lambda_1, \ldots, \lambda_n) \in L_2(\mathbb{R}^\infty, d\rho(\lambda)) = L_2(l_\infty\left(\left(\frac{1}{N_k}\right)_{k=1}^{\infty}\right),$ $d\rho(\lambda))$ and take any $\varepsilon > 0$ $(\varepsilon \le \|u_n\|_{L_2(\mathbb{R}^\infty, d\rho(\lambda))})$.
Choose K such that there exists a (generally speaking not cylindrical) function $\bar{u}(\lambda_1, \ldots, \lambda_n, \ldots)$ which equals zero outside the set

$$S_K = \left\{\lambda \in l_\infty\left(\left(\frac{1}{N_k}\right)_{k=1}^{\infty}\right) \mid \|\lambda\|_{l_\infty(\frac{1}{N_k})} \le K\right\}$$

and for which $\|u_n - \bar{u}\|_{L_2(\mathbb{R}^\infty, d\rho(\lambda))} < \varepsilon$. Then choose $m \ge n$ and $P(\lambda_1, \ldots, \lambda_m)$ is such a way that

$$1 - \exp\Big(-\Big(\sum_{k=m+1}^{\infty} 1/k^2\Big)K\Big) \le \varepsilon / \|u_n\|_{L_2(\mathbb{R}^\infty, d\rho(\lambda))}$$

and

$$\|u_n(\lambda_1, \ldots, \lambda_n) - P(\lambda_1, \ldots, \lambda_m)\exp\Big(-\sum_{k=1}^{m} \lambda_k^2 / k^2 N_k\Big)\|_{L_2(\mathbb{R}^\infty, d\rho(\lambda))} \le \varepsilon.$$

We have the following estimate:

$$\|u_n(\lambda_1,\dots,\lambda_n)-P(\lambda_1,\dots,\lambda_m)\exp(-\|\lambda\|^2_{H_-})\|_{L_2(\mathbb{R}^\infty,d\rho(\lambda))}\le$$

$$\le\|u_n(\lambda_1,\dots,\lambda_n)-P(\lambda_2,\dots,\lambda_m)\exp(-\sum_{k=1}^m\lambda_k^2/k^2N_k)\|_{L_2(\mathbb{R}^\infty,d\rho(\lambda))}+$$

$$+\|P(\lambda_1,\dots,\lambda_m)\exp(-\sum_{k=1}^\infty\lambda_k^2/k^2N_k)\,(1-\exp(-\sum_{k=m+1}^\infty\lambda_k^2/k^2N_k)\|_{L_2(\mathbb{R}^\infty,d\rho(\lambda))}\le$$

$$\le\varepsilon+\|P(\lambda_1,\dots,\lambda_m)\exp(-\sum_{k=1}^m\lambda_k^2/k^2N_k)\|_{L_2(\mathbb{R}^\infty\setminus S_\varkappa,d\rho(\lambda))}+$$

$$+\|P(\lambda_1,\dots,\lambda_m)\exp(-\sum_{k=1}^m\lambda_k^2/k^2N_k)\,(1-\exp(-\sum_{k=m+1}^\infty\lambda_k^2/k^2N_k)\|_{L_2(S_\varkappa,d\rho(\lambda))}\le$$

$$\le\varepsilon+\frac{\varepsilon}{\|u_n\|_{L_2(\mathbb{R}^\infty,d\rho(\lambda))}}(\|u_n\|_{L_2(\mathbb{R}^\infty,d\rho(\lambda))}+\varepsilon)+$$

$$+\|P(\lambda_1,\dots,\lambda_m)\exp(-\sum_{k=1}^m\lambda_k^2/k^2N_k)-\bar{u}(\lambda_1,\dots,\lambda_n,\dots)\|_{L_2(\mathbb{R}^\infty\setminus S_k,d\rho(\lambda))}\le$$

$$\le\varepsilon+\varepsilon+\varepsilon^2/\|u_n\|_{L_2(\mathbb{R}^\infty,d\rho(\lambda))}+2\varepsilon\le5\varepsilon.$$

So the lemma is proved and it follows that $e^{-\|\lambda\|^2_{H_-}}$ is cyclic. □

Because the existence of a vacuum for the family $(A_k)_{k=1}^\infty$ of CSO is equivalent to the existence of a cyclic vector for the family, Definition 3 is equivalent to the following one.

DEFINITION 7. A family $(A_k)_{k=1}^\infty$ of CSO has simple joint spectrum if there exists a joint cyclic vector of the family.

It should be noted that equivalence of the Definitions 3 and 7 means that for any probability measure $\rho(\cdot)$ on $(\mathbb{R}^\infty,\mathbf{B}(\mathbb{R}^\infty))$ there is an equivalent measure $\rho_1(\cdot)$ such that all moments

$$S_\alpha^{\rho_1(\cdot)}=\int_{\mathbb{R}^\infty}\lambda_1^{\alpha_1}\cdots\lambda_n^{\alpha_n}\cdots d\rho_1(\lambda)\quad(\alpha=(\alpha_1,\dots,\alpha_n,\dots)\in\mathbb{N}_0^\infty)$$

exist and the corresponding moment problem has a unique solution.

Remark 5. For general families $(A_\alpha)_{\alpha\in\Lambda}$ of CSO with simple joint spectrum, the spectral theorem in the form of operators of multiplication also holds. Now however, the existence of a vacuum does not imply the existence of a cyclic vector (see A.V. Kosyak, Yu.S. Samoĭlenko [2]). We give an example.

Example 7. In the space $H=L_2(S'(\mathbb{R}^1),d\mu(\xi))$ $(\xi\in S'(\mathbb{R}^1))$ where $S'(\mathbb{R}^1)=\bigcup_{n=0}^\infty H_{-n}=\bigcup_{n=0}^\infty W_2^{-n}(\mathbb{R}^1,\frac{dx}{(1+|x|^2)^n})$ is the Schwartz space of generalized functions on the real line and $W_2^{-n}(\mathbb{R}^1,\frac{dx}{(1+|x|^2)^n})$ the space dual to the Sobolev space with polynomial weight, consider the family $(A_\phi)_{\phi\in S(\mathbb{R}^1)}$ of CSO of multiplication by the linear functions (ξ,ϕ). Let, further, $\xi_n\in S'(\mathbb{R}^1)\setminus H_{-n}$ $(H_{-n}=W_2^{-n}(\mathbb{R}^1,\frac{dx}{(1+|x|^2)^n}))$ be a fixed sequence of generalized functions.

Take an atomic probability measure $d\mu_0(\xi) : \mu_0(\xi_n) = c_n$ $(n=1,2,\ldots)$, $\sum_{n=1}^{\infty} c_n = 1$ and assume that the family $(A_\phi)_{\phi \in S(\mathbb{R}^1)}$ on $L_2(S'(\mathbb{R}^1), d\mu_0(\xi))$ has a cyclic vector $\Omega_1(\xi)$. As $\Omega_1(\xi) \in \bigcap_{\phi \in S(\mathbb{R}^1)} \mathrm{D}(A_\phi)$, we have $\|A_\phi \Omega_1\|^2 = \int_{S'(\mathbb{R}^1)} |(\xi,\phi)\,\Omega_1(\xi)|^2 \, d\mu_0(\xi) = \sum_{n=1}^{\infty} |(\xi_n,\phi)|^2 \, |\Omega_1(\xi_n)|^2 c_n < \infty$ for all $\phi \in S(\mathbb{R}^1)$. But for any sequence of generalized functions $\xi_n \in S'(\mathbb{R}^1) \setminus H_{-n}$, there exists a trial function $\phi \in S(\mathbb{R}^1)$ such that the sequence of numbers $(|(\xi_n,\phi)|)_{n=1}^{\infty}$ increases faster than any given sequence $(a_n)_{n=1}^{\infty}$. Consequently, the last inequality holds only if $\Omega_1(\xi_n) = 0$ $(n \geq n_0)$. But this contradicts the fact that $\mathrm{CSL}\{A_{\phi_1}^{\alpha_1} \cdots A_{\phi_n}^{\alpha_n} \cdots \Omega_1 \mid \phi_1, \ldots, \phi_n, \ldots \in S(\mathbb{R}^1), \alpha \in \mathbb{N}_0^{\infty}\} = H$.

The fact that the family $(A_\phi)_{\phi \in S(\mathbb{R}^1)}$ of CSO on $L_2(S'(\mathbb{R}^1), d\mu_0(\xi))$ does not have a cyclic vector can be reformulated as follows: on $S'(\mathbb{R}^1)$, there are measures $d\mu(\xi)$ with the property that any equivalent measure does not have the moments $M_n(\phi_1, \ldots, \phi_n) = \int_{S'(\mathbb{R}^1)} (\xi,\phi_1) \cdots (\xi,\phi_n) \, d\mu(\xi)$ $(\phi_1, \ldots, \phi_n \in S(\mathbb{R}^1))$. In particular, all measures equivalent to the measure $d\mu_0(\xi)$ do not even have the first moment.

This statement is in correspondence with the results in R.L. Dobrushin and R.A. Minlos [1]. It shows that the existence of the moments for linear random functions implies the continuity of the corresponding moment forms.

1.4. Spectral theorem for a countable collection of CSO in terms of multiplication operators.

If a self-adjoint operator A_k on H has a simple spectrum with vacuum Ω (i.e. with such a vector that $\mathrm{CLS}\{E(\Delta)\Omega\}_\Delta = H (\Delta \in \mathrm{B}(\mathbb{R}^1))$, then the operator measure $E_k(\cdot)$ and the scalar probability measure $\rho_k(\cdot) = (E_k(\cdot)\Omega, \Omega)$ are equivalent. The operators with multiple spectrum do not have a vacuum vector. Nevertheless, there exists a vector Ω (a vector of maximal spectral type), such that the measures $E_k(\cdot)$ and $\rho_k(\cdot) = (E_k(\cdot)\Omega, \Omega)$ are equivalent (see A.I. Plesner [1]).

The situation for a finite or even for a countable collection of CSO is similar, namely we have the following proposition.

PROPOSITION 3. For any countable collection $(A_k)_{k=1}^{\infty}$ of CSO, there exists a vector Ω such that the projection-valued measure $E(\cdot)$ on $(\mathbb{R}^\infty, \mathrm{B}(\mathbb{R}^\infty))$ is equivalent to the scalar measure $\rho(\cdot) = (E(\cdot)\Omega, \Omega)$.

Proof. Without loss of generality, we may assume that all operators in the collection $(A_k)_{k=1}^{\infty}$ are bounded. (The change of variables $\lambda_k \mapsto \mathrm{arctg}\,\lambda_k$, $k = 1,2,\ldots$ transforms the collection of unbounded operators into a collection $(A_k)_{k=1}^{\infty}$ of bounded operators with spectrum $\sigma(A_1, \ldots, A_n) \subset \left[-\frac{\pi}{2}, \frac{\pi}{2}\right]^\infty \subset \mathbb{R}^\infty$. Thus also the property of being a vector of maximal spectral type remains.) Further, let $(e_i)_{i=1}^{\infty}$ be an orthonormal basis in H. Now, choose a sequence of vectors Ω_j and subspaces H_j as follows. Set $\Omega_1 = e_1$, and $H_1 = \text{CLS}\{E(\Delta)\Omega_1\}_\Delta$. Among the vectors $e_k \ominus H_1 (k=2,...)$, choose the first nonzero vector and denote it by Ω_2. (If all the vectors $e_k \ominus H_1, k=2,...,$ are zero then the collection $(A_k)_{k=1}^{\infty}$ has simple joint spectrum.) Set $H_2 = \text{CLS}\{E(\Delta)\Omega_2\}_\Delta$, and so on. Suppose that the number of vectors $\Omega_1, \dots, \Omega_m$ and corresponding subspaces $H_1, \dots, H_m$ $(H_k = \text{CLS}\{E(\Delta)\Omega_k\}_\Delta ; k=1, \dots, m)$ is finite. We choose $\Omega = \sum_{j=1}^{m} \Omega_j$ and show that Ω is a vector of maximal spectral type. Namely, we have that $H = \oplus \sum_{k=1}^{m} H_k$, with H_k an invariant subspace for the collection, that the vectors Ω_k are cyclic vectors for $(A_j)_{j=1}^{\infty}$ in H_k, $E(\cdot) = E_1(\cdot) + \cdots + E_m(\cdot)$ $(E_k(\cdot) = P_{H_k} E(\cdot) P_{H_k})$, and we have that the projection-valued measures $E_k(\cdot)$ are equivalent to the scalar measures $\rho_k(\cdot) = (E_k(\cdot)\Omega_k, \Omega_k) = (E(\cdot)\Omega_k, \Omega_k)$. Consequently, the operator measure $E(\cdot) = \sum_{k=1}^{m} E_k(\cdot)$. is equivalent to the scalar measure $(E(\cdot)\Omega, \Omega) = \sum_{k=1}^{m} (E(\cdot)\Omega_k, \Omega_k) = \sum_{k=1}^{m} \rho_k(\cdot)$. Now, if the number of vectors $\Omega_1, \dots, \Omega_n, \dots$ is infinite, then $\Omega = \sum_{k=1}^{\infty} 2^{-k} \Omega_k$ is a vector of maximal spectral type. □

Note that because the cyclic vectors of the collection are dense in the cyclic subspaces, the set of vectors of maximal spectral type is dense in H.

Now, using Proposition 3, we construct a vector Ω_1 in H of maximal spectral type. Denote $H_1 = \text{CLS}\{E(\Delta)\Omega_1\}_\Delta$. Then let Ω_2 be a vector of maximal spectral type in $H \ominus H_1$, and denote $H_2 = \text{CLS}\{E(\Delta)\Omega_2\}_\Delta$, and so on. The subspaces H_k are cyclic invariant subspaces of the collection of operators and $H = \oplus \sum_{k=1}^{m} H_k$ (m can be infinite). Because of the choice of Ω_{k-1} and Ω_k, all the operator measures $E_k(\cdot) = P_{H_k} E(\cdot) P_{H_k}$ are absolutely continuous with respect to the operator measures $E_{k-1}(\cdot) = P_{H_{k-1}} E(\cdot) P_{H_{k-1}}$.

If we now use the spectral theorem 4 in the form of multiplication operators for the collection $(A_j \restriction H_k)_{j=1}^{\infty}$ $(k=1, \dots, m)$ of CSO with a simple joint spectrum, we get the following theorem.

THEOREM 6. (spectral theorem in terms of multiplication operators for a countable collection of CSO). For any countable collection $(A_k)_{k=1}^{\infty}$ of CSO on H, there exists a sequence of probability measures $\rho_1(\cdot) \gg \rho_2(\cdot) \gg \cdots$ ($\gg$ means that $\rho_{k+1}(\cdot)$ is absolutely continuous with respect to the measure $\rho_k(\cdot)$ defined on $(\mathbb{R}^\infty, \mathrm{B}(\mathbb{R}^\infty))$) and there exists a unitary transformation $U : H \to \oplus \sum_{j=1}^{m} L_2(\mathbb{R}^\infty, d\rho_j(\lambda))$ such that for all $k = 1,2,...$

$$A_k = U^* \lambda_k U \tag{15}$$

(λ_k is the operator of multiplication by the k-th independent variable in

$\oplus \sum_{j=1}^{m} L_2(\mathbb{R}^\infty, d\rho_j(\lambda)))$.

Using this theorem, we can consider different collections of CSO, for which the collection of measures $(\rho_j(\cdot))_{j=1}^m$ could be chosen to be Gaussian, Markov, invariant with respect to a group G of transformations, etc. Depending on each case, such a collection of CSO is called Gaussian, Markov or G-invariant, etc.

Remark 6. The spectral theorem in terms of multiplication operators also holds for general families $(A_\alpha)_{\alpha\in\Lambda}$ of CSO.

1.5. Unitary invariants of families of CSO.

Recall that two collections $(A_k)_{k=1}^\infty$ and $(\tilde{A}_k)_{k=1}^\infty$ of CSO on H and $\tilde{H}$ are unitarily equivalent if there exists a unitary operator $U : H \to \tilde{H}$ such that for all $k = 1,2,...$

$$A_k = U^* \tilde{A}_k U.$$

A characteristic of the collection $(A_k)_{k=1}^\infty$ is called a unitary invariant if it is the same for all unitarily equivalent collections $(\tilde{A}_k)_{k=1}^\infty$.

For a collection $(A_k)_{k=1}^\infty$, its joint spectrum is a simple example of a unitary invariant because the support of the measures $E(\cdot)$ and $U^* E(\cdot) U$ coincide.

However, the spectrum is a poor invariant. A collection of operators $(\lambda_k)_{k=1}^\infty$ on $L_2(\mathbb{R}^\infty, dg(\lambda))$ with continuous spectrum and a collection of operators that has a complete system of eigenvectors and a countable everywhere dense set of sequences in $\mathbb{R}^\infty$ as its eigenvalues, are very different, although for both the spectrum is $\mathbb{R}^\infty$.

We can find more detailed invariants if we first decompose the spectral measures in some natural way and then consider the support.

For example, separate the pure point measures $\rho_{p.p}(\cdot)$ on $(\mathbb{R}^\infty, \mathbf{B}(\mathbb{R}^\infty))$ and the continuous measures $\rho_c(\cdot)$ (i.e. measures with $\rho_c(\lambda) = 0$ for all $\lambda \in \mathbb{R}^\infty$. The unique decomposition of any measure $\rho(\cdot)$ into a point and a continuous component: $\rho(\cdot) = \rho_{p.p}(\cdot) + \rho_c(\cdot)$ implies a decomposition of the spectrum as

$$\sigma(A_1, \ldots, A_n, ...) = \sigma_{p.p}(A_1, \ldots, A_n, ...) \cup \sigma_c(A_1, \ldots, A_n, ...),$$

where σ_{pp} and σ_c are the supports for the corresponding $\rho_{pp}(\cdot)$ and $\rho_c(\cdot)$. Here, both $\sigma_{p.p}(A_1, \ldots, A_n, ...)$ and $\sigma_c(A_1, \ldots, A_n, ...)$ are unitary invariants of the collection.

Note, that if the joint spectrum of a collection $(A_k)_{k=1}^n$, $n = 1,2,...$ is discrete, it does not imply that the spectrum $\sigma(A_1, \ldots, A_n, ...)$ of the entire collection $(A_k)_{k=1}^\infty$ will be discrete.

Example 8. Let $H = L_2(\{0,1\}^\infty, d\rho(\lambda))$, where $d\rho(\lambda) = \bigotimes_{k=1}^{\infty} d\rho_k(\lambda_k)$ $(\rho_k(1) = \rho_k(0) = ½)$. The joint spectrum of the collection $(\lambda_k)_{k=1}^\infty$ of operators with a discrete spectrum (projections) is continuous.

Undoubtedly, a more detailed unitary invariant of a collection is its spectral type, i.e. a set of probability measures on $(\mathbb{R}^\infty, \mathbf{B}(\mathbb{R}^\infty))$ which are equivalent to the joint resolution of the identity $E(\cdot)$ of the collection. For a family $(A_k)_{k=1}^\infty$ of operators with a simple joint spectrum, Theorem 3 implies that this is a complete set of unitary invariants. Indeed we have that the collections $(A_k)_{k=1}^\infty$ and $(\tilde{A}_k)_{k=1}^\infty$ of CSO with a simple joint spectrum are unitarily equivalent if and only if they are of the same spectral type.

Before considering the question of a complete system of unitary invariants of a collection of operators with a multiple joint spectrum, we prove the following lemmas.

LEMMA 5. A bounded operator T which commutes with the joint resolution of the identity of the CSO $(\lambda_k)_{k=1}^\infty$ on $L_2(\mathbb{R}^\infty, d\rho(\lambda))$ is an operator of multiplication by a complex-valued measurable essentially bounded function $T(\lambda)$ on $L_2(\mathbb{R}^\infty, d\rho(\lambda))$.

Proof. Let T be a bounded operator on $L_2(\mathbb{R}^\infty, d\rho(\lambda))$ which commutes with the resolution of the identity $E(\cdot)$ for the CSO of multiplication by independent variables. Consider the total set of characteristic functions $h(\Delta) = E(\Delta)\cdot 1$. Because the operator T commutes with $E(\Delta)$, it follows that

$$T\,h(\Delta) = T\,E(\Delta)\cdot 1 = E(\Delta)\,T(\lambda) = h(\Delta)\,T(\lambda)$$

where $T(\lambda) = T\cdot 1$, i.e.

$$T\,h(\Delta) = T(\lambda)\,h(\Delta).$$

As the operator T is bounded, the measurable function $T(\lambda)$ turns out to be essentially bounded: ess sup $|T(\lambda)| < \infty$. Extending T from the dense set $M = \mathrm{LS}\{h(\Delta)\}_\Delta$ to the whole space $L_2(\mathbb{R}^\infty, d\rho(\lambda))$ we find that the operator T is an operator of multiplication by a measurable essentially bounded function $T(\lambda)$. ▯

DEFINITION 8. A bounded operator $S : H \to H$ with the property that for two collections $(A_k)_{k=1}^\infty$ and $(\tilde{A}_k)_{k=1}^\infty$ of CSO on H

$$\tilde{A}_k S = S A_k,$$

is called an intertwining operator.

The following lemma is analogous to the well-known Schur lemma in representation theory.

LEMMA 6. If the spectra of the collections $(A_k)_{k=1}^\infty$ and $(\tilde{A}_k)_{k=1}^\infty$ on H do not intersect, i.e.

$$\sigma(A_1, \ldots, A_n, \ldots) \cap \sigma(\tilde{A}_1, \ldots, \tilde{A}_n, \ldots) = \varnothing$$

then any intertwining operator equals zero.

Proof. This follows from the fact that for all finite collections $(A_k)_{k=1}^n$ and $(\tilde{A}_k)_{k=1}^n$ such that $\sigma(A_1, \ldots, A_n) \cap \sigma(\tilde{A}_1, \ldots, \tilde{A}_n) = \varnothing$, S is still an intertwining operator. So $S = 0$. ▯

Now, we go back to the questions of a complete collection of unitary invariants for a family of CSO.

THEOREM 7. The sequence of spectral types of the measures $\rho_1(\cdot) \gg \rho_2(\cdot) \gg \cdots \gg \rho_k(\cdot) \gg \cdots$ on $(\mathbb{R}^\infty, \mathrm{B}(\mathbb{R}^\infty))$ constructed in Theorem 6 is a complete system of unitary invariants of the collection.

Proof. Of course, if the measures $\rho_k(\cdot)$ and $\mu_k(\cdot)$ on $(\mathbb{R}^\infty, \mathrm{B}(\mathbb{R}^\infty))$ are unitarily equivalent for all $k = 1, \ldots, m$ then the collections of the operators $(\lambda_j)_{j=1}^\infty$ on $\oplus \sum_{k=1}^{m} L_2(\mathbb{R}^\infty, d\rho_k(\lambda))$ and on $\oplus \sum_{k=1}^{m} L_2(\mathbb{R}^\infty, d\mu_k(\lambda))$ are unitarily equivalent.

Now we prove the converse. Let the collections $(A_k)_{k=1}^\infty$ on H and $(\tilde{A}_k)_{k=1}^\infty$ on $\tilde{H}$ be unitarily equivalent, i.e. $A_k = U^* \tilde{A}_k U (k=1,2,...)$. Then the decompositions of the identity are equivalent, $E(\cdot) = U^* \tilde{E}(\cdot) U$, and consequently the spectral types of the measures $E(\cdot)$ and $\tilde{E}(\cdot)$ are equal. Now, the spectral types of the operator measures $E(\cdot)\restriction H \ominus \mathrm{CLS}\{E(\Delta)\Omega_1\}_\Delta$ and $\tilde{E}(\cdot)\tilde{H} \ominus \mathrm{CLS}\{\tilde{E}(\Delta)\tilde{\Omega}_1\}_\Delta$ are the same (here Ω_1 is a vector of the maximal spectral type of the collection $(A_k)_{k=1}^\infty$ and $\tilde{\Omega}_1 = U\Omega_1$ is a vector of the maximal spectral type of the collection $(\tilde{A}_k)_{k=1}^\infty$). Now we use induction. □

The concrete criteria for the measure equivalence of product measures, Gaussian measures and other measure classes allows to decide whether the collections are unitarily equivalent or not.

We give, without proof, several known theorems concerning the absolute continuity of measures on $(\mathbb{R}^\infty, \mathrm{B}(\mathbb{R}^\infty))$. We use the following notations: $\mu(\cdot) \ll \nu(\cdot)$ means that the measure $\mu(\cdot)$ is absolutely continuous with respect to the measure $\nu(\cdot)$; $\mu(\cdot) \sim \nu(\cdot)$ means that the measures $\mu(\cdot)$ and $\nu(\cdot)$ are equivalent, and $\mu(\cdot) \perp \nu(\cdot)$ indicates that $\mu(\cdot)$ and $\nu(\cdot)$ are singular (orthogonal).

PROPOSITION 4. If the measures $\mu(\cdot)$ and $\nu(\cdot)$ are such that $\mu(\cdot) \ll \nu(\cdot)$, then for their finite dimensional projections $\mu_n(\cdot)$ and $\nu_n(\cdot)$ we have: $\mu_n(\cdot) \ll \nu_n(\cdot)$ for all $n \in \mathbb{N}$, $p_n(\lambda_1, \ldots, \lambda_n) = \dfrac{d\mu_n}{d\nu_n}(\lambda_1, \ldots, \lambda_n)$ and for almost all $\lambda \in \mathbb{R}^\infty$ with respect to the measure $\nu(\cdot)$ there exists

$$p(\lambda) = \lim_{n \to \infty} p_n(\lambda_1, \ldots, \lambda_n). \tag{16}$$

THEOREM 8. A measure $\mu(\cdot)$ is absolutely continuous with respect to a measure $\nu(\cdot)$ if and only if the function $p(\lambda)$, defined by (16) satisfies

$$\int_{\mathbb{R}^\infty} p(\lambda)\, d\nu(\lambda) = 1. \tag{17}$$

If (17) holds, then

$$p(\lambda)=\frac{d\mu}{d\nu}(\lambda) \quad (\operatorname{mod} \nu(\cdot)). \tag{18}$$

We also give criteria for equivalence of product measures and Gaussian measures.

Let $\mu(\cdot)=\bigotimes_{k=1}^{\infty}\mu_k(\cdot)$ and $\nu(\cdot)=\bigotimes_{k=1}^{\infty}\nu_k(\cdot)$ be product measures and let the measures $\mu_k(\cdot)$ and $\nu_k(\cdot)$ $(k=1,2,...)$ be equivalent.

Theorem 9. The product measures $\mu(\cdot)$ and $\nu(\cdot)$ are either equivalent or orthogonal. Here $\mu(\cdot)\sim\nu(\cdot)$ if and only if

$$\prod_{k=1}^{\infty}\int_{\mathbb{R}^1}\sqrt{\frac{d\mu_k}{d\nu_k}(\lambda_k)}\, d\nu_k(\lambda_k)>0.$$

If $\mu(\cdot)\sim\nu(\cdot)$ then

$$\frac{d\mu}{d\nu}(\lambda)=\prod_{k=1}^{\infty}\frac{d\mu_k}{d\nu_k}(\lambda). \tag{19}$$

The Gajek-Feldman theorem states that Gaussian measures on a Hilbert space are either orthogonal or equivalent. It follows from Lemma 3 that the same is also true for the space $\mathbb{R}^{\infty}$.

Consider two Gaussian measures $g_{B_1}(\cdot)$ and $g_{B_2}(\cdot)$ on $(\mathbb{R}^{\infty}, \mathbf{B}(\mathbb{R}^{\infty}))$ with zero means. Choose $l_2\,((N_k^{-1})_{k=1}^{\infty})$ in such a way that the matrices B_1 and B_2 correspond to nuclear operators on $l_2\,((N_k^{-1})_{k=1}^{\infty})$. So we have $g_{B_1}(l_2\,((N_k^{-1})_{k=1}^{\infty}))=g_{B_2}(l_2\,((N_k^{-1})_{k=1}^{\infty}))=1$. In this situation we can apply the absolute continuity theorems to the Gaussian measures on a Hilbert space. In particular, we have the following.

THEOREM 10. The measures $g_{B_1}(\cdot)$ and $g_{B_2}(\cdot)$ on $l_2\,((N_k^{-1})_{k=1}^{\infty})$ are equivalent if and only if there exists a symmetric Hilbert-Schmidt operator S such that $I+S$ is invertible and such that

$$B_2=B_1+B_1^{1/2}\,S\,B_1^{1/2}.$$

Remark 7. Theorem 7 also holds for a general collection $(A_\alpha)_{\alpha\in\Lambda}$ of CSO.

1.6. A joint domain of a family of CSO. Joint analytic, entire and bounded vectors.

If $\mathbf{D}(A_k)$ and $\mathbf{D}(A_j)$ are the corresponding domains of the operators A_k and A_j then generally speaking, the closure of $\mathbf{D}(A_j)\cap\mathbf{D}(A_k)$ is not equal to H. More then that, it even could be equal to zero.

However, if the operators commute, this cannot happen. Even for a countable collection $(A_k)_{k=1}^{\infty}$ of CSO, the closure of the intersection of their domains equals to H. We prove a more general statement.

THEOREM 11. For any countable collection $(A_k)_{k=1}^{\infty}$ of CSO, there exists a topologically dense nuclear linear topological space Φ in H such that:
1) A_k maps Φ into Φ continuously for all $k = 1,2,\dots$;
2) Φ is a core for all $A_k (k=1,2,\dots)$.

Proof. Suppose that the operators $E(\Delta)$, $\Delta \in \mathrm{B}(\mathbb{R}^{\infty})$ have a common cyclic vector Ω (the collection of operators has a simple joint spectrum). From Theorem 4 it follows that in this case the space H is isomorphic to the space $L_2(\mathbb{R}^{\infty}, d\rho(\lambda))$. Under this isomorphism the operators A_k correspond to the operators of multiplication by $\lambda_k (k=1,2,\dots)$. Then we can set

$$\Phi = \lim_{\substack{\rightarrow \\ m,N\to\infty}} \Phi_{m,N}$$

where $\lim\limits_{\rightarrow}$ is the inductive limit and $\Phi_{m,N} = \{\phi \in H \mid \phi = \sum\limits_{|\alpha| \le m} a_{\alpha} A^{\alpha} \Omega_1\}$ (Ω_1 is a cyclic vector for the operators $(A_k)_{k=1}^{\infty}$; $|\alpha| = \alpha_1 + \cdots + \alpha_N$; $m = 0,1,2,\dots$, $N = 1,2,\dots$; a_{α}-arbitrary numbers). For the operators $(\lambda_k)_{k=1}^{\infty}$ on $L_2(\mathbb{R}^{\infty}, d\rho(\lambda))$, the space Φ is a set of cylindrical polynomials multiplied by the fixed function $\Omega_1(\lambda)$, with the topology brought about by the inductive limit of the corresponding finite dimensional spaces. In particular, if the set of cylindrical polynomials is dense in $L_2(\mathbb{R}^{\infty}, d\rho(\lambda))$, then we can choose it to be Φ, endowed with the inductive limit topology. The theorem now follows from the use of the foregoing construction and Theorem 6. □

DEFINITION 9. A nuclear topological space, topologically dense in H, such that all the operators $A_k(k=1,2,\dots)$ continuously map Φ into Φ is called a rigging.

We can now state Theorem 11 in a different way.

THEOREM 12. A countable collection of commuting self-adjoint operators always admits a rigging.

Example 9. The construction of a rigging is particularly simple for the operators λ_k on $L_2(\{0,1\}^{\infty}, d\rho(\lambda))$ (see Example 5). In this case, the subspaces of cylindrical functions are $L_2(X_n, d\rho_n(\lambda_1, \dots, \lambda_n))$ ($X_n = \{0,1\}^n$, $\rho_n(\cdot)$ is a projection of the measure $\rho(\cdot)$ on X_n). They are finite dimensional and $\Phi = \lim\limits_{\rightarrow} L_2(X_n, d\rho_n)$ is a nuclear rigging of the space $L_2(\{0,1\}^{\infty}, d\rho(\lambda))$.

In the particular case, that $\rho(\cdot) = \bigotimes\limits_{k=1}^{\infty} \rho_k(\cdot)$ $(\rho_k(1)=\rho_k(0)=\frac{1}{2})$ is the Haar measure of the group $\{0,1\}^{\infty} = \mathbb{Z}_2^{\infty}$, $L_2(\{0,1\}^{\infty}, d\rho(\lambda))$ is isomorphic to $L_2([0,1], dx)$. Under this isomorphism the operators λ_k on $L_2(\mathbb{Z}_2^{\infty}, d\rho(\lambda))$ correspond to the operators N_k defined by $(N_k f)(x) = (N_k f)(0, i_1 i_2 \cdots) = i_k f(x)$ (here $0, i_1 i_2 \cdots$ is a binary decomposition of $x \in [0,1]$). The corresponding rigging Φ is the rigging with Walsh finite series. The dual space Φ' is a set of formal infinite Walsh series. Here the joint spectrum of the system of operators λ_k, each with discrete spectrum, is continuous. The corresponding common generalized eigenfunctions are the δ-functions in the points that are binary irrational. $\square$

It follows from the construction of Φ in the proof of the theorem that all the vectors ϕ from Φ belong to the domain of all operators $A_1^{\alpha_1} \cdots A_n^{\alpha_n} \cdots$, where $(\alpha_1, \alpha_2, \ldots, \alpha_n, \ldots)$ is a finite multiindex.

DEFINITION 10. If for all finite multiindices $\alpha = (\alpha_1, \ldots, \alpha_n, \ldots) \in \mathbb{N}_0^\infty$, a vector $\phi \in H$ is such that $\phi \in \mathbf{D}(A_1^{\alpha_1} \cdots A_n^{\alpha_n} \cdots)$, then ϕ is called a joint infinitely differentiable vector of the collection $(A_k)_{k=1}^\infty$, and we write $\phi \in H^\infty(A_1, \ldots, A_n, \ldots)$.

DEFINITION 11. A vector $\phi \in H$ is called a joint analytic vector of the collection $(A_k)_{k=1}^\infty$ if for all $n = 1, 2, \ldots$ there exists $s_n > 0$ such that

$$\sum_{k=0}^{\infty} \frac{1}{k!} \sum_{|\alpha|=k} \|A_1^{\alpha_1} \cdots A_n^{\alpha_n} \phi\| s_n^k < \infty$$

$$(|\alpha| = \alpha_1 + \cdots + \alpha_n).$$

We denote the set of joint analytic vectors of the collection by $H^\omega(A_1, \ldots, A_n, \ldots)$.

DEFINITION 12. A vector $\phi \in H$ is called a joint entire vector of the collection $(A_k)_{k=1}^\infty$ if for all $n = 1, 2, \ldots$ and $s > 0$

$$\sum_{k=0}^{\infty} \frac{1}{k!} \sum_{|\alpha|=k} \|A_1^{\alpha_1} \cdots A_n^{\alpha_n} \phi\| s^k < \infty.$$

The set of joint entire vectors will be denoted by $H^C(A_1, \ldots, A_n, \ldots)$.

DEFINITION 13. A vector $\phi \in H$ is called a joint bounded vector of the collection $(A_k)_{k=1}^\infty$ if for all $n = 1, 2, \ldots$ there exists $s_n > 0$ such that

$$\|A_1^{\alpha_1} \cdots A_n^{\alpha_n} \phi\| \leq s_n^{|\alpha|}.$$

The set of these vectors is denoted by $H^B(A_1, \ldots, A_n, \ldots)$. We have the following inclusions:

$$H^\infty(A_1, \ldots, A_n, \ldots) \supseteq H^\omega(A_1, \ldots, A_n, \ldots) \supseteq$$

$$\supseteq H^C(A_1, \ldots, A_n, \ldots) \supseteq H^B(A_1, \ldots, A_n, \ldots). \tag{20}$$

This follows immediately from

$$H^\infty(A_1, \ldots, A_n) \supseteq H^\omega(A_1, \ldots, A_n) \supseteq$$

$$\supseteq H^C(A_1, \ldots, A_n) \supseteq H^B(A_1, \ldots, A_n) \tag{21}$$

for a finite family $(A_k)_{k=1}^n$ of CSO, where each of the sets are respectively defined as:

$$\phi \in \mathbf{D}(A_1^{\alpha_1} \cdots A_n^{\alpha_n}) \quad \text{for all} \quad (\alpha_1, \ldots, \alpha_n) \in \mathbb{N}^n;$$

$$\exists s > 0 : \sum_{k=0}^{\infty} \frac{1}{k!} \sum_{|\alpha|=k} \|A_1^{\alpha_1} \cdots A_n^{\alpha_n} \phi\| s^k < \infty ;$$

$$\forall s > 0 \sum_{k=0}^{\infty} \frac{1}{k!} \sum_{|\alpha|=k} \|A_1^{\alpha_1} \cdots A_n^{\alpha_n} \phi\| s^k < \infty ;$$

$$\exists s > 0 : \|A_1^{\alpha_1} \cdots A_n^{\alpha_n} \phi\| \le s^{|\alpha|} .$$

THEOREM 13. For any countable collection $(A_k)_{k=1}^{\infty}$ of CSO

$$\bigcap_{k=1}^{\infty} H^{\#}(A_k) = H^{\#}(A_1, \ldots, A_n, \ldots) \tag{22}$$

(here # stands for ∞, ω, C or B).

Proof. Show, for example, that

$$\bigcap_{k=1}^{\infty} H^{\omega}(A_k) = H^{\omega}(A_1, \ldots, A_n, \ldots). \tag{23}$$

Because of Definition 11, it is sufficient to show that for all $n = 2,3,\ldots$

$$\bigcap_{k=1}^{n} H^{\omega}(A_k) = H^{\omega}(A_1, \ldots, A_n). \tag{23'}$$

The theorem can be proved using results of R. Goodman [1]. Here we give a direct proof. Without loss of generality, assume that $n = 2$ and that the operators A_1 and A_2 have simple joint spectrum, i.e. there exists a vector $\Omega \in H$ such that $\mathrm{CLS}\{E_1(\Delta_1) E_2(\Delta_2)\Omega\} = H$, where $E_1(\cdot)$ and $E_2(\cdot)$ are decompositions of the identity for the operators A_1 and A_2, $\Delta_1, \Delta_2 \in \mathbf{B}(\mathbb{R}^1)$ and CLS means closed linear span. Represent the operators A_1, A_2 as operators of multiplication by the independent variables λ_1, λ_2 in $L_2(\mathbb{R}^2, d\mu(\lambda_1, \lambda_2))$, where $\mu(\cdot)$ is a probability measure on $\mathbb{R}^2$. The proof is based on the following lemma.

LEMMA 7. Let x be an operator of multiplication by the independent variables in $L_2(\mathbb{R}^1, d\nu(x))$, where $\nu(\cdot)$ is a probability measure on $(\mathbb{R}^1, \mathbf{B}(\mathbb{R}^1))$. If $\exp(\varepsilon|x|)$ is the multiplication operator by the function $\mathbb{R}^1 \ni x \mapsto \exp(\varepsilon|x|) \in \mathbb{R}^1$ in $L_2(\mathbb{R}^1, d\nu(x))$, $\varepsilon > 0$, and if $\mathbf{D}(\exp(\varepsilon|x|))$ is the domain of the operator $\exp(\varepsilon|x|)$, then $H^{\omega}(x) = \bigcup_{\varepsilon>0} \mathbf{D}(\exp(\varepsilon|x|))$.

Proof. Let $\phi(x) \in H^{\omega}(x)$. Then there exists an $s > 0$ such that

$$\sum_{n=0}^{\infty} \frac{1}{n!} \|x^n \phi(x)\| s^n < \infty. \tag{24}$$

For $0 < \varepsilon \le s$, the series $\sum_{n=0}^{\infty} \frac{(\varepsilon|x|)^n}{n!} \phi(x)$ converges absolutely in $L_2(\mathbb{R}^1, d\nu(x))$ to the function $\exp(\varepsilon|x|)\phi(x)$. So, $\exp(\varepsilon|x|)\phi(x) \in L_2(\mathbb{R}^1, d\nu(x))$ for $0 < \varepsilon \le s$, i.e. $\phi(x) \in \mathbf{D}(\exp(\varepsilon|x|))$ $(0 < \varepsilon \le s)$.

Now, let $\phi \in \bigcup_{\varepsilon>0} \mathrm{D}(\exp(\varepsilon|x|))$. Then there exist an $\varepsilon>0$ such that $\exp(\varepsilon|x|)\phi(x) \in L_2(\mathbb{R}^1, d\nu(x))$. We make an estimate for the series in (24). To do this, we determine the maximum of the function $h_{n,\varepsilon}(x) = x^n \exp(-\varepsilon x)$ for $n = 0,1,\ldots,\ \varepsilon>0,\ x \geq 0$. Clearly, $\max_x (h_{n,\varepsilon}(x)) = h_{n,\varepsilon}\left(\frac{n}{\varepsilon}\right) = \left(\frac{n}{e\varepsilon}\right)^n$. Thus, we have

$$|x|^n \exp(-\varepsilon|x|) \leq \left(\frac{n}{e\varepsilon}\right)^n$$

or

$$|x|^n \leq \left(\frac{n}{e\varepsilon}\right)^n \exp(\varepsilon|x|). \tag{25}$$

Using (25) for the series (24) we get the estimate

$$\sum_{n=0}^{\infty} \frac{1}{n!} \|x^n \phi\| s^n \leq \sum_{n=0}^{\infty} \frac{1}{n!} \left(\frac{n}{e\varepsilon}\right)^n \|\exp(\varepsilon|x|)\phi\| s^n =$$

$$= \|\exp(\varepsilon|x|)\phi\| \sum_{n=0}^{\infty} \frac{1}{n!} \left(\frac{n}{e\varepsilon}\right)^n s^n. \tag{26}$$

The series (26) converges for $0 < s < \lim_{n\to\infty} \frac{(n+1)!\,(e\varepsilon)^{n+1}\, n^n}{n!\,(e\varepsilon)^n\,(n+1)^{n+1}} =$

$$= \lim_{n\to\infty} \frac{e\varepsilon\, n^n}{(n+1)^n} = e\,\varepsilon \lim_{n\to\infty} \left(\frac{n}{n+1}\right)^n = \varepsilon.$$

So the series in (24) converges for $0 < s \leq \varepsilon$, i.e. $\phi \in H^\omega(x)$. □

LEMMA 8. Let λ_1, λ_2 be operators of multiplication on independent variables in $L_2(\mathbb{R}^2, d\mu(\lambda_1,\lambda_2))$, where $\mu(\cdot)$ is a probability measure on $(\mathbb{R}^2, \mathrm{B}(\mathbb{R}^2))$. If $\exp(\varepsilon(|\lambda_1|+|\lambda_2|))$ is the operator of multiplication by the function $\mathbb{R}^2 \ni (\lambda_1,\lambda_2) \mapsto \exp(\varepsilon(|\lambda_1|+|\lambda_2|)) \in \mathbb{R}^1$ $(\varepsilon>0)$ and $\mathrm{D}(\exp(\varepsilon(|\lambda_1|+|\lambda_2|)))$ is the domain of the operator $\exp(\varepsilon(|\lambda_1|+|\lambda_2|))$, then

$$H^\omega(\lambda_1,\lambda_2) = \bigcup_{\varepsilon>0} \mathrm{D}(e^{\varepsilon(|\lambda_1|+|\lambda_2|)}). \tag{27}$$

Proof. It is not difficult to check, that to define a joint analytic vector of a collection $(A_k)_{k=1}^\infty$ of commuting operators, we could require that either of the series

$$\sum_{n=0}^{\infty} \frac{1}{n!} \max_{|\alpha|=n} \|A_1^{\alpha_1} \cdots A_m^{\alpha_m} \phi\| s^n < \infty,$$

$$\sum_{n=0}^{\infty} \frac{1}{n!} \sum_{|\alpha|=n} \|A_1^{\alpha_1} \cdots A_m^{\alpha_m} \phi\| s^n < \infty \tag{28}$$

be convergent. Let $\phi \in H^\omega(\lambda_1,\lambda_2)$. Then there exists an $s>0$ such that

$$\sum_{n=0}^{\infty} \frac{1}{n!} \sum_{1 \le i_1, \ldots, i_n \le 2} \|\lambda_{i_1} \cdots \lambda_{i_n} \phi\| s^n \le \infty.$$

For $0 < \varepsilon \le s$, the series $\sum_{n=0}^{\infty} \frac{1}{n!} [\varepsilon(|\lambda_1| + |\lambda_2|)]^n \phi(\lambda_1, \lambda_2)$ converges absolutely in $L_2(\mathbb{R}^2, d\mu(\lambda_1, \lambda_2))$ to the function $\exp(\varepsilon(|\lambda_1|+|\lambda_2|))\, \phi(\lambda_1, \lambda_2)$. So, for $0 < \varepsilon \le s$, $\exp(\varepsilon(|\lambda_1|+|\lambda_2|))\, \phi(\lambda_1, \lambda_2) \in L_2(\mathbb{R}^2, d\mu(\lambda_1, \lambda_2))$.

Now, let $\phi \in \bigcup_{\varepsilon>0} D(\exp(\varepsilon(|\lambda_1|+|\lambda_2|)))$. Then, there exists an $\varepsilon > 0$ such that $\exp(\varepsilon(|\lambda_1|+|\lambda_2|))\, \phi(\lambda_1, \lambda_2) \in L_2(\mathbb{R}^2, d\mu(\lambda_1, \lambda_2))$. Estimate the series in (28). To do that, we use the inequality, analogous to (25):

$$|\lambda_1|^{\alpha_1} |\lambda_2|^{\alpha_2} \le \left[\frac{\alpha_1}{e\varepsilon}\right]^{\alpha_1} \left[\frac{\alpha_2}{e\varepsilon}\right]^{\alpha_2} \exp(\varepsilon(|\lambda_1|+|\lambda_2|)) \tag{29}$$

$(\alpha_1, \alpha_2 = 0, 1, \ldots; \varepsilon > 0)$.

Taking into account (29), for the series (28) we get the estimate:

$$\sum_{n=0}^{\infty} \frac{1}{n!} \sum_{|\alpha|=n} \|\lambda_1^{\alpha_1} \lambda_2^{\alpha_2} \phi\| s^n \le \sum_{n=0}^{\infty} \sum_{|\alpha|=n} \frac{1}{\alpha_1!\, \alpha_2!} \|\lambda_1^{\alpha_1} \lambda_2^{\alpha_2} \phi\| s^n \le$$

$$\le \sum_{n=0}^{\infty} \sum_{|\alpha|=n} \frac{1}{\alpha_1!\, \alpha_2!} \left[\frac{\alpha_1}{e\varepsilon}\right]^{\alpha_1} \left[\frac{\alpha_2}{e\varepsilon}\right]^{\alpha_2} \|\exp(\varepsilon(|\lambda_1|+|\lambda_2|))\phi\| s^{\alpha_1+\alpha_2} =$$

$$= \|\exp((|\lambda_1|+|\lambda_2|))\phi\| \prod_{k=1}^{2} \sum_{\alpha_k=0}^{\infty} \frac{1}{\alpha_k} \left[\frac{\alpha_k}{e\varepsilon}\right]^{\alpha_k} s^{\alpha_k}. \tag{30}$$

Similarly to (26) the series in (30) converge for $0 < s < \varepsilon$, i.e. $\phi \in H^{\omega}(\lambda_1, \lambda_2)$. □

Using Lemmas 7 and 8, rewrite (23') in the form

$$\bigcup_{\varepsilon>0} D(\exp(\varepsilon(|\lambda_1|+|\lambda_2|))) = (\bigcup_{\varepsilon_1>0} D(\exp(\varepsilon_1 |\lambda_1|))) \cap (\bigcup_{\varepsilon_2>0} D(\exp(\varepsilon_2 |\lambda_2|))).$$

It is enough to show that

$$\bigcup_{\varepsilon>0} D(\exp(\varepsilon(|\lambda_1|+|\lambda_2|))) \supseteq (\bigcup_{\varepsilon_1>0} D(\exp(\varepsilon_1 |\lambda_1|))) \cap (\bigcup_{\varepsilon_2>0} D(\exp(\varepsilon_2 |\lambda_2|))).$$

Let $\phi \in L_2(\mathbb{R}^2, d\mu(\lambda_1, \lambda_2))$ be such that

$$\phi \in (\bigcup_{\varepsilon_1>0} D(\exp(\varepsilon_1 |\lambda_1|))) \cap (\bigcup_{\varepsilon_2>0} D(\exp(\varepsilon_2 |\lambda_2|))).$$

Then there exists $\varepsilon_1 > 0$ and $\varepsilon_2 > 0$ such that

$$\int_{\mathbb{R}^2} |\exp(\varepsilon_1 |\lambda_1|)\, \phi(\lambda_1, \lambda_2)|^2 \, d\mu(\lambda_1, \lambda_2) < \infty$$

$$\int_{\mathbb{R}^2} |\exp(\varepsilon_2 |\lambda_2|)\, \phi(\lambda_1, \lambda_2)|^2 \, d\mu(\lambda_1, \lambda_2) < \infty.$$

Because $\exp(2\varepsilon_1 |\lambda_1|) + \exp(2\varepsilon_2 |\lambda_2|) \geq 2\exp(\varepsilon_1 |\lambda_1| + |\varepsilon_2 |\lambda_2|)$ we have $\int_{\mathbb{R}^2} |\exp(\varepsilon(|\lambda_1| + |\lambda_2|))\, \phi(\lambda_1, \lambda_2)|^2 \, d\mu(\lambda_1, \lambda_2) < \infty$ where $\varepsilon = \frac{1}{2} \min(\varepsilon_1, \varepsilon_2)$. So (23) has been proved. By direct verification one can prove the remaining equalities in (22). □

Now we prove a statement more general then the statement that $\bigcap_{k=1}^{\infty} \mathbf{D}(A_k)$ is dense.

THEOREM 14. We have

$$\overline{H^B(A_1, \ldots, A_n, \ldots)} = \overline{\bigcap_{k=1}^{\infty} H^B(A_k)} = H. \tag{31}$$

Proof. Let us find a dense set in H consisting of vectors such that for all $n = 1,2,\ldots$

$$\|A_1^{\alpha_1} \cdots A_n^{\alpha_n} \phi\| \leq s_{n,\phi}^{|\alpha|} \quad (\alpha_1, \ldots, \alpha_n = 0,1,\ldots;\ |\alpha| = \alpha_1 + \cdots + \alpha_n).$$

For that we use Lemma 3 and choose a real Hilbert space $l_2((N_k^{-1})_{k=1}^{\infty}) = \{\lambda \in \mathbb{R}^{\infty}\ \|\lambda\|^2_{l_2((N_k^{-1})_{k=1}^{\infty})} = \sum_{k=1}^{\infty} \frac{\lambda_k^2}{N_k} < \infty\}$ in $\mathbb{R}^{\infty}$ with full spectral measure. Because of the σ-additivity of the spectral measure on $l_2((N_k^{-1})_{k=1}^{\infty})$ (here $S_R(l_2(N_k^{-1})_{k=1}^{\infty})$ is a sphere of radius R in $l_2((N_k^{-1})_{k=1}^{\infty})$), the vectors from $\bigcup_{R=1}^{\infty} E(S_R(l_2((N_k^{-1})_{k=1}^{\infty}))H$ are dense in H. If $\phi \in E(S_R(l_2((N_k^{-1})_{k=1}^{\infty})))H$, then for all $n = 1,2,\ldots$ we have $\|A_1^{\alpha_1} \cdots A_n^{\alpha_n} \phi\| \leq (R\, C_n)^{|\alpha|}\, \|\phi\|$. □

Remark. For general families $(A_\alpha)_{\alpha \in \Lambda}$ of CSO, $\overline{\bigcap_{\alpha \in \Lambda} \mathbf{D}(A_\alpha)}$ need not coincide with the whole H. We give example of a family of CSO such that $\bigcap_{\alpha \in \Lambda} \mathbf{D}(A_\alpha) = \{0\}$ (see A.V. Kosyak and Yu.S. Samoĭlenko [2]).

Example 10. Let $L_0([0,1], dx)$ be the set of all measurable almost everywhere finite functions $\alpha(\cdot) : [0,1] \to \mathbb{R}^1$, let A_α be the operator of multiplication by the function $\alpha(\cdot) \in L_0([0,1], dx)$ in $H = L_2([0,1], dx)$. Then $\mathbf{D}(A_\alpha) = \{f(\cdot) \in H \mid \int_0^1 |\alpha(x) f(x)|^2 \, dx < \infty\}$.

PROPOSITION 5.

$$\bigcap_{\alpha \in L_0([0,1],dx)} \mathbf{D}(A_\alpha) = \{0\}.$$

Indeed, if $\int_0^1 |f_0(x)|^2 \, dx = \varepsilon > 0$, then choosing $0 = x_0 < x_1 < \cdots < x_k < \cdots < 1$ such that $\int_{x_{k-1}}^{x_k} |f_0(x)|^2 \, dx = \frac{\varepsilon}{2^k}$ $(k = 1,2,\ldots)$ and letting $\alpha_0(x) = 2^k (x_{k-1} \leq x < x_k)$ we get

$$\int_0^1 |\alpha_0(x) f_0(x)|^2 \, dx = \varepsilon \sum_{k=1}^{\infty} 2^k = \infty.$$

Because $\alpha_0(\cdot) \in L_0([0,1], dx)$, we have

$$f_0(\cdot) \notin \mathbf{D}(A_{\alpha_0}) \supset \bigcap_{\alpha \in L_0([0,1],dx)} \mathbf{D}(A_\alpha).$$

Note that the intersection of the domains of the family of CSO with empty joint spectrum considered in Yu.M. Berezanskiĭ [9], also contains only the zero vector.

1.7. Functions of a countable family of CSO. Essentially infinite-dimensional functions.

Let A_j be a self-adjoint operator on H, $E_j(\cdot)$ be its resolution of the identity, and $\rho_j(\cdot)$ its spectral measure. Denote by $L_0(\mathbb{R}^1, d\rho_j(\lambda_j))$ the totality of all the Borel functions $\mathbb{R}^1 \ni \lambda_j \mapsto \phi(\lambda_j) \in \mathbb{C}^1$ which are finite almost everywhere with respect to $\rho_j(\cdot)$. For every $\phi(\cdot) \in L_0(\mathbb{R}^1, d\rho_j(\lambda_j))$ we define the function $\phi(A_j)$ of the operator A_j on H by

$$\phi(A_j) = \int_{\mathbb{R}^1} \phi(\lambda_j) \, dE_j(\lambda_j) \tag{32}$$

$$\mathbf{D}(\phi(A_j)) = \{f \in H \mid \int_{\mathbb{R}^1} |\phi(\lambda_j)|^2 \, d(E_j(\lambda_j) f, f) < \infty\}.$$

The function $\phi(\cdot)$ completely determines whether the operator $\phi(A_j)$ belongs to a particular class of operators, and also determines some other characteristics of $\phi(A_j)$. For instance, the operator $\phi(A_j)$ is self-adjoint if and only if $\phi(\lambda_j) \in \mathbb{R}^1 (\mathrm{mod}\, \rho_j(\cdot))$. In general, $\phi(A_j)$ will be a normal operator on H. Now, the boundedness of $\phi(A_j)$ is equivalent to the property that $\phi(\cdot) \in L_\infty(\mathbb{R}^1, d\rho_j(\lambda_j))$ and in this case

$$\|\phi(A_j)\| = \operatorname{ess\,sup}_{\lambda_j \in \mathbb{R}^1} |\phi(\lambda_j)| = \|\phi(\cdot)\|_{L_\infty(\mathbb{R}^1, d\rho_j(\lambda_j))}.$$

For a bounded operator A_j, formula (32) defines a (unique) $*$-homomorphism of the C^*-algebras: $C(\sigma(A_j)) \ni \phi(\cdot) \mapsto \phi(A_j) \in L(H)$ such that if $q(\lambda_j) = \lambda_j$, then $q(A_j) = A_j$.

Consider a family $(A_k)_{k=1}^{\infty}$ of CSO on H. Let $E(\cdot)$ be the resolution of the identity and $\rho(\cdot)$ - the spectral measure of the family.

DEFINITION 14. For a function $\phi \in L_0(\mathbb{R}^\infty, d\rho(\lambda))$ (i.e. ϕ is $\mathbf{B}(\mathbb{R}^\infty)$ measurable and finite almost everywhere with respect to $\rho(\cdot)$) we define the function $\phi(A_1, \ldots, A_n, \ldots)$ by setting

$$\phi(A_1, \ldots, A_n, \ldots) = \int_{\mathbb{R}^\infty} \phi(\lambda_1, \ldots, \lambda_n, \ldots) \, dE(\lambda_1, \ldots, \lambda_n, \ldots), \tag{33}$$

$$\mathbf{D}(\phi(A_1,\ldots,A_n,\ldots)) = \{f \in H \mid \int_{\mathbb{R}^\infty} |\phi(\lambda_1,\ldots,\lambda_n,\ldots)|^2 \, d(E(\lambda_1,\ldots,\lambda_n,\ldots)f,f) < \infty\}.$$

As in the case of a single operator, the properties of the function $\phi(\lambda_1,\ldots,\lambda_n,\ldots)$ determine whether the operator $\phi(A_1,\ldots,A_n,\ldots)$ belongs to a particular class of operators on H. Indeed, the operator $\phi(A_1,\ldots,A_n,\ldots)$ is normal for an arbitrary function $\phi(\cdot)$, self-adjoint if and only if $\phi(\lambda_1,\ldots,\lambda_n,\ldots) \in \mathbb{R}^1$ (mod $\rho(\cdot)$), bounded if and only if $\phi(\cdot) \in L_\infty(\mathbb{R}^\infty, d\rho(\lambda))$.

If the operators of the collection $A = (A_j)_{j=1}^\infty$ are bounded and $\|A_j\| \le C_j$ ($j = 1,2,\ldots$), then we can consider a C^*-algebra $\mathbf{U}$ which is the closure in the C^*-norm of $\lim\limits_{\rightarrow} C(K_1 \times \cdots \times K_n)$ (with the inclusions

$$j_{n+1}^n : C(K_1 \times \cdots \times K_n) \ni f(\lambda_1,\ldots,\lambda_n) \mapsto f(\lambda_1,\ldots,\lambda_n) \in C(K_1 \times \ldots \times K_{n+1})).$$

Consider the set $K_1 \times \cdots \times K_n \times \cdots$, which is compact in the product topology.

PROPOSITION 6. $\mathbf{U}$ is $*$-isomorphic to the C^*-algebra of continuous functions on the compact set $K_1 \times \cdots \times K_n \times \cdots$.

Proof. Elements of $\lim\limits_{\rightarrow} C(K_1 \times \cdots \times K_n \times \cdots) \subset \mathbb{R}^\infty$. By the Stone-Weierstrass theorem, $\lim\limits_{\rightarrow} C(K_1 \times \cdots \times K_n)$ is dense in $C(K_1 \times \cdots \times K_n \times \cdots)$, and consequently the elements of $\mathbf{U}$ can be identified with the continuous functions $\phi(x_1,\ldots,x_n,\ldots)$ on the compact set $K_1 \times \cdots \times K_n \times \cdots$. They "weakly depend on the far-away variables", or "they are almost cylindrical" in the sense that there exists on $K_1 \times \cdots \times K_n \times \cdots$ a sequence of continuous cylindrical functions $\phi_n(x_1,\ldots,x_n)$ uniformly convergent on $K_1 \times \cdots \times K_n \times \cdots$ to $\phi(\lambda_1,\ldots,\lambda_n,\ldots)$.

PROPOSITION 7. There exists a unique $*$-homomorphism of the C^*-algebras: $\mathbf{U} \ni \phi(\lambda_1,\ldots,\lambda_n,\ldots) \mapsto \phi(A_1,\ldots,A_n,\ldots) \in L(H)$ such that for $q_j(\lambda_1,\ldots,\lambda_n,\ldots) = \lambda_j$ we have $q_j(A_1,\ldots,A_n,\ldots) = A_j$. This homomorphism is defined by formula (33).

Proof. For cylindrical functions, the uniqueness of the $*$-homomorphism defined by (33) follows from the theorem on the functional calculus for finite collections of CSO. Because $\|\phi(\lambda_1,\ldots,\lambda_n)\|_U = \|\phi(A_1,\ldots,A)\|_{L(H)}$, this $*$-homomorphism can be extended to the whole C^*-algebra $\mathbf{U}$. The limiting process under the integral sign in (33) shows that this extension is also defined by (33). □

So, one can easily construct a functional calculus for both "almost cylindrical" functions and collections of unbounded CSO. Formula (33) extends this functional calculus to all $\phi(\cdot) \in L_0(\mathbb{R}^\infty, d\rho(\lambda))$.

However, among the functions $\phi(\cdot) \in L_0(\mathbb{R}^\infty, d\rho(\lambda))$ of a family $\mathbf{A} = (A_k)_{k=1}^\infty$ of CSO there could exist functions for which there are no finite-dimensional approximations. To be more precise, let $\mathbf{B}^n(\mathbb{R}^\infty)$ be the σ-algebra generated by the cylindrical sets of the form $\mathbb{R}^n \times \Delta \times \mathbb{R}^\infty$ ($\Delta \in \mathbf{B}(\mathbb{R} \mapsto p)$, $p = 1,2,\ldots$), let $\mathbf{B}^n_{\mu(\cdot)}(\mathbb{R}^\infty)$ be the completion of the σ-algebra $\mathbf{B}^n(\mathbb{R}^\infty)$ with respect to the measure $\mu(\cdot)$ on $(\mathbb{R}^\infty, \mathbf{B}(\mathbb{R}^\infty))$ and

$\mathbf{B}^{\infty}_{\mu(\cdot)} = \overline{\bigcap_{n=1}^{\infty} \mathbf{B}^{n}_{\mu(\cdot)}(\mathbb{R}^{\infty}) \cap \mathbf{B}(\mathbb{R}^{\infty})}.$

DEFINITION 15. A function $\phi(\lambda_1, \ldots, \lambda_n, \ldots) \in L_0(\mathbb{R}^{\infty}, d\mu(\lambda))$ is called essentially infinite-dimensional if it is $\mathbf{B}^{\infty}_{\mu(\cdot)}$-measurable.

Values of an essentially infinite-dimensional function do not depend on the first n coordinates $\lambda_1, \ldots, \lambda_n$ $(n = 1, 2, \ldots)$.

The existence of $\mathbf{B}^{\infty}_{\mu(\cdot)}$-measurable functions different from a constant depends on the measure $\mu(\cdot)$ on $(\mathbb{R}^{\infty}, \mathbf{B}(\mathbb{R}^{\infty}))$. Thus, on the σ-algebra $\mathbf{B}^{\infty}_{\omega(\cdot)}$ the product-measures $d\omega(\lambda) = \bigotimes_{k=1}^{\infty} d\omega_k(\lambda_k)$ assume only the values 0 and 1, and so there are no $\mathbf{B}^{\infty}_{\omega(\cdot)}$-measurable functions on $L_0(\mathbb{R}^{\infty}, \bigotimes_{k=1}^{\infty} d\omega_k(\lambda_k))$ different from a constant. In the book of A.V. Skorokhod [3] one gives a criterion that the measure $\mu(\cdot)$ would assume only the values 0 and 1 on $\mathbf{B}^{\infty}_{\mu(\cdot)}$ (i.e. would be tail-trivial), and so there would not be any essentially infinite-dimensional functions different from a constant in $\mathbf{B}^{\infty}_{\mu(\cdot)}$. The existence of nontrivial essentially infinite-dimensional functions of a collection $A = (A_j)_{j=1}^{\infty}$ of CSO depends on the spectral measure of this collection.

Example 11. Essentially infinite-dimensional functions.

Suppose that $\mu(\{\lambda \in \mathbb{R}^{\infty} \mid \overline{\lim_{n\to\infty}} |\lambda_n| \text{ exists and is finite}\}) = 1$, then

$$\phi(\lambda_1, \ldots, \lambda_n, \ldots) = \overline{\lim_{n\to\infty}} |\lambda_n| \in L_0(\mathbb{R}^{\infty}, d\mu(\lambda)) \tag{34}$$

is $\mathbf{B}^{\infty}_{\mu(\cdot)}$-measurable. However, for a rather large class of measures $\omega_A(\cdot) = \omega(A\cdot)$ this function is constant almost everywhere (the matrix $A = (a_{ij})_{i,j=1}^{\infty}$ has on each row and each column only a finite number of elements different from zero, $d\omega(\lambda) = \bigotimes_{k=1}^{\infty} d\omega_k(\lambda_k)$ is a product measure with the zero first moments and finite second moments) (see V.V. Buldigin [1, §4.2]). So, for a collection $(A_k)_{k=1}^{\infty}$ of CSO such that its spectral measure is equivalent to $\omega_A(\cdot)$, we have that $\overline{\lim_{n\to\infty}} |A_n| = cI$.

Also the functions

$$\lim_{n\to\infty} \frac{\lambda_1 + \cdots + \lambda_n}{B_n} \quad (B_n > 0,\ B_n \underset{n\to\infty}{\to} \infty) \tag{35}$$

are essentially infinite-dimensional. For a collection $(A_k)_{k=1}^{\infty}$ of CSO such that $\lim_{n\to\infty} \frac{\lambda_1 + \cdots + \lambda_n}{B_n}$ exists and finite for almost all $\lambda \in \mathbb{R}^{\infty}$, using the spectral measure $\mu(\cdot)$ of this collection, one can define a self-adjoint operator $\lim_{n\to\infty} \frac{A_1 + \cdots + A_n}{B_n}$. Whether it is trivial or not essentially depends on the values of the measure $\mu(\cdot)$ on $\mathbf{B}^{\infty}_{\mu(\cdot)}$ and on the values of the function $\lim_{n\to\infty} \frac{\lambda_1 + \cdots + \lambda_n}{B_n}$ on the set of full measure.

□

Remark 9. Calculation of distributions of functionals of random processes (generalized random processes) is, in essence, the calculation of the spectral measure of functions of a family of CSO of multiplication by a continuum of independent variables.

1.8. Countable families of CSO connected through linear relations.

So far we considered spectral problems of countable families not connected by any algebraic relations. So suppose that it is additionally known that a family is linearly dependent.

If $A_1, \dots, A_n$ is a finite family of CSO (for simplicity, we can consider them to be bounded) connected by a linear relation

$$\sum_{k=1}^{n} c_k A_k = 0 \tag{36}$$

and its discrete joint spectrum is $\sigma(A_1, \dots, A_n) \subset \mathbb{R}^n$, then by applying (36) to a common eigenvector $e_{\lambda_1, \dots, \lambda_n}$ $((\lambda_1, \dots, \lambda_n) \in \sigma(A_1, \dots, A_n))$ of the CSO $(A_k)_{k=1}^n : A_k e_{\lambda_1, \dots, \lambda_n} = \lambda_k e_{\lambda_1, \dots, \lambda_n}$, we get $\sum_{k=1}^{n} c_k \lambda_k = 0$, i.e. the points of the joint spectrum of such a family of CSO will be connected by the same relations as the operators. The spectrum of such a set of operators is no longer an arbitrary countable subset of $\mathbb{R}^n$.

Let now $(A_k)_{k=1}^n$ be a finite family of CSO, but let it not be assumed any longer that they are bounded and that their joint spectrum $\sigma(A_1, \dots, A_n)$ is discrete. Then, because of the commutativity there exists $\Phi \subset H$ such that $A_k \Phi \subset \Phi$ $(k = 1, \dots, n)$, Φ is a rigging for $(A_k)_{k=1}^n$ and Φ is a domain of essential self-adjointness for all $\sum_{k=1}^{n} c_k A_k$ $(c_k \in \mathbb{R}^1, k = 1, \dots, n)$. Suppose now, that for all $u \in \Phi$

$$\left(\sum_{k=1}^{n} c_k A_k\right) u = 0. \tag{37}$$

Then for the set of points $(\lambda_1, \dots, \lambda_n) \in \sigma(A_1, \dots, A_n)$ of full spectral measure, we have

$$\sum_{k=1}^{n} c_k \lambda_k = 0.$$

The proof is given in Yu.M. Berezanskiĭ [10, 11] by using the decomposition theory with respect to the common generalized eigenvectors of the CSO.

In the case of countable collections of CSO, connected by a finite or a countable set of functions, which include only a finite number of operators, every relation defines a closed subset in $\mathbb{R}^\infty$ of full spectral measure, and, consequently, every point of the spectrum satisfies the same relation. We summarize the above results in the following theorem.

THEOREM 15. The spectrum of a collection $(A_k)_{k=1}^{\infty}$ of CSO connected by a finite or a countable set of linear relations

$$\sum_{k=1}^{\infty} c_k^{(n)} A_k = 0 \quad (n = 1,2,...)$$

such that for all n there is only a finite number of $(c_k^{(n)})_{k=1}^{\infty}$ different from zero, is a closed subset of $\mathbb{R}^{\infty}$ of the points $\lambda = (\lambda_1, \ldots, \lambda_n, ...)$ with the property that

$$\sum_{k=1}^{\infty} c_k^{(n)} \lambda_k = 0 \quad (n = 1,2,...).$$

Nevertheless, when considering a concrete example of a family of CSO, we will not be finding the joint spectrum of this family of operators but rather go from a family of CSO with linear constraints to a linearly independent family, the spectrum of which already can be any measurable set in a corresponding space of sequences.

Example 12. In what follows, consider a countable collection $(A_j^k)_{k=0;j=0,1,\ldots,2^k-1}^{\infty}$, connected by the relations

$$A_j^k = A_{2j}^{k+1} + A_{2j+1}^{k+1} \quad (k = 0,1,...;\ j = 0,1,\ldots, 2^k - 1).$$

It is convenient to set this family in the following scheme

$$\begin{array}{ccccccc} & & & A_0^0 & & & \\ & A_0^1 & & & & A_1^1 & \\ A_0^2 & & A_1^2 & & A_2^2 & & A_3^2 \end{array}$$

Here, the sum of any two "lower" operators gives the "upper".

Construct a linearly independent collection of the operators, such that any operator of the collection $(A_j^k)_{k,j}$ can be linearly expressed in terms of the linearly independent collection. We define $(\mathbf{A}_{i_1,\ldots,i_n\cdots})_{i_1\cdots i_n\cdots\in\mathbb{Z}_{2,0}^{\infty}}$ ($\mathbb{Z}_{2,0}^{\infty}$ is a set of sequences of finite binary indexes) as:

$$\mathbf{A}_{000\ldots} = A_0^0$$

$$\mathbf{A}_{100\ldots} = A_0^1 - A_1^1$$

$$\mathbf{A}_{010\ldots} = A_0^2 - A_1^2 + A_2^2 - A_3^2$$

$$\mathbf{A}_{110\ldots} = A_0^2 - A_1^2 - A_2^2 - A_3^2$$

....

Note, that this formulas coincide with the formulas, expressing Walsh functions $(u_{i_1\cdots i_n\cdots})_{i_1\cdots i_n\cdots\in\mathbb{Z}_{2,0}^{\infty}}$ as linear combinations of the functions

$$h_j^k(x)=\begin{cases}1, & x\in\left[\dfrac{j}{2^k},\dfrac{j+1}{2^k}\right],\\[2mm] 0, & x\notin\left[\dfrac{j}{2^k},\dfrac{j+1}{2^k}\right]\end{cases}$$

which are defined on $(0,1]$ (see, for example, S. Kaczmarz and H. Steinhaus [1]).

The spectral measure of this linearly independent family of CSO is contained in $\mathbb{R}^{\mathbb{Z}^{\infty}_{2,0}}$.

Comments to Chapter 1.

1. Foundations of measure theory in $\mathbb{R}^\infty$ are given, for example, in A.N. Kolmogorov [1]; P. Halmos [1]; Y. Umemura [1]; G.E. Shilov and Fan Dyk Tin [1], I.I. Gikhman, A.V. Skorokhod and M.I. Yadrenko [1] etc.

For an approach to study finite and infinite collections of commuting normal operators using a joint resolution of the identity see A.I. Plesner and V.A. Rohlin [1]; A.I. Plesner [1]; Yu.M. Berezanskiĭ [5,6,7,8,9]; L. Gårding [3]; M.Sh. Birman and M.Z. Solomyak [1] etc. Theorem 1 is proved in a more general setting in Yu.M. Berezanskiĭ [9].

2. To define the spectrum of a collection, and to prove that it is nonempty, we followed the monograph of Yu.M. Berezanskiĭ [9]. Subsequent study of the spectrum of a countable collection of CSO follows the same scheme of studying the spectrum of a self-adjoint operator as in N.I. Akhiezer and I.M. Glazman [1] with necessary modifications for a countable collection.

3. Here, we give some results of the papers V.I. Kolomytsev and Yu.S. Samoĭlenko [2,3].

4. The proofs follow the book of N.I. Akhiezer and I.M. Glazman [1].

5. The treatment of the question of unitary invariants of a countable collection of CSO is given along the lines of the book A.I. Plesner [1]. For theorems on absolute continuity of the measures on $\mathbb{R}^\infty$ see, for example, the monograph A.V. Skorokhod [3] and bibliography given there.

6. The section contains the extension of the results obtained in V.I. Kolomytsev, Yu.S. Samoĭlenko [2]; R. Goodman [1]; A.V. Kosyak [1].

7. The essentially infinite-dimensional functions are treated in Yu.S. Samoĭlenko, G.F. Us [1,2].

8. Only the simplest families of CSO, connected by a linear relations, are considered. In essence, the example given is for general families of CSO, connected by algebraic and differential relations (see C. Ionescu Tulcea [1], G. Maltese [1,2], Yu.M. Berezanskiĭ [10,11]).

Chapter 2.
UNITARY REPRESENTATIONS OF INDUCTIVE LIMITS OF COMMUTATIVE LOCALLY COMPACT GROUPS

This section could also be called: families of commuting unitary operators. But here we will be mainly concerned with commutative groups of unitary operators, the study of which is reduced to the study of countable collections of commuting self-adjoint operators. Regardless the fact that these groups of unitary operators are not representations of locally compact groups, they have a number of properties that make them like representations of finite-dimensional commutative Lie groups.

2.1. Inductive limits of locally compact groups.

While studying functions of infinitely many variables and spaces of these functions, there arises the need to study different symmetry groups (which are, generally, not locally compact) of infinite-dimensional spaces and their representations. Many natural symmetry groups are inductive limits of locally compact topological groups or they contain dense subgroups of this type.

Let $G_\alpha(\alpha\in\Lambda)$ be a set, of locally compact topological groups partially ordered with respect to inclusions such that for any ordered pair of indices $\alpha\leq\beta$ the monomorphisms $j^\alpha_\beta : G_\alpha \hookrightarrow G_\beta$ are continuous and the topology on G_α coincides with the topology on G_α induced by G_β.

DEFINITION 1. The group $G=\bigcup_{\alpha\in\Lambda} G_\alpha$ with the topology induced by the inductive limit of topological groups $G_\alpha(\alpha\in\Lambda)$ is called the inductive limit of the groups $G_\alpha(\alpha\in\Lambda)$ and denoted by $G=j-\lim\limits_{\rightarrow} G_\alpha$ (here $j=(j^\alpha_\beta)$).

If the meaning of the inclusions $j=(j^\alpha_\beta)$ is clear from the context, then sometimes the index j will be left out from the notation of the inductive limit of groups. In what follows we will mainly take Λ to be the set of positive integers $\mathbb{N}$.

Now, let all G_n be connected Lie groups and G_n be a Lie subgroup of G_{n+1} for all $n=1,2,\ldots$. Then we can construct a chain of Lie algebras $(\mathfrak{g}_n)_{n\in\mathbb{N}}$ ordered with respect to inclusion, corresponding to the chain of the groups $G_1\hookrightarrow G_2\hookrightarrow\cdots\hookrightarrow G_n\hookrightarrow\cdots$: $\mathfrak{g}_n$ is a Lie subalgebra of $\mathfrak{g}_{n+1}$ for all $n\in\mathbb{N}$ and the operator $j^n_{n+1} : G_n\hookrightarrow G_{n+1}$ induces the inclusion operator $J^n_{n+1} : \mathfrak{g}_n\hookrightarrow\mathfrak{g}_{n+1}$.

DEFINITION 2. The Lie algebra $\mathfrak{g}=J-\lim\limits_{\rightarrow}\mathfrak{g}_n$ $(J=(J^n_{n+1}),\, n\in\mathbb{N})$ will be called the Lie algebra of the group $G=j-\lim\limits_{\rightarrow} G_n$.

If $(X_k)_{k=1}^{\dim\mathfrak{g}_n}$ is a basis in the Lie algebra $\mathfrak{g}_n(n\in\mathbb{N})$, then sequentially extending it to a basis of $\mathfrak{g}_{n+1}$, $\mathfrak{g}_{n+2}$ and so on, we will construct a basis of the, generally speaking, infinite-dimensional Lie algebra g.

We will give two examples of inductive limits of Lie groups and their Lie algebras that will be used to illustrate the material that follows.

Let G be a connected locally compact Lie group, g be its Lie algebra, $(X_k)_{k=1}^n$ be a basis in g satisfying the commutation relations:

$$[X_l, X_k] = \sum_{m=1}^{n} c_{lk}^m X_m.$$

DEFINITION 3. A group of functions defined on a manifold X with values in a group G is called a group of G-currents on X.

The choice of either continuous or smooth, or finite or piecewise-constant mappings $X \to G$ determines the corresponding groups of continuous, smooth, finite, or step G-currents on X. Amongst the groups of G-currents, we consider two groups which are inductive limits of locally compact groups.

Example 1. The group G_0^∞ of finite G-currents on a countable set: $G_0^\infty = i-\underset{\rightarrow}{\lim}\, G^n$, where the inclusions $i_{n+1}^n: \quad G^n \hookrightarrow G^{n+1}$ are defined as follows: $i_{n+1}^n(g_1, \ldots, g_n) = (g_1, \ldots, g_n, e) \in G^{n+1}$ (e the identity element of G).

PROPOSITION 1. The Lie algebra of this group is $\mathrm{g}_0^\infty = J-\underset{\rightarrow}{\lim}\, \mathrm{g}^n$ with the corresponding inclusions in the algebra being $J_{n+1}^n(X_{i_1}, \ldots, X_{i_n}) = (X_{i_1}, \ldots, X_{i_n}, 0) \in \mathrm{g}$. The basis in g_0^∞ can be chosen to be $X_k^{(j)} = (0, \ldots, \underbrace{0, X_k, 0}_{j\text{-th place}},\ldots)$ and the commutation relations in this basis are:

$$[X_l^{(j)}, X_k^{(i)}] = \delta_{ij} \sum_{m=1}^{n} c_{lk}^m X_m.$$

In particular, we mention the additive group $\mathbb{R}_0^\infty = i-\underset{\rightarrow}{\lim}\, \mathbb{R}^n$ of all finite sequences of real numbers with the topology defined by the inductive limit of the spaces $\mathbb{R}^n$.

The Lie algebra of the group $\mathbb{R}_0^\infty = i-\underset{\rightarrow}{\lim}\, \mathbb{R}^n$ is a countable-dimensional commutative Lie algebra with the basis $(X^{(j)})_{j=1}^\infty$ in which $[X^{(j)}, X^{(i)}] = 0$ $(i,j = 1,2,\ldots)$. □

Example 2. The group $G_{\pi(2)}^\infty$ of step G-currents on the interval $[0,1]$ with discontinuities in the binary-rational points: $G_{\pi(2)}^\infty = j-\underset{\rightarrow}{\lim}\, G^{2^n}$, where the inclusions $j_{n+1}^n : G^{2^n} \hookrightarrow G^{2^{n+1}}$ are defined as follows:

$$j_{n+1}^n\,(g_1, g_2, \ldots, g_{2^n}) = (g_1, g_1, g_2, g_2, \ldots, g_{2^n}, g_{2^n}) \in G^{2^{n+1}}.$$

The vectors $X_k^l(n) = (\overbrace{0, \ldots, \underbrace{0, X_l, 0}_{k\text{-th place}}, \ldots, 0}^{2^n})$ form a basis in the Lie algebra $\mathbf{g}^{2^n}$ and the inclusions are $J^n_{n+1}\, X^l_k(n) = X^l_{2k}(n+1) + X^l_{2k+1}(n+1)$. We have the following proposition.

PROPOSITION 2. In $\mathbf{g}^{\infty}_{\pi(2)}$ one can choose a basis $(X^l_{i_1,\ldots,i_k,\ldots})^n_{l=1}$ $(i_1,\ldots,i_k,\ldots,\in \mathbb{Z}_2$ where starting with some number $i_{N+1} = \cdots = 0)$ similar to the Walsh basis for all step functions defined on $[0,1]$ with discontinuities in the binary rational points. There we have the relations:

$$[X^l_{i_1}\cdots_{i_k}\cdots, X^m_{j_1}\cdots_{j_k}\cdots] = \sum_{p=1}^{n} c^p_{lm}\, X^p_{i_1+j_1}\cdots_{i_k+j_k}\cdots$$

($i_k + j_k$ is addition in $\mathbb{Z}_2$).

In Section 6 of this chapter we will come back to the group $\mathbb{R}^{\infty}_{\pi(2)}$ and study its unitary representations. □

We give two more examples, showing that the class of topological groups, which are inductive limits of locally compact groups, is rather rich.

Example 3. The class of groups, which are inductive limits of locally compact groups, contains all locally finite groups. In particular, it contains the group $S_\infty = i - \varinjlim S_n$, which is the group of permutations of a countable set, such that each permutation transforms only a finite number of points of the set. □

Example 4. Take a countable-dimensional complex (real) Hilbert space H, with orthonormal basis $(e_k)^{\infty}_{k=1}$. Denote by $SU(n)$, $U(n)$, $U(p, n-p)$, $GL(n, \mathbb{C})$, $(SO(n), O(n)$, $O(p, n-p)$, $GL(n, \mathbb{R}))$ the corresponding groups of unitary (orthogonal), of J-unitary (J-orthogonal), and of invertible operators on the space $H_n = LS\{e_1, \ldots, e_n\}$. Consider these operators as operators on H leaving the vectors $e_{n+1}, e_{n+2},\ldots$ invariant. Then we can define the groups $U(\infty) = \varinjlim U(n)$, $SU(\infty) = \varinjlim SU(n)$, $SU(p,\infty) = \varinjlim SU(p, n-p)$, $GL(\infty, \mathbb{C}) = \varinjlim GL(n, \mathbb{C})$ $(O(\infty) = \varinjlim O(n)$, $SO(\infty) = \varinjlim SO(n)$, $O(p,\infty) = \varinjlim O(p, n-p)$, $GL(\infty, \mathbb{R}) = \varinjlim GL(n, \mathbb{R}))$ which are the groups of unitary, of J-unitary (orthogonal, J-orthogonal), or of invertible operators on H that can be represented as the direct sum of a finite-dimensional operator and the identity. Their Lie algebras are the corresponding Lie algebras of the finite-dimensional operators on H. □

In general a topological group $G = \varinjlim G_n$ does not have all remarkable properties enjoyed by the locally compact groups. First of all, there is no invariant σ-additive measure on $(G, \mathbf{B}(G))$, where $\mathbf{B}(G)$ is a Borel σ-algebra of the subsets of G. Indeed, we have the following proposition (see A. Weil [1]).

PROPOSITION 3. There is an invariant σ-finite Borel measure on a topological group G if and only if G is locally compact.

In particular, there is no invariant σ-finite measure on the group $\mathbb{R}^{\infty}_0$. Moreover, there is not even a nontrivial Borel σ-finite measure $\omega(\cdot)$ that is $\mathbb{R}^{\infty}_0$-quasi-invariant (i.e.

$\omega_t(\cdot) = \omega(\cdot + t) \sim \omega(\cdot)$ for all $t \in I\!R_0^\infty$) (see Chapter 3).

Nevertheless, it will be shown that commutative groups which are inductive limits of locally compact groups, satisfy a duality law, the Naĭmark theorem, and have some other properties in common with the commutative locally compact groups and their representations.

Some properties of noncommutative locally compact groups remain true in groups which are inductive limits of locally compact groups. We study such groups in Part II.

In the class of arbitrary topological groups (even commutative) there exist groups with properties quite different from those satisfied by locally compact groups (see, for example, Remarks 1 and 3 below).

2.2. Character theory. The duality law for inductive limits of commutative locally compact groups.

A character $\chi(\cdot)$ of a topological abelian group G is a continuous homomorphism of G into the group S^1 of complex numbers with modulus equal to 1.

On the set of characters $\hat{G}$ define the topology of uniform convergence on compact sets in G, i.e. sets of the form

$$U(K, \varepsilon, \chi_0(\cdot)) = \{\chi(\cdot) \in \hat{G} \mid |\chi(g) - \chi_0(g)| < \varepsilon \ \forall g \in K\}$$

(here $\chi_0(\cdot)$ is a fixed character, $\varepsilon > 0$ and K is a compact subset of G) are open in $\hat{G}$ and the collection of such open sets establishes a basis for the topology in $\hat{G}$.

The pointwise product of two characters is again a character and with this multiplication $\hat{G}$ becomes a commutative topological group, which is called the character group. If G is a locally compact group, then $\hat{G}$ is also a locally compact group and we have Pontryagin's duality law: $\hat{\hat{G}} = G$.

Here, we study the group of characters of the group $G = \lim\limits_{\rightarrow} G_n$ (with G_n locally compact commutative topological groups for all n), and prove a duality law for this class of groups.

Because G_n is a locally compact abelian group, its character group $\hat{G}_n$ is also locally compact. Moreover, the monomorphism $j_{n+1}^n : G_n \hookrightarrow G_{n+1}$ determines an epimorphism $p_n^{n+1} : \hat{G}_{n+1} \to \hat{G}_n$ by restricting each character $\chi(\cdot) \in G_{n+1}$ to the subgroup G_n. In this way, there arises the projective system $p - \lim\limits_{\leftarrow} \hat{G}_n (p = (p_n^{n+1})_{n=1}^\infty)$ of locally compact groups. We will call $\chi_n(\cdot) \in \hat{G}_n$ and $\chi_{n+k}(\cdot) \in \hat{G}_{n+k}$ equivalent if $\chi_{n+k}(\cdot) \restriction_{G_n} = p_n^{n+1} \cdots p_{n+k-1}^{n+k} \chi_{n+k}(\cdot) = \chi_n(\cdot)$ and set the group $\lim\limits_{\leftarrow} \hat{G}_n$ to be the group $\prod\limits_{n=1}^\infty \hat{G}_n$ factored with respect to this equivalence relations. The topology in $p - \lim\limits_{\leftarrow} \hat{G}_n$ is the projective limit topology generated by the topological spaces $\hat{G}_n$ i.e. the weakest topology in

$\lim\limits_{\leftarrow} \hat{G}_n$ such that all the mappings $p_n : p-\lim\limits_{\leftarrow} \hat{G}_n \to \hat{G}_n$ $(n=1,2,...)$ are continuous.

Example 5. Let us describe $\lim\limits_{\leftarrow} \mathbb{R}^n$. Because $\mathbb{R}^n = \hat{\mathbb{R}}_n \ni (\lambda_1, \ldots, \lambda_n)$ and $\chi_{(\lambda_1,\ldots,\lambda_n)}(t_1, \ldots, t_n) = \exp(i \sum\limits_{k=1}^{n} t_k \lambda_k)$, the inclusions $i_{n+1}^n : \mathbb{R}^n \hookrightarrow \mathbb{R}^{n+1}$ given by the formula $i_{n+1}^n(t_1, \ldots, t_n) = (t_1, \ldots, t_n, 0) \in \mathbb{R}^{n+1}$ define the projections $p_n^{n+1}(\lambda_1, \ldots, \lambda_{n+1}) = (\lambda_1, \ldots, \lambda_n) \in \mathbb{R}^n$ and, consequently the elements $\hat{\mathbb{R}}_0^{\infty} \ni (\lambda_1, \ldots, \lambda_n, \ldots)$ are all sequences of real numbers. The weakest topology in $\mathbb{R}^\infty$ in which all the mappings $p_n(\lambda_1, \ldots, \lambda_n, \ldots) = (\lambda_1, \ldots, \lambda_n) \in \mathbb{R}^n$ are continuous, is the product topology in $\mathbb{R}^\infty$, i.e. we have $i-\lim\limits_{\rightarrow}\hat{}\ \mathbb{R}^n = \hat{\mathbb{R}}_0^{\infty} = \mathbb{R}^\infty$. ▯

The following example is a generalization of the previous one.

Example 6. Consider the topological groups $\sum\limits_{k=1}^{\infty} G_k = i-\lim\limits_{\rightarrow} G_1 \times \cdots \times G_n$ and $\prod\limits_{k=1}^{\infty} G_k = G_1 \times \cdots \times G_n \times \cdots$, where the G_k are locally compact abelian groups $(k=1,2,...)$.

PROPOSITION 4. The groups $\sum\limits_{k=1}^{\infty} G_k$ and $\prod\limits_{k=1}^{\infty} \hat{G}_k$ are dual, i.e. $\sum\limits_{k=1}^{\infty}\hat{}\ G_k = \prod\limits_{k=1}^{\infty} \hat{G}_k$ and $\prod\limits_{k=1}^{\infty}\hat{}\ G_k = \sum\limits_{k=1}^{\infty} \hat{G}_k$. ▯

Let us prove a more general assertion.

THEOREM 1. The topological groups $j-\lim\limits_{\rightarrow} G_n$ and $p-\lim\limits_{\leftarrow} \hat{G}_n$ are dual, i.e. $j-\lim\limits_{\rightarrow}\hat{}\ G_n = p-\lim\limits_{\leftarrow} \hat{G}_n$ and $\rho-\lim\limits_{\rightarrow}\hat{}\ G_n = j-\lim\limits_{\rightarrow} G_n$.

Proof. Because the projections on G_n of the open sets in $G = j-\lim\limits_{\rightarrow} G_n$ are open in G_n according to the definition of the inductive limit topology, for any continuous character $\chi(\cdot)$ in $\hat{G}$, its restriction $\chi(\cdot)\restriction_{G_n} = \chi_n(\cdot)$ is a character in G_n, i.e. $\chi_n(\cdot) \in \hat{G}_n$ $(n=1,2,...)$. So for any continuous character $\chi(\cdot)$ of the group G there is a sequence $\chi_n(\cdot) \in \hat{G}_n$ such that $p_n^{n+1}\ \chi_{n+1}(\cdot) = \chi_n(\cdot)$ $(n=1,2,...)$.

Conversely, for any such sequence $(\chi_n(\cdot))_{n=1}^{\infty}$ $(p_n^{n+1}\ \chi_{n+1}(\cdot) = \chi_n(\cdot))$ there is a homomorphism $\chi(\cdot)$ from G into S^1 which is continuous as the preimage in S^1 of any open set in G is the union of an increasing sequence of open sets in G_n imbedded in G and these sets are open in $G = j-\lim\limits_{\rightarrow} G_n$. So there is a one-to-one correspondence between the characters of the group G and the elements of $p-\lim\limits_{\leftarrow} \hat{G}_n$. The mapping $p(\chi(\cdot)) = (\chi(\cdot)\restriction_{G_n})_{n=1}^{\infty} = (\chi_n(\cdot))_{n=1}^{\infty}$ is an algebraic isomorphism between the groups $\hat{G}$ and $p-\lim\limits_{\leftarrow} \hat{G}_n$.

We prove that this mapping is a homeomorphism. Because the compacts K in G are the images of the compacts K_n in G_n under $j_n : G_n \hookrightarrow G$ $(n=1,2,...)$ the topology in $\hat{G}$ of uniform convergence on compact sets in G is generated by the sets

$$U(K, \varepsilon, 1) = \{\chi(\cdot) \in \hat{G} \mid |\chi(j_n g_n) - 1| < \varepsilon \; \forall g_n \in K_n\}$$

as well as shifts, unions and finite intersections of these sets. But the topology in $p-\underset{\leftarrow}{\lim}\, \hat{G}_n$ is defined by the preimages under $p_n : p-\underset{\leftarrow}{\lim}\, \hat{G}_n \to \hat{G}_n$ of the open sets in $\hat{G}_n$ which are also given by the sets

$$U_n(K_n, \varepsilon, 1) = \{\chi_n(\cdot) \in \hat{G}_n \mid |\chi_n(g_n) - 1| < \varepsilon \; \forall g_n \in K_n\}$$

i.e. by the sets

$$U(K, \varepsilon, 1) = \{\chi(\cdot) \in p-\underset{\leftarrow}{\lim}\, \hat{G}_n \mid |\chi(j_n g_n) - 1| =$$

$$= |p_n \chi(g_n) - 1| < \varepsilon \; \forall g_n \in K_n\}.$$

So $\hat{G} = j-\underset{\rightarrow}{\lim}\hat{\ } G_n = p-\underset{\leftarrow}{\lim}\, \hat{G}_n$. Analogously one proves the equality $\hat{\hat{G}} = p-\underset{\leftarrow}{\lim}\hat{\ } \hat{G}_n = j-\underset{\rightarrow}{\lim}\, G_n = G$. □

Remark 1. For arbitrary topological groups the duality principle does not hold in general. Thus, for a linear topological space (abelian topological group) which does not have nonzero continuous linear functionals, there is only the identity character.

2.3. Unitary representations of inductive limits of commutative locally compact groups.

The main point of this section is to develop a spectral theory for a family $(U_g)_{g \in G}$ of unitary operators on a separable Hilbert space such that the mappings $G \ni g \mapsto U_g \in \mathbf{U}(H)$ are strongly continuous homomorphisms of the group G into the group $\mathbf{U}(H)$ of unitary operators on H. This is the same thing as studying the unitary representations of the topological group $G = j-\underset{\rightarrow}{\lim}\, G_n$.

For a commutative locally compact group G_n we have an analogue of the spectral theorem in terms of a resolution of the identity.

THEOREM 2. (Naĭmark). For every unitary representation of an abelian locally compact group G_n on a separable Hilbert space H there is a unique orthogonal resolution of the identity $E_n(\cdot)$ on $(\hat{G}_n, \mathbf{B}(\hat{G}_n))$ such that

$$U_{g_n} = \int_{\hat{G}_n} \chi_n(g_n)\, dE_n(\chi_n). \tag{1}$$

The integral in (1) converges in strong sense.

If for G_n we take the locally compact group $\mathbb{R}^n$, then Theorem 2 is the classical theorem of Stone.

THEOREM 3 (Stone). Every strongly continuous n-parameter group of unitary operators $U_{(t_1,\ldots,t_n)}$ $((t_1,\ldots,t_n)\in \mathbb{R}^n)$ has the form

$$U_{(t_1,\ldots,t_n)} = \exp(i\sum_{k=1}^{n} t_k A_k) = \int_{\mathbb{R}^n} \exp(i\sum_{k=1}^{n} t_k\lambda_k)\, dE_n(\lambda). \tag{2}$$

The integral in (2) converges in strong sense.

Here, $(A_k)_{k=1}^n$ is a finite collection of CSO and $E_n(\cdot)$ is its resolution of the identity, (see Chapter 1, 1.1).

Before we look at the analogue of Theorem 2 for the group $G = j-\lim\limits_{\rightarrow} G_n$, we consider the unitary representations of $\mathbb{R}_0^\infty$ i.e. the mappings $\mathbb{R}_0^\infty \ni t = (t_1,\ldots,t_n,\ldots) \mapsto U_{(t_1,\ldots,t_n,\ldots)} \in \mathrm{U}(H)$ such that

1) $U_{t^{(1)}+t^{(2)}} = U_{t^{(1)}}U_{t^{(2)}}\ \forall t^{(1)}, t^{(2)} \in \mathbb{R}_0^\infty$;

2) U_t is strongly continuous with respect to $t\in \mathbb{R}_0^\infty$.

Similar to Theorem 3 we have the following

THEOREM 4. For a unitary representation of $\mathbb{R}_0^\infty$ there corresponds a unique resolution of the identity $E(\cdot)$ on $(\mathbb{R}^\infty, \mathrm{B}(\mathbb{R}^\infty))$ such that we have the representation

$$U_t = \int_{\mathbb{R}^\infty} \exp(i(t,\lambda))\, dE(\lambda). \tag{3}$$

The integral in (3) converges in strong sense.

Proof. We have a direct proof of this theorem by using Theorem 3 for $n = 1,2,\ldots$. Since the corresponding unitary representations $\mathbb{R}^n \ni (t_1,\ldots,t_n) \mapsto U_{(t_1,\ldots,t_n,0,\ldots)}$ are strongly continuous, we have

$$U_{(t_1,\ldots,t_n,0,\ldots)} = \int_{\mathbb{R}^n} \exp(i\sum_{k=1}^{n} t_k\lambda_k)\, dE_n(\lambda_1,\ldots,\lambda_n). \tag{4}$$

Because the measures $dE_n(\lambda_1,\ldots,\lambda_n)$ in (4) are uniquely determined, the orthogonal operator measures $dE_n(\lambda_1,\ldots,\lambda_n)$ on $(\mathbb{R}^n, \mathrm{B}(\mathbb{R}^n))$ $(n=1,2,\ldots)$ are compatible to each other:

$$\int_{\mathbb{R}^1_{n+1}} dE_n(\lambda_1,\ldots,\lambda_n,\lambda_{n+1}) = dE_n(\lambda_1,\ldots,\lambda_n).$$

As in the proof of Theorem 1 of Chapter 1, 1.1, one proves that there exists a unique measure $E(\cdot)$ on $(\mathbb{R}^\infty, \mathrm{B}(\mathbb{R}^\infty))$ such that $\int_{\mathbb{R}^1_{n+1}\times\mathbb{R}^1_{n+2}\times\cdots} dE(\lambda_1,\ldots,\lambda_n,\ldots) = dE_n(\lambda_1,\ldots,\lambda_n)$ and so we have the representation (3).

We give another proof of this theorem using the fact that the one-parameter groups of the unitary operators $U_{(0,\ldots,0,t_k,0,\ldots)}$ are strongly continuous, with respect to $t_k \in \mathbb{R}^1$. From Stone's theorem it follows that there exists a self-adjoint operator A_k on H such that

$$U_{(0,\ldots,0,t_k,0,\ldots)} = \exp(i\, t_k A_k) \quad (t_k \in \mathbb{R}^1,\ k=1,2,\ldots).$$

Because the strongly continuous one-parameter groups $U_{t_k} = \exp(i\, t_k A_k)$ and $U_{t_j} = \exp(i\, t_j A_j)$ commute, the self-adjoint operators A_k and A_j also commute. So the collection $(A_k)_{k=1}^{\infty}$ is a collection of CSO. Applying Theorem 1 of Chapter 1 to this collection we determine a resolution of the identity which is an orthogonal operator-valued measure on $(\mathbb{R}^{\infty}, \mathrm{B}(\mathbb{R}^{\infty}))$. The formulas for the functional calculus for a family of CSO (Chapter 7, 7.1) yield the equality

$$U_{(t_1,\ldots,t_n,\ldots)} = \exp(i \sum_{k=1}^{n} t_k A_k) = \int_{\mathbb{R}^{\infty}} \exp(i \sum_{k=1}^{n} t_k \lambda_k)\, dE(\lambda).$$

□

Remark 2. The collection $(A_k)_{k=1}^{\infty}$ of CSO is a representation of a basis in the commutative Lie algebra g_0^{∞} of the group $\mathbb{R}_0^{\infty}$. The corresponding representation of the Lie algebra g_0^{∞} can be extended to a representation of the group $\mathbb{R}_0^{\infty} \ni (t_1,\ldots,t_n,\ldots)$. Here, the resolution of the identity for the unitary representation $\mathbb{R}_0^{\infty} \ni t \mapsto U_t$ and the corresponding collection $(A_k)_{k=1}^{\infty}$ of CSO are the same. □

PROPOSITION 5. The joint spectrum of the countable family $(A_k)_{k=1}^{\infty}$ of CSO is simple if and only if the corresponding representation $\mathbb{R}_0^{\infty} \ni t \mapsto U_t = \exp(i \sum_{k=1}^{\infty} t_k A_k)$ is cyclic, i.e. if there exists a vector $\Omega \in H$ such that $\underset{t \in \mathbb{R}_0^{\infty}}{\mathrm{CLS}} \{U_t \Omega\} = H$.

The proof follows directly from Theorem 4 of Chapter 1 that allows to realize the Hilbert space H as $L_2(\mathbb{R}^{\infty}, d\mu(\lambda))$ and the operators $(A_k)_{k=1}^{\infty}$ as multiplication operators $(\lambda_k)_{k=1}^{\infty}$. Since $\mu(\cdot)$ is a probability measure on $(\mathbb{R}^{\infty}, \mathrm{B}(\mathbb{R}^{\infty}))$, $\underset{t \in \mathbb{R}_0^{\infty}}{\mathrm{CLS}} \{\exp(i \sum_{k=1}^{\infty} t_k \lambda_k)\} = H$, and consequently $\Omega = 1$ is a cyclic vector of the representation. □

Example 7. Consider a $\mathbb{R}_0^{\infty}$-quasiinvariant probability measure $\rho(\cdot)$ on $(\mathbb{R}^{\infty}, \mathrm{B}(\mathbb{R}^{\infty}))$ (i.e. $\rho(\cdot)$ is such that $\forall t \in \mathbb{R}_0^{\infty}$ the measures $\rho_t(\cdot) = \rho(\cdot + t)$ and $\rho(\cdot)$ are equivalent). Consider the unitary representation $\mathbb{R}_0^{\infty} \ni t \mapsto U_t \in \mathrm{U}(L_2(\mathbb{R}^{\infty}, d\rho(x)))$ given by $(U_t f)(x) = \sqrt{\frac{d\rho_t}{d\rho}(x)}\, f(x+t)$.

This is a strongly continuous unitary representation with spectral properties essentially dependent on the measure $\rho(\cdot)$. Even for Gaussian measures it need not be cyclic. The spectral measure of this representation is the spectral measure of the corresponding collection $(D_k)_{k=1}^{\infty}$ of CSO (see below Chapter 3). □

Now we formulate the spectral theorem in terms of a resolution of the identity for a unitary representation of the commutative group $G = j - \underset{\rightarrow}{\lim}\, G_n$.

THEOREM 5. (Spectral theorem in terms of a resolution of the identity for a unitary representation of the group $G = j - \underset{\rightarrow}{\lim}\, G_n$). To every unitary representation of the inductive limit $G = j-\lim\limits_{\rightarrow} G_n$ of commutative locally compact groups G_n $(n=1,2,\ldots)$ there corresponds a unique resolution of the identity $E(\cdot)$ on $(\hat{G}, \mathbf{B}(\hat{G}))$.

Conversely, if there exists a resolution of the identity $E(\cdot)$ on $(\hat{G}, \mathbf{B}(\hat{G}))$, then the corresponding representation has the form:

$$G \ni g \mapsto U_g = \int_{\hat{G}} \chi(g)\, dE(\chi)\ (\chi \in \hat{G}). \tag{5}$$

The convergence of the integral in (5) is understood in strong sense.

We can prove this theorem in the same way as we proved Theorem 4, i.e. by showing that the orthogonal operator measures $E_n(\cdot)$ on $(\hat{G}_n, \mathbf{B}(\hat{G}_n))$ agree with each other, and that there is an unique resolution of the identity $E(\cdot)$ on $(\hat{G}, \mathbf{B}(\hat{G}))$. Because all $\hat{G}_n$ are locally compact for such a sequence of probability measures on $(\hat{G}_n, \mathbf{B}(\hat{G}_n))_{n=1}^{\infty}$ we can apply Kolmogorov's theorem. □

Remark 3. For an arbitrary topological group, Theorem 5 does not hold. In particular, the resolution of the identity for the unitary representation on $L_2(\mathbb{R}^{\infty}, dg(x))$ of the abelian group $l_2 \ni t = (t_1, t_2, \ldots) \mapsto (U_t f)(x) = \sqrt{\frac{dg_t}{dg}}(x)\, f(x+t)$ is not concentrated on the set of continuous characters of the group l_2 (here $dg(x) = dg_I(x)$).

For the spectral measure of unitary representation in H, $l_2 \ni t \mapsto U_t$, to be concentrated on the set of continuous characters of the group l_2, i.e. on l_2, the representation has to be continuous with respect to a stronger topology for the group (see V.V. Sazonov [1]). This condition is obtained by using equivalent terms of existence of a rigging in the article Yu.M. Berezanskiĭ, Yu.G. Kondrat'ev and I.M. Gali [1].

2.4. The Spectral theorem for representations in terms of multiplication operators.

In the same way as the spectral theorem for a collection $(A_k)_{k=1}^{\infty}$ of CSO can be formulated in terms of a resolution of the identity (Chapter 1, 1.1) as well as in terms of multiplication operators (Chapter 1, 1.3), the Spectral Theorem for a unitary representation of $\mathbb{R}_0^{\infty}$ can be formulated both in terms of a resolution of the identity and in terms of multiplication operators.

THEOREM 6. (the Spectral Theorem for the operators of a unitary representation of the group $\mathbb{R}_0^{\infty}$ in terms of multiplication operators). For any unitary representation $\mathbb{R}_0^{\infty} \ni t \mapsto U_t$ in H there exists a sequence of probability measures on $(\mathbb{R}^{\infty}, \mathbf{B}(\mathbb{R}^{\infty}))$, $\rho_1(\cdot) \gg \rho_2(\cdot) \gg \cdots$, and a unitary transformation $U : H \to \oplus \sum_k L_2(\mathbb{R}^{\infty}, d\rho_k(\lambda))$, such that for all $t \in \mathbb{R}_0^{\infty}$

$$U_t = U^* \exp(i(t,\lambda))\, U \tag{6}$$

$(\exp(i(t,\lambda))$ is the operator of multiplication by the function $e^{i(t,\lambda)}$ $(\lambda \in \mathbb{R}^{\infty})$ in

$\oplus \sum_k L_2(\mathbb{R}^\infty, d\rho_k(\lambda)))$.

This theorem can be proved by using Theorem 6 (Chapter 1), i.e. the CSO $(A_k)_{k=1}^\infty$ corresponding to a unitary representation of the group $\mathbb{R}_0^\infty$ (see Remark 2) are realized in a unitarily equivalent way as operators of multiplication by $(\lambda_k)_{k=1}^\infty$ in the space $\oplus \sum_k L_2(\mathbb{R}^\infty, d\rho_k(\lambda))$, $(\rho_1(\cdot) \gg \rho_2(\cdot) \gg \cdots)$. Because the representation operators $(U_t)_{t \in \mathbb{R}_0^\infty}$ are functions of the CSO $(A_k)_{k=1}^\infty$, viz. $U_t = \exp(i \sum_{k=1}^\infty t_k A_k)$, they will transform into operators of multiplication by the function $\exp(i \sum_{k=1}^\infty t_k \lambda_k) = \exp(i(t, \lambda))$, in the space $\oplus \sum_k L_2(\mathbb{R}^\infty, d\rho_k(\lambda))$. □

Also the Spectral Theorem for a unitary representation of the group $G = j-\lim_{\rightarrow} G_n$ can be formulated in terms of multiplication operators.

THEOREM 7. (the Spectral Theorem for the unitary representation operators of the inductive limit $G = j-\lim_{\rightarrow} G_n$ of locally compact abelian groups G_n in terms of multiplication operators). For any unitary representation of $G = j-\lim_{\rightarrow} G_n \ni g \mapsto U_g$ on H there exists a sequence of probability measures $\rho_1(\cdot) \gg \rho_2(\cdot) \gg \cdots$ on $(\hat{G}, \mathbf{B}(\hat{G}))$ and a unitary transformation $U : H \to \oplus \sum_k L_2(\hat{G}, d\rho_k(\chi))$ such that for all $g \in G$

$$U_g = U^* \chi(g) U$$

($\chi(g)$ is the operator of multiplication by the function $\chi(g)$ $(\chi(\cdot) \in \hat{G})$ on $\oplus \sum_k L_2(\hat{G}, d\rho_k(\chi)))$.

Proof. If the representation $\hat{G} \ni g \mapsto U_g$ is cyclic on H and Ω is a cyclic vector of this representation, then the isometric correspondence between the dense sets $(\sum_k c_k U_{g_k})\Omega$ in H and $\sum_k c_k \chi(g_k)$ in $L_2(\hat{G}, d(E(\chi)\Omega, \Omega))$ defines the wanted unitary operator $U : H \to L_2(\hat{G}, d(E(\chi)\Omega, \Omega))$.

If the representation is not cyclic, then the proof follows the proof of Theorem 6 of Chapter 1 and is based on the following proposition (which is the analogue of Proposition 3 of Chapter 1).

PROPOSITION 6. For any unitary representation $G \ni g \mapsto U_g$ there exists a vector of maximal spectral type, i.e. a vector Ω such that the projection-valued measure $E(\cdot)$ on $(\hat{G}, \mathbf{B}(\hat{G}))$ is equivalent to the scalar measure $\rho(\cdot) = (E(\cdot)\Omega, \Omega)$.

As before, the representations of $G = j-\lim_{\rightarrow} G_n$ are studied up to unitary equivalence.

DEFINITION 4. Two unitary representation of G, U_g on H and V_g on $\tilde{H}$ are called unitarily equivalent if there exists a unitary operator $U : H \to \tilde{H}$ such that $\forall g \in G$

$$U_g = U^* V_g U.$$

DEFINITION 5. A unitary invariant of a representation is a characteristic that is the same for the entire class of unitarily equivalent representations.

Before looking at the unitary invariants of the representations of the group $G = j - \lim\limits_{\to} G_n$, we prove the following theorem (analogous to Theorem 7 of Chapter 1).

THEOREM 8. The sequence of spectral types of the measures $\rho_1(\cdot) \gg \rho_2(\cdot) \gg \cdots$ on $(\mathbb{R}^\infty, \mathbf{B}(\mathbb{R}^\infty))$ constructed in Theorem 6 is a complete collection of unitary invariants of a unitary representation of $\mathbb{R}^\infty$, i.e. two unitary representations U_t and V_t of $\mathbb{R}^\infty$ are unitarily equivalent if and only if the corresponding spectral types of the measures $\rho_1(\cdot) \gg \rho_2(\cdot) \gg \cdots$ and $\tilde{\rho}_1(\cdot) \gg \tilde{\rho}_2(\cdot) \gg \cdots$ are the same.

Proof. We prove that two unitary representations $\mathbb{R}_0^\infty \ni t \mapsto U_t$ on H and V_t on H are unitarily equivalent if and only if the corresponding collections of CSO $(A_k)_{k=1}^\infty$ on H and $(B_k)_{k=1}^\infty$ on $\tilde{H}$ are unitarily equivalent (there exists a unitary operator $U : H \to \tilde{H}$ such that for all $k = 1,2,\ldots$ $A_k = U^* B_k U$). Indeed, the unitary equivalence of the collections $(A_k)_{k=1}^\infty$ and $(B_k)_{k=1}^\infty$ of CSO implies the unitary equivalence of the representations $\mathbb{R}_0^\infty \ni t \mapsto U_t = \exp(i \sum\limits_{k=1}^\infty t_k A_k)$ on H and $V_t = \exp(i \sum\limits_{k=1}^\infty t_k B_k)$ on $\tilde{H}$ with $\exp(i \sum\limits_{k=1}^\infty t_k A_k) = U^* \exp(i \sum\limits_{k=1}^\infty t_k B_k) U$ for all $t \in \mathbb{R}_0^\infty$ because these unitary operators are the same, for example, on the dense set of joint entire vectors of the collection $(A_k)_{k=1}^\infty$ of CSO in H. Conversely, as the resolutions of the identity $E(\cdot)$ and $F(\cdot)$ are uniquely defined on $(\mathbb{R}^\infty, \mathbf{B}(\mathbb{R}^\infty))$ by their corresponding representations U_t on H and V_t on $\tilde{H}$, for any $\Delta \in \mathbf{B}(\mathbb{R}^\infty)$ we have $U^* F(\Delta) U = E(\Delta)$. Consequently, the collections $(A_k)_{k=1}^\infty$ on H and $(B_k)_{k=1}^\infty$ on $\tilde{H}$ of CSO are unitarily equivalent.

So the unitary invariants of the collection $(A_k)_{k=1}^\infty$ of CSO and the unitary invariants of the representation $\mathbb{R}_0^\infty \ni t \mapsto U_t = \exp(i \sum\limits_{k=1}^\infty t_k A_k)$ are the same. By Theorem 7 of Chapter 1, the sequence of spectral types of the measures $\rho_1(\cdot) \gg \rho_2(\cdot) \gg \cdots$ is a complete collection of unitary invariants of the CSO $(A_k)_{k=1}^\infty$ and so the same is true for the corresponding representation of $\mathbb{R}_0^\infty$. ▯

A generalization of Theorem 8 for unitary representations of indutive limits G of locally abelian groups is also true.

THEOREM 9. The sequence of spectral types of the measures $\rho_1(\cdot) \gg \rho_2(\cdot) \gg \cdots$ on $(\hat{G}, \mathbf{B}(\hat{G}))$ constructed in Theorem 7 is a complete collection of unitary invariants of the unitary representation of the group $G = j - \lim\limits_{\to} G_n$ (G_n is a locally compact abelian group for all $n = 1,2,\ldots$).

Proof. First suppose that the representation $G \ni g \mapsto U_g$ is cyclic in H, i.e. for a vector Ω we have $\underset{g \in G}{\mathrm{CLS}} \{U_g \Omega\} = H$. Then this representation is unitarily equivalent to the representation $G \ni g \mapsto \chi(g)$ on $L_2(\hat{G}, d(E(\chi)\Omega, \Omega)) = L_2(\hat{G}, d\rho(\chi))$. The property of being cyclic is, of course, a unitary invariant of the representation and so for a representation with a simple spectrum the theorem could be restated as follows: for two cyclic unitary representations of the group G, $G \ni g \mapsto U_g$ on H and $G \ni g \mapsto V_g$ on H to be unitarily equivalent it is necessary and sufficient that their decompositions of the identity be equivalent. But if Ω is a cyclic vector of the representation U_g on H, if $E(\cdot)$ is its resolution of the identity and if there exists a unitary operator $U : H \to \tilde{H}$ such that $U_g = U^* V_g U$ for all $g \in G$, then $U\Omega = \tilde{\Omega}$ is cyclic for the representation V_g and $F(\cdot) = U E(\cdot) U^*$ is its resolution of the identity. So, $\rho(\cdot) = (E(\cdot)\Omega, \Omega)_H = (U E(\cdot) U^* U\Omega, U\Omega) = (F(\cdot)\tilde{\Omega}, \tilde{\Omega})_{\tilde{H}} = \tilde{\rho}(\cdot)$. Conversely, if the measures $\rho(\cdot)$ and $\tilde{\rho}(\cdot)$ on $(\hat{G}, \mathbf{B}(\hat{G}))$ are equivalent, then the unitary operator $U : L_2(\hat{G}, d\rho(\chi)) \to L_2(\hat{G}, d\tilde{\rho}(\chi))$ given by

$$(U f)(\chi) = \sqrt{\frac{d\tilde{\rho}}{d\rho}(\chi)}\, f(\chi)$$

defines a unitary equivalence of the representations $U_g = \chi(g)$ on $L_2(\hat{G}, d\rho(\chi))$ and $V_g = \chi(g)$ on $L_2(\hat{G}, d\tilde{\rho}(\chi))$.

Consider the general case. By Theorem 7, any unitary representation $G \ni g \mapsto U_g$ on H is unitarily equivalent to the representation $G \ni g \mapsto \chi(g)$ on $\oplus \sum_k L_2(\hat{G}, d\rho_k(\chi))$, where $\rho_1(\cdot) \gg \rho_2(\cdot) \gg \cdots$, $\rho_1(\cdot) = (E(\cdot)\Omega, \Omega)$ (Ω is a vector of maximal spectral type), $\rho_2(\cdot) = (E_1(\cdot)\Omega_1, \Omega_1)$ ($E_1(\cdot)$ is a spectral measure for $U_g \restriction_{H \ominus \mathrm{CLS}\{U_g\Omega | g \in G\}}$, Ω_1 is its vector of maximal spectral type) and so on. Then, if the representation $G \ni g \mapsto V_g$ is unitarily equivalent to the representation $G \ni g \mapsto \chi(g)$ on $\oplus \sum_k L_2(\hat{G}, d\tilde{\rho}_k(\chi))$, where $\tilde{\rho}_1(\cdot) \gg \tilde{\rho}_2(\cdot) \gg \cdots$ and $\tilde{\rho}_k(\cdot) \sim \rho_k(\cdot)$ $(k = 1, 2, \ldots)$, the unitary equivalence of V_g and U_g is checked directly.

Conversely, if $U : H \to \tilde{H}$ is a unitary operator such that $U_g = U^* V_g U$ for all $g \in G$, then by setting $\tilde{\Omega}_k = U\Omega_k \in \tilde{H}$ we get by induction
$\rho_k(\cdot) = (E_k(\cdot)\Omega_k, \Omega_k)_H = (F_k(\cdot)\tilde{\Omega}_k, \tilde{\Omega}_k)_{\tilde{H}} = \tilde{\rho}_k(\cdot)$. □

2.5. Gårding domains and entire vectors for a representation.

The operators of a unitary representation of an inductive limit of abelian locally compact Lie groups $G = j - \varinjlim G_n$ are, of course, defined on the entire representation space H. However, the self-adjoint operators which generate the corresponding representation of the commutative Lie algebra $\mathrm{g} = \varinjlim \mathrm{g}_n$, are, generally speaking not bounded.

For a locally compact Lie group G there is a standard procedure due to Gårding to construct a general invariant dense domain D_G for a representation of the Lie algebra $\mathfrak{g}_G$ using the unitary representation of the corresponding Lie group $G \ni g \mapsto U_g$. We review briefly Gårding's construction of D_G for the representation U_G. Let $C_0^\infty(G)$ be the set of all infinitely differentiable functions on the Lie group G with compact support. As shown by L. Gårding [1] the set D_G of finite linear combinations of the vectors

$$\phi_f = \int_G f(g)\, U(g)\, \phi\, dg\,, \quad \phi \in H\,, \quad f \in C_0^\infty(G) \tag{7}$$

is dense everywhere in H. This set is contained in the domain of all generators of one-parameter subgroups of G and these generators map D_G into itself, so D_G consists of infinitely differentiable vectors for the representation U_g. Recall that a vector $x \in H$ is called infinitely differentiable (analytic) for U if the mapping $G \ni g \mapsto U(g)x \in H$ belongs to the class C^∞ (analytic). We denote the space of infinitely differentiable (analytic) vectors for a representation U by $H^\infty(U)$ $(H^\omega(U))$. For a representation U one can introduce a continuous scale of spaces of analytic vectors $H_t^\omega(U)$, $0 < t < \infty$ such that $H^\omega(U) = \bigcup_{t>0} H_t^\omega(U)$. This set $H^c(U) = \bigcap_{t>0} H_t^\omega(U)$ is called the space of entire vectors of the representation U.

Following G. Hegerfeldt [2], a Gårding domain for a strongly continuous unitary representation of the group G is a domain D which is invariant with respect to the operators of the representation and the generators of all one-parameter subgroups, and which further is a core for these generators.

E. Nelson [1] has strengthened Gårding's result by showing that every strongly continuous unitary representation of a Lie group on a separable Hilbert space has an everywhere dense set of analytic vectors. R. Goodman [3] has shown that the set of entire vectors for an irreducible representation of a nilpotent Lie group is everywhere dense in H.

For the groups (even commutative) which are not locally compact the Gårding domain does not always exist (see below Remark 6). Nevertheless M. Reed [1], P. Richter [1], J. Simon [2] and G. Hegerfeldt [2] have constructed Gårding domains which consist of analytic or entire vectors for unitary representations of some infinite-dimensional groups.

However their constructions differ from (7).

The construction of D for unitary representations of groups which are not locally compact analogous to the Gårding construction, needs a modification because now there is no invariant measure on the group. In addition, the space $C_0^\infty(G)$ for a group G which is not locally compact, is not studied in that much detail as in the locally compact case. The best studied spaces of functions of infinitely many variables are spaces of cylindrical functions, but in the case of infinitely many variables a cylindrical function cannot have compact support.

Using (7), we give below a construction of the Gårding domain for the strongly continuous unitary representations of the inductive limits of commutative locally compact Lie groups $G = \lim\limits_{\rightarrow} \mathbb{R}^n$ which are the simplest non-locally compact groups. Here, instead of an invariant measure on G and the space $C_0^\infty(G)$ we use a quasi-invariant measure (Gaussian measure with a correlation operator dependent on the representation) and the space A(G) of entire functions of minimal type and with order of growth not greater than two (see Yu.G. Kondrat'ev, Yu.S. Samoĭlenko [2], Yu.M. Berezanskiĭ [9]). The constructed Gårding domain consists of entire vectors for the representation operators of the Lie algebra. It has a nuclear topology such that the representation operators are continuous, and it also has some other properties stated, see Theorem 11.

First we give the construction of the Gårding domain for unitary representations of $\mathbb{R}^1$ and then generalize it for the infinite dimensional case. This construction uses the Gaussian measure on the group and the properties of the nuclear space $\mathbb{Z}_0^2(\mathbb{R}^1)$ which consists of restrictions to $\mathbb{R}$ of entire functions of minimal type with order of growth not greater than two. Then this construction is generalized for the representations of the group $\mathbb{R}_0^\infty = \bigcup\limits_{n=1}^{\infty} \mathbb{R}^n \times (0,0,\ldots)$.

In the case of a unitary strongly continuous representation of the additive group $\mathbb{R}^1$ on a separable Hilbert space, the Gårding domain D is defined to be the set of all finite linear combinations of vectors of the form $\phi_f = \int\limits_{\mathbb{R}^1} f(t)\, U(t)\, \phi\, dt,\ f \in C_0^\infty(\mathbb{R}^1),\ \phi \in H,$ where dt is the Lebesgue measure on $\mathbb{R}^1$.

We show what kind of function space $N(\mathbb{R}^1)$ under the integral sign in (7) can be choosen in order

a) to change the Lebesgue measure in (7) into the Gaussian measure;

b) to include in $N(\mathbb{R}^1)$ the identity (so that in what follows one can consider cylindrical functions);

c) that the constructed set of vectors is invariant with respect to the representation operators and can be endowed with a nuclear topology in which the representation operators must be continuous.

If the representation is cyclic, then for D to be dense in H, it is sufficient to fix a cyclic vector $\phi_0 \in H$ in (7). Assuming that the representation is cyclic and taking into account a) - c) we find that the domain D becomes $D = \{ \int\limits_{\mathbb{R}^1} a(t)\, U(t)\, \phi_0\, dg(t) \mid a \in N(\mathbb{R}^1)\}$.

From the way the representation operators act on the vectors of D we get

$$U_s(\int\limits_{\mathbb{R}^1} a(t)\, U_t\, \phi_0\, dg(t)) = \int\limits_{\mathbb{R}^1} a(t-s)\, \frac{dg(t-s)}{dg(t)}\, U(t)\, \phi_0\, dg(t).$$

So for D to be invariant with respect to the representation operators we have to choose $N(\mathbb{R}^1)$ such that the operators $T_s (s \in \mathbb{R}^1)$

$$(T_s a)(t) = a(t-s)\frac{dg(t-s)}{dg(t)} = a(t-s)\exp(2ts-s^2) \tag{9}$$

map D into itself. Letting $a(t)=1$ in (9) we are led to the conclusion that $N(\mathbb{R}^1)$ must include functions with exponential growth. We will show below that a suitable space is the space $\mathbb{Z}_0^2(\mathbb{R}^1)$ defined as follows, is a suitable space.

Let $\mathbb{Z}_0^2(\mathbb{C}^1)$ be the set of entire functions $u(z)$, $\mathbb{C}^1 \ni z \mapsto u(z) \in \mathbb{C}^1$ which satisfy the estimate $|u(z)| \le C_{u,\varepsilon}\exp(\varepsilon |z|^2)$ $(z \in \mathbb{C}^1)$ for an arbitrary $\varepsilon > 0$. Then $\mathbb{Z}_0^2(\mathbb{R}^1) = \mathbb{Z}_0^2(\mathbb{C}^1)\restriction_{\mathbb{R}^1}$.

We will need another description of $\mathbb{Z}_0^2(\mathbb{R}^1)$. Let $L_2(\mathbb{R}^1, dg)$ be the space of functions with summable square with respect to the measure $dg(t) = \frac{1}{\sqrt{\pi}} e^{-t^2}\, dt$. The Hermite polynomials $(h_k)_{k=0}^{\infty}$

$$h_k(t) = \frac{(-1)^k}{(2^k k!)^{1/2}} e^{t^2} \frac{d^k}{dt^k}(e^{-t^2})$$

form an orthonormal basis in $L_2(\mathbb{R}^1, dg)$. To every function $u \in L_2(\mathbb{R}^1, dg)$ there corresponds the sequence $(u_k)_{k=0}^{\infty}$ of Fourier coefficients with respect to the basis $(h_k)_{k=0}^{\infty}$.

For every $l = 1,2,\ldots$ introduce the Hilbert space

$$A_l(\mathbb{R}^1) = \{u \in L_2(\mathbb{R}^1, dg) \mid \sum_{k=0}^{\infty} |u_k|^2 l^k < \infty\}.$$

Obviously, $A_1(\mathbb{R}^1) = L_2(\mathbb{R}^1, dg)$, $A_{l+1}(\mathbb{R}^1) \subset A_l(\mathbb{R}^1)$. The set $\bigcap_{l=1}^{\infty} A_l(\mathbb{R}^1)$ contains all possible linear combinations of the Hermite polynomials and consequently is dense in every $A_l(\mathbb{R}^1)$. The space $\mathrm{A}(\mathbb{R}^1) = \varprojlim A_l(\mathbb{R}^1)$ is nuclear.

Introduce also the space

$$\mathrm{B}(\mathbb{R}^1) = \{\sum_{k=0}^{\infty} u_k \frac{t^k}{(2^k k!)^{1/2}} \;\Big|\; \sum_{k=0}^{\infty} |u_k|^2 m^k < \infty,\ m = 1,2,\ldots,\ t \in \mathbb{R}^1\}.$$

It turns out (see Yu.G. Kondrat'ev and Yu.S. Samoĭlenko [2]) that

$$\mathbb{Z}_0^2(\mathbb{R}^1) = \mathrm{A}(\mathbb{R}^1) = \mathrm{B}(\mathbb{R}^1).$$

Now, the Hilbert space H can be realized as a space $L_2(\mathbb{R}^1, d\mu)$, where μ is a probability measure. With such a realization the operators U_t correspond to the multiplication operators in $L_2(\mathbb{R}^1, d\mu)$ by the functions $\mathbb{R}^1 \ni x \mapsto e^{itx} \in \mathbb{C}^1$, $t \in \mathbb{R}^1$ and the cyclic vector ϕ_0 correspond to the function identically equal to one. The set $D = \{\int_{\mathbb{R}^1} a(t)e^{itx}\, dg(x) \mid a \in \mathrm{A}(\mathbb{R}^1)\} \subset L_2(\mathbb{R}^1, d\mu)$ is linear. Consider the mapping F_0

$$\mathrm{A}(\mathbb{R}^1) \ni a \mapsto (F_0 a)(x) = \int_{\mathbb{R}^1} a(t) e^{itx} dg(t) \in L_2(\mathbb{R}^1, d\mu).$$

From the inequality

$$\|F_0 a\|^2_{L_2(\mathbb{R}^1, d\mu)} = \int_{\mathbb{R}^1} \left| \int_{\mathbb{R}^1} a(t) e^{itx} dg(t) \right|^2 d\mu(x) \leq$$

$$\leq \left(\int_{\mathbb{R}^1} | a(t) | \, dg(t) \right)^2 \leq \|a\|^2_{L_2(\mathbb{R}^1, dg)} = \|a\|^2_{A_1(\mathbb{R}^1)} \tag{10}$$

it follows that $D \subset L_2(\mathbb{R}^1, d\mu)$ and that F_0 is continuous.

The mapping F_0 induces the topology of the space $\mathrm{A}(\mathbb{R}^1)$ on D. If F_0 is not a bijection from $\mathrm{A}(\mathbb{R}^1)$ onto D, then the topology on D is the same as the factor-topology on the space $\mathrm{A}(\mathbb{R}^1) / \mathrm{Ker} F_0$.

THEOREM 10. The following statements hold:

1) D is a nuclear linear topological space;

2) D is topologically densely imbedded in H;

3) $U_t \in L(D \to D) \; \forall t \in \mathbb{R}^1$;

4) D is contained in the domains of the generators of one-parameter subgroups of the group $\mathbb{R}^1$ and these generators map D into itself;

5) D consists of the entire vectors for the generators of one-parameter subgroups of the group $\mathbb{R}^1$;

6) the function $\mathbb{R}^1 \ni f \mapsto U_t f \in D$ is continuous for all $f \in D$.

Proof. 1. D is a nuclear space as a factor space of the nuclear space $\mathrm{A}(\mathbb{R}^1)$.

2. We will show that D is topologically densely imbedded in $L_2(\mathbb{R}^1, d\mu)$. From the relations (N.Ya. Vilenkin [2, ch. XI, §5])

$$\int_{\mathbb{R}^1} h_k(t) e^{itx} dg(t) = \frac{(ix)^k}{(2^k k!)^{1/2}} e^{x^2/4}$$

it follows that

$$D = \mathrm{A}(\mathbb{R}^1) e^{-x^2/4}. \tag{11}$$

Taking into consideration that A contains all possible polynomials, the proof that D is dense in $L_2(\mathbb{R}^1, d\mu)$ is a consequence of Lemma 1.

LEMMA 1. Let $\rho(\cdot)$ be an arbitrary measure on $(\mathbb{R}^1, \mathbf{B}(\mathbb{R}^1))$, let $f \in L_2(\mathbb{R}^1, d\rho)$, $f \neq 0 (\mathrm{mod}\, \rho)$ and let there exists $\delta > 0$ such that $e^{\delta |t|} f(t) \in L_2(\mathbb{R}^1, d\rho)$. Then $\mathrm{CLS}\{t^n f(t)\}_{n=0}^{\infty} = L_2(\mathbb{R}^1, d\rho)$.

Proof. Let $h \in L_2(\mathbb{R}^1, d\rho)$ be a function such that $\int_{\mathbb{R}^1} t^n h(t) f(t) \, d\rho(t) = 0$, $n = 0, 1, \ldots$, we will show that $h = 0$ in $L_2(\mathbb{R}^1, d\rho)$ using the properties of the Fourier transform of sign changing measures. Recall that a sign changing measure ω on the σ-algebra $\mathbf{R}$ of subsets of the space R is a finite absolutely continuous function defined on the sets

$$\mathbf{R} \ni \Delta \mapsto \omega(\Delta) \in \mathbb{R}^1.$$

For sign changing measures, the Jordan decomposition $\omega = \omega_+ - \omega_-$ holds, where ω_+, ω_- are measures. Obviously, $|\omega| = \omega_+ + \omega_-$ is also a measure.

Define the Fourier transform of a sign changing measure ω on the σ-algebra $\mathbf{B}(\mathbb{R}^1)$ of Borel sets, to be the integral

$$g(\lambda) = \int_{\mathbb{R}^1} e^{-it\lambda}\, d\omega(t) = \int_{\mathbb{R}^1} e^{-it\lambda}\, d\omega_+(t) - \int_{\mathbb{R}^1} e^{-it\lambda}\, d\omega_-(t), \quad \lambda \in \mathbb{R}^1.$$

The Fourier transform of a sign changing measure has the uniqueness property: if $g(\lambda) \equiv 0$ then $\omega(\Delta) = 0 \ \forall \Delta \in \mathbf{B}(\mathbb{R}^1)$. If the sign changing measure ω is such that for some $\delta > 0 \int_{\mathbb{R}} e^{\delta|t|}\, d\omega(t) < \infty$ then $g(\lambda)$ can be analytically extended to the strip $|\operatorname{Im}\lambda| < \delta$.

Assume that f and h are real valued. If not we consider the complex sign changing measures

$$\mathbf{B}(\mathbb{R}^1) \ni \Delta \mapsto \omega(\Delta) = \omega_1(\Delta) + i\omega_2(\Delta) \in \mathbb{C}^1,$$

where ω_1, ω_2 are real sign changing measures. Consider the integral $\omega(\Delta) = \int_\Delta f(t)\, h(t)\, d\rho(t), \ \Delta \in \mathbf{B}(\mathbb{R}^1)$.

It is clear that ω is a sign changing measure and

$$\int_{\mathbb{R}^1} e^{\delta|t|}\, d\,|\omega(t)| = \int_{\mathbb{R}^1} e^{\delta(t)}\, |f(t)\, h(t)|\, d\rho(t) < \infty.$$

So the Fourier transform of the sign changing measure ω

$$g(\lambda) = \int_{\mathbb{R}^1} e^{-i\lambda t}\, d\omega(t), \quad \lambda \in \mathbb{R}^1$$

can be analytically extended to the strip $|\operatorname{Im}\lambda| < \infty$. Because

$$g^{(n)}(0) = \int_{\mathbb{R}^1} (-it)^n\, d\omega(t) = \int_{\mathbb{R}^1} (-it)^n f(t)\, h(t)\, d\rho(t) = 0$$

$$n = 0, 1, \ldots$$

we have $g(\cdot) \equiv 0$. Thus, it follows from the uniqueness that $\omega(\Delta) = \int_\Delta f(t)\, h(t)\, d\rho(t) = 0$ $\forall \Delta \in \mathbf{B}(\mathbb{R}^1)$ and so $f(t)\, h(t) = 0 \pmod{\rho}$ and $h(t) = 0 \pmod{\rho}$.

The inequality (10) shows that the imbedding of D into $L_2(\mathbb{R}^1, d\mu)$ is topological.

3. We show that $U_t \in L(D \to D)$. From (8) it follows that property 3 is equivalent with the operators (9) $(T_s\, a)(t) = a(t-s)\, e^{2ts - s^2}$ being continuous.

LEMMA 2. We have $T_s \in L(\mathbf{A}(\mathbb{R}^1) \to \mathbf{A}(\mathbb{R}^1))$.

Proof. Denote by $L(l_1, l_2)$ $(\sigma_2(l_1, l_2))$ the space of continuous (Hilbert-Schmidt) operators from $\mathbf{A}_{l_1}(\mathbb{R}^1)$ into $\mathbf{A}_{l_2}(\mathbb{R}^1)$. Because the space $\mathbf{A}(\mathbb{R}^1)$ has the projective limit topology, to prove the lemma it is sufficient to show that $\forall s \in \mathbb{R}^1$ $l_1 \in \mathbb{N}, \exists l_2 \in \mathbb{N} : T_s \in L(l_2, l_1)$.

It follows from the inclusion $\sigma_2(l_2, l_1) \subset L(l_2, l_1)$ that it is sufficient to show that $T_s \in \sigma_2(l_2, l_1)$. Note, that $\|T\|_{L(l_2,l_1)} \le \|T\|_{\sigma_2(l_2,l_1)}$.

An easy calculation shows that for $T \in \sigma_2(l_2, l_1)$

$$\|T\|^2_{\sigma_2(l_2,l_1)} = \sum_{k,m=0}^{\infty} \frac{|T_{k,m}|^2}{l_2^k\, l_1^m}, \quad \text{where } T_{k,m} = (T\, h_m, h_k)_{L_2(\mathbb{R}^1, dg)}.$$

Let us calculate

$$T_{k,m}(s) = (T_s\, h_m, h_k)_{L_2(\mathbb{R}^1,dg)} = \int_{\mathbb{R}^1} e^{-s^2+2ts}\, h_m(t-s) \times$$

$$\times h_k(t)\, dg(t) = \frac{1}{\sqrt{\pi}} \int_{\mathbb{R}^1} h_m(t-s)\, h_k(t)\, e^{-s^2+2ts-t^2}\, dt =$$

$$= \frac{1}{\sqrt{\pi}} \int_{\mathbb{R}^1} h_m(t-s)\, h_k(t)\, e^{-(t-s)^2}\, dt = \frac{1}{\sqrt{\pi}} \int_{\mathbb{R}^1} h_m(t)\, h_k(t+s)\, e^{-t^2}\, dt =$$

$$= \frac{1}{(2^{(m+k)}\, m!\, k!)^{1/2}} \int_{\mathbb{R}^1} H_m(t)\, H_k(t+s)\, dg(t)$$

where $H_k(t) = (2^k\, k!)^{1/2}\, h_k(t) = (-1)^k\, e^{t^2} \dfrac{d^k}{dt^k}(e^{-t^2})$.

Noting that

$$H_k(t+s) = (-1)^k\, e^{(t+s)^2} \frac{d^k}{dt^k}(e^{-(t+s)^2}) =$$

$$= (-1)^k\, e^{(t+s)^2} \frac{d^k}{dt^k}(e^{-t^2}\, e^{-s^2-2ts}) = (-1)^k\, e^{(t+s)^2} \sum_{n=0}^{k} C_k^n \frac{d^n}{dt^n}(e^{-t^2}) \times$$

$$\times \frac{d^{k-n}}{dt^{k-n}}(e^{-s^2-2ts}) = (-1)^k\, e^{s^2-2ts} \sum_{n=0}^{k} C_k^n\, e^{t^2} \frac{d^n}{dt^n}(e^{-t^2}\,(-1)^{k-n} \times$$

$$\times (2s)^{k-n}\, e^{-s^2-2ts}) = \sum_{n=0}^{k} C_k^n (2s)^{k-n}\, H_n(t)$$

we have

$$T_{k,m}(s) = \frac{2}{(2^{m+k}\, k!\, m!)^{1/2}} \sum_{n=0}^{k} C_k^n (2s)^{k-n} \int_{\mathbb{R}^1} H_m(t)\, H_n(t)\, dg(t) =$$

$$= \frac{1}{(2^{m+k}\,k\,!\,m\,!)^{1/2}} \sum_{n=0}^{k} C_k^n (2s)^{k-n}\,(2^m\,m\,!)\,\delta_{m,n}.$$

So

$$T_{k,m}(s) = \begin{cases} 0 & m > k\,, \\ \dfrac{1}{(2^{m+k}\,m\,!\,k\,!)^{1/2}}\, C_k^m\,(2s^2)^{k-m}\,2^m\,m\,! & m \le k\,, \end{cases} \tag{12}$$

and

$$\|T_s\|^2_{\sigma_2(l_2,l_1)} = \sum_{0\le m\le k} \frac{|T_{k,m}(s)|^2}{l_1^m\,l_2^k} = \sum_{n=0}^{\infty}\sum_{m=0}^{\infty} \frac{|T_{n+m,m}(s)|^2}{l_1^m\,l_2^{n+m}} =$$

$$= \sum_{n=0}^{\infty}\sum_{m=0}^{\infty} C_{n+m}^m \frac{1}{n!}(2s^2)^n \frac{1}{l_1^m\,l_2^{n+m}} = \sum_{n=0}^{\infty}\left[\frac{2s^2}{l_2}\right]^n \frac{1}{(n\,!)^2} \sum_{m=0}^{\infty} \frac{(n+m)!}{m!} \frac{1}{(l_1\,l_2)^m} =$$

$$= \sum_{n=0}^{\infty}\left[\frac{2s^2}{l_2}\right]^n \frac{1}{(n\,!)^2} \frac{1}{n\,!\,(1-\frac{1}{l_1\,l_2})^{n+1}} = \left[\frac{l_1\,l_2}{l_1\,l_2-1}\right] \sum_{n=0}^{\infty}\left[\frac{2s^2\,l_1}{l_1\,l_2-1}\right]^n \frac{1}{(n\,!)^3} \le$$

$$\le \left[\frac{l_1\,l_2}{l_1\,l_2-1}\right] \sum_{n=0}^{\infty}\left[\frac{2s^2\,l_1}{l_1\,l_2-1}\right]^n \frac{1}{n\,!} = \left[\frac{l_1\,l_2}{l_1\,l_2-1}\right] e^{\frac{2s^2 l_1}{l_1 l_2 - 1}} < \infty.$$

We used the series

$$\frac{1}{n\,!(1-z)^{n+1}} = \sum_{m=0}^{\infty} \frac{(n+m)!}{m\,!} z^m \quad \text{for} \quad |z| < 1$$

obtained by n times differentiating the series $\dfrac{1}{1-z} = \sum_{m=0}^{\infty} z^m\,,\ |z| < 1.$

4. Here we show that the domains of the generators of one-parameter subgroups of the group $\mathbb{R}^1$ contain D and the generators map D into itself.

Consider the subgroup $\{e^{itx} \mid t \in \mathbb{R}^1\}$. Its generator is a self-adjoint operator A of multiplication in $L_2(\mathbb{R}^1, d\mu)$ by the independent variable

$$(A\,f)\,(t) = t\,f(t)\,,\quad f \in D(A)\,,$$

$$D(A) = \{f \in L_2(\mathbb{R}^1, d\mu) \mid \|A\,f\|^2 = \int_{\mathbb{R}^1} |t\,f(t)|^2\,d\mu(t) < \infty\}.$$

Other generators have the form $k\,A$, where $k \in \mathbb{R}^1$. It easily follows from (11) that $D \subset D(A)$ and $A(D) \subset D(A)$.

5. To prove that D consists of entire vectors for the generators of the one-parameter subgroups of the group $\mathbb{R}^1$ it is sufficient to consider a single generator A. The set $H^c(A)$ of entire vectors for the operator A equals $\bigcap_{\varepsilon>0} D(e^{\varepsilon|x|})$, where $D(e^{\varepsilon|x|})$ is the domain of

the operator of multiplication in $L_2(\mathbb{R}^1, d\mu)$ by the function $\mathbb{R}^1 \ni x \mapsto e^{\varepsilon|x|} \in \mathbb{R}^1$. Using (11), we get $D \subset \bigcap_{\varepsilon>0} D(e^{\varepsilon|x|})$.

6. We show that the function $\mathbb{R}^1 \ni y \mapsto u_y f \in D \ \forall f \in D$ is continuous. Taking into consideration the topology of D and using (9) it is sufficient to show that the function $\mathbb{R}^1 \ni s \mapsto T_s a \in \mathrm{A}(\mathbb{R}^1)$ is continuous for all $a \in \mathrm{A}(\mathbb{R}^1)$. But we have the following lemma.

LEMMA 3. $\forall l_1 \in \mathbb{N} \ \exists l_2 \in \mathbb{N}$ such that $\lim_{s\to 0} \|T_s - T_0\|_{L(l_2,l_1)} = 0$.

Proof. As in the proof of Lemma 2, it is sufficient to show that $\lim_{s\to 0} \|T_s - T_0\|_{\sigma_2(l_2,l_1)} = 0$. Using (12) we have

$$((T_s - T_0)h_m, h_k) = (T_s h_m, h_k) - (h_m, h_k) = T_{k,m}(s) - \delta_{k,m} =$$

$$= \begin{cases} 0 & m \geq k \\ T_{k,m}(s), & m < k. \end{cases}$$

Consequently

$$\|T_s - T_0\|^2_{\sigma_2(l_2,l_1)} = \sum_{0\leq m<k} \frac{|T_{k,m}(s)|^2}{l_1^m l_2^k} = \left[\frac{l_1 l_2}{l_1 l_2 - 1}\right] \sum_{n=1}^{\infty} \left[\frac{2s^2 l_1}{l_1 l_2 - 1}\right]^n \frac{1}{(n!)^3} \leq$$

$$\leq \left[\frac{l_1 l_2}{l_1 l_2 - 1}\right] \sum_{n=1}^{\infty} \left[\frac{2s^2 l_1}{l_1 l_2 - 1}\right]^n \frac{1}{n!} = \left[\frac{l_1 l_2}{l_1 l_2 - 1}\right] \left[\exp\left[\frac{2s^2 l_1}{l_1 l_2 - 1}\right] - 1\right] \to 0, \ s \to 0.$$

Remark 4. If the initial representation is not cyclic, then there exists a countable sequence of measures $(\mu_n(\cdot))_{n=1}^{\infty}$ on $\mathbb{R}^1$ such that H is unitarily equivalent to $\oplus \sum_{n=1}^{\infty} L_2(\mathbb{R}^1, d\mu_n)$, and the operator $U(t)$ restricted to each $L_2(\mathbb{R}^1, d\mu_n)$ equals the multiplication operator e^{itx}. Then we construct the Gårding domain D_n with the properties 1) - 6) on every $L_2(\mathbb{R}^1, d\mu_n)$ and define $D = \oplus \sum_{n=1}^{\infty} D_n$.

□

Now consider a strongly continuous unitary representation of the group of finite real number sequences $t = (t_1, \ldots, t_n, \ldots) \in \mathbb{R}_0^{\infty}$ on a separable Hilbert space $H : \mathbb{R}_0^{\infty} \ni t \mapsto U_t \in \mathrm{U}(H)$. Suppose that the representation is cyclic. Then the space H is isometric to the space $L_2(\mathbb{R}^{\infty}, d\mu)$, where μ is some probability measure on the measurable space $(\mathbb{R}^{\infty}, \mathrm{B}(\mathbb{R}^{\infty}))$.

The operators U_t correspond to the multiplication operators $e^{i(t,x)}$ on $L_2(\mathbb{R}^{\infty}, d\mu)$ and a cyclic vector corresponds to the identity function. From Lemma 3 of Chapter 1 it follows that for any probability measure μ on $\mathbb{R}^{\infty}$ there is a Hilbert space of full μ-measure, i.e. there is $l_2(a_n) \subset \mathbb{R}^{\infty}$, $a_n > 0$, $n = 1,2,\ldots$ such that $\mu(l_2(a_n)) = 1$.

On the dual space $l_2\left[\frac{1}{a_n}\right]$ define the Gaussian measure

$dg_\alpha(t) = \bigotimes_{k=1}^{\infty} \sqrt{\frac{\alpha_k}{\pi}} \exp(-\alpha k\, t_k^2)\, dt_k$ such that the measure $g_\alpha(l_2\left[\frac{1}{a_n}\right]) = 1$. It follows from A.N. Kolmogorov, A.Ya. Khinchin criterion that it is sufficient to choose the sequence $(\alpha_k)_{k=1}^{\infty}$ such that $\sum_{k=1}^{\infty} (a_k \alpha_k)^{-1} < \infty$.

We will construct the Gårding space as in the one-dimensional case.

$$D = \{ \int_{\mathbb{R}^\infty} a(t)\, e^{i(\lambda,t)}\, dg_\alpha(t) \mid a \in \mathrm{A}^\alpha(\mathbb{R}^\infty)\} ,$$

where $\mathrm{A}^\alpha(\mathbb{R}^\infty)$ is the infinite tensor product of $\mathrm{A}^\alpha(\mathbb{R}^1)$. More exactly, denote by $\mathbb{N}_0^\infty$ the set of natural number sequences, which starting from a certain entry (dependent on the sequence) take the value one.

Let $\tau \in \mathbb{N}_0^\infty$, $\tau = (t_k)_{k=1}^\infty$. Denote $A_\tau^\alpha(\mathbb{R}^\infty) = \bigotimes_{k=1}^{\infty} A_{\tau_k}^{\alpha_k}(\mathbb{R}^1)$, $\mathrm{A}^\alpha(\mathbb{R}^\infty) = \underset{\tau \in \mathbb{N}_0^\infty}{\mathrm{prlim}}\, A_\tau^\alpha(\mathbb{R}^\infty)$ where $A_{\tau_k}^{\alpha_k}(\mathbb{R}^1) = \{f^{\alpha_k}(\cdot) = f(\sqrt{\alpha_k}\,\cdot) \mid f \in A_{\tau_k}(\mathbb{R}^1)\}$.

The space $\mathrm{A}(\mathbb{R}^\infty)$ is a nuclear space which consists of cylindrical functions (see Yu.M. Berezanskiĭ [9], Yu.G. Kondrat'ev, Yu.S. Samoĭlenko [2]). The space $\mathrm{A}(\mathbb{R}^\infty)$ can be described as in the one-dimensional case. Denote by $\mathbb{Z}_0^2(\mathbb{C}^p)$ the space of entire functions $\mathbb{C}^p \ni z \mapsto u(z) \in \mathbb{C}^1$ satisfying the estimate $|u(z)| \le C_{u,\varepsilon} \exp(\varepsilon \sum_{n=1}^{p} |z_n|^2)$ for any $\varepsilon > 0$. Define $\mathbb{Z}_0^2(\mathbb{R}^p) = \mathbb{Z}_0^2(\mathbb{C}^p)\restriction_{\mathbb{R}^p}$, $\mathbb{Z}_0^2(\mathbb{R}_0^\infty) = j - \underset{\rightarrow}{\lim}\, \mathbb{Z}_0^2(\mathbb{R}^p)$, where the inclusion $j_{p+1}^p : \mathbb{Z}_0^2(\mathbb{R}^p) \to \mathbb{Z}_0^2(\mathbb{R}^{p+1})$ is defined to be the multiplication by the identity. Then $\mathrm{A}^\alpha(\mathbb{R}^\infty) = \mathbb{Z}_0^2(\mathbb{R}_0^\infty)$.

Let us introduce a topology on D. To do it consider the mapping F_0:

$$\mathrm{A}^\alpha(\mathbb{R}^\infty) \ni a \mapsto (F_0 a)(t) = \int_{\mathbb{R}^1} a(x)\, e^{i(t,x)}\, dg_\alpha(x) \in L_2(\mathbb{R}^\infty, d\mu).$$

From the inequality

$$\|F_0 a\|^2_{L_2(\mathbb{R}^\infty, d\mu)} = \int_{\mathbb{R}^\infty} \Big| \int_{\mathbb{R}^\infty} a(x)\, e^{i(t,x)}\, dg_\alpha(x)\Big|^2 d\mu(t) \le$$

$$\le \Big(\int_{\mathbb{R}^\infty} |a(x)|\, dg_\alpha(x)\Big)^2 \le \|a\|^2_{L_2(\mathbb{R}^\infty, dg_\alpha)} \|1\|^2_{L_2(\mathbb{R}^\infty, dg_\alpha)} = \|a\|^2_{A_1^\alpha(\mathbb{R}^\infty)} \tag{13}$$

it follows that $D \subset L_2(\mathbb{R}^\infty, d\mu)$.

The mapping F_0 induces the topology of the space $\mathrm{A}^\alpha(\mathbb{R}^\infty)$ on D. If F_0 is not one-to-one, then the topology on D equals the topology of the factor space $\mathrm{A}^\alpha(\mathbb{R}^\infty) / \mathrm{Ker} F_0$.

THEOREM 11. The following statements hold:

1) D is a nuclear topological space;

2) D is topologically densely imbedded in H;

3) $U_t \in L(D \to D)$, $t \in \mathbb{R}_0^\infty$;

4) D is contained in the domains of the generators of all one-parameter subgroups of the group $\mathbb{R}_0^\infty$, and these generators map D into itself;

5) D consists of the entire vectors for the generators of all one-parameter subgroups of the group $\mathbb{R}_0^\infty$;

6) the function $\mathbb{R}_0^\infty \ni t \mapsto U_t f \in D$ is continuous $\forall f \in D$.

Proof.

1. D is nuclear being the factor space $A^\alpha(\mathbb{R}^\infty) / \mathrm{Ker} F_0$ of the nuclear space $A^\alpha(\mathbb{R}^\infty)$.

2. To prove that $D \subset L_2(\mathbb{R}^\infty, d\mu)$ is dense, we prove a lemma similar to Lemma 1.

Introduce the necessary notations. Let I be the set of nonnegative integers, I_0^∞ the totality of nonnegative integer sequences which starting form a certain entry (depending on the sequence) assume the values zero. For $\beta \in I_0^\infty$ by $\nu(\beta)$ denote the minimal $m = 1, 2, \ldots$ such that $\beta_{m+1} = \beta_{m+2} = \cdots = 0$.

Let $\nu(\beta) = p$, $\beta = (\beta_1, \beta_2, \ldots, \beta_0, 0, \ldots) \in I_0^\infty$, $(x_1, x_2, \ldots) \in \mathbb{R}^\infty$, the symbols $|\beta|$, x^β, D^β are defined to be

$$|\beta| = \sum_{i=1}^{p} |\beta_i|, \; x^\beta = x_1^{\beta_1} x_2^{\beta_2} \cdots x_p^{\beta_p}, \; D^\beta = \frac{\partial^{|\beta|}}{\partial x_1^{\beta_1} \partial x_2^{\beta_2} \cdots \partial x_p^{\beta_p}}.$$

LEMMA 4. Let μ be an arbitrary probability measure on $\mathbb{R}^\infty$, let $\mathbb{R}^\infty \supset H_-$ be a Hilbert space of full measure, f a function in $L_2(\mathbb{R}^\infty, d\mu)$, $f \neq 0 (\mathrm{mod}\, \mu)$ such that $\exists \delta > 0 \exp(\delta \|x\|_{H_-}) f(x) \in L_2(\mathbb{R}^\infty, d\mu)$. Then $\mathrm{CLS}\{x^\beta f(x) \mid \beta \in I_0^\infty\} = L_2(\mathbb{R}^\infty, d\mu)$.

Proof. Let $h \in L_2(\mathbb{R}^\infty, d\mu)$ be such that $\int_{\mathbb{R}^\infty} x^\beta f(x) \overline{h(x)}\, d\mu(x) = 0$, $\beta \in I_0^\infty$. We prove that $h = 0$ in $L_2(\mathbb{R}^\infty, d\mu)$. The proof follows that of Lemma 1.

Let ω be a finite sign changing measure on $(\mathbb{R}^\infty, \mathbf{B}(\mathbb{R}^\infty))$ where $\mathbb{R}^\infty = \mathbb{R}^1 \times \mathbb{R}^1 \times \ldots$, $\mathbf{B}(\mathbb{R}^\infty)$ is a σ-algebra generated by cylindrical sets with he Borel base in $\mathbb{R}^\infty$. The Jordan decomposition holds for ω, $\omega = \omega_+ - \omega_-$, where ω_+ and ω_- are measures. Clearly $|\omega| = \omega_+ + \omega_-$ is also a measure.

Define the Fourier transform of a sign changing measure to be the integral $g(t) = \int_{\mathbb{R}^\infty} e^{-i(t,x)}\, d\omega(x) = \int_{\mathbb{R}^\infty} e^{-i(t,x)}\, d\omega_+(x) - \int_{\mathbb{R}^\infty} e^{-i(t,x)}\, d\omega_-(x)$. The Fourier transform of a sign changing measure has the uniqueness property: if $g(t) \equiv 0$ then $\omega(\Delta) = 0$ $\forall \Delta \in \mathbf{B}(\mathbb{R}^\infty)$.

If a sign changing measure ω is such that for some $\delta > 0$ $\int_{\mathbb{R}^\infty} \exp(\delta \|x\|_{H_-})\, d|\omega(x)| < \infty$ then for any fixed n, the function $g(t)$, considered as a function of n variables, can be analytically extended into the domain $\{z = x + iy \in \mathbb{C}^n \mid \|y\|_{H_+} < \delta\}$, where H_+ is the space dual to H_- with respect to $l_2 \subset \mathbb{R}^\infty$.

Suppose that the functions f and h take real values. If not we consider the complex-valued measures $\mathbf{B}(\mathbb{R}^\infty) \ni \Delta \mapsto \omega(\Delta) = \omega_1(\Delta) + i\,\omega_2(\Delta) \in \mathbb{C}^1$, where ω_1, ω_2 are real sign changing measures. Set $\omega(\Delta) = \int_\Delta f(x)\,h(x)\,d\mu(x)$. Obviously ω is a finite sign changing measure for which $\int_{\mathbb{R}^\infty} \exp(\delta \|x\|_{H_-})\,d\,|\omega(x)| = \int_{\mathbb{R}^\infty} \exp(\delta\|x\|_{H_-})\,|f(x)\,h(x)|\,d\mu(x) < \infty$.

Thus for any fixed $n = 1,2,\ldots$ the function $g(t)$ considered as a function of n variables can be analytically extended into the domain $\{z = x+iy \in \mathbb{C}^n \mid \|y\|_{H_+} < \delta\}$. As $(D^\beta g)(z)|_{z=0} = \int_{\mathbb{R}^\infty} (-ix)^\beta f(x)\,h(x)\,d\mu(x) = 0$, $\beta \in I_0^\infty$, $g(\cdot) \equiv 0$. So, from the uniqueness property it follows that $\omega(\Delta) = 0\ \forall \Delta \in \mathbf{B}(\mathbb{R}^\infty)$. So $f(x)\,h(x) = 0$ and $h(x) = 0$ almost everywhere with respect to μ. $\square$

Taking into consideration that the description of the space

$$D = \mathbf{A}^\alpha(\mathbb{R}^\infty)\exp\left(-\frac{1}{4}\sum_{k=1}^\infty \frac{x_k^2}{\alpha_k}\right) \tag{14}$$

is the same as in one-dimensional case for $D = \mathbf{A}(\mathbb{R}^1)\exp\left(-\frac{x^2}{4}\right)$ this concludes the density proof of D. From (13) it follows that the imbedding $D \subset L_2(\mathbb{R}^\infty, d\mu)$ is topological.

3. We prove that $U_s \in L(D \to D)$. For this it is necessary to prove that the operator T_s^α, $s = (s_1, s_2, \ldots, s_p, 0, \ldots) \in \mathbb{R}_0^\infty$, $\alpha = (\alpha_k)_{k=1}^\infty$

$$\mathbf{A}^\alpha(\mathbb{R}^\infty) \ni a(t) \mapsto (T_s^\alpha a)(t) = \exp\left(\sum_{k=1}^p \alpha_k(2t_ks_k - s_k^2)\right) a(t-s) \in \mathbf{A}^\alpha(\mathbb{R}^\infty)$$

is continuous.

Because

$$\mathbf{A}^\alpha(\mathbb{R}^\infty) = \left\{\sum_{\beta\in I_0^\infty} u_\beta h_\beta^\alpha \mid \sum_{\beta \in I_0^\infty} |u_\beta|^2 \tau^\beta < \infty,\ \tau \in \mathbb{N}_0^\infty\right\}$$

where $h_\beta^\alpha = h_{\beta_1}^{\alpha_1}(t_1) \cdots h_{\beta_p}^{\alpha_p}(t_p)$, $\beta = (\beta_1, \ldots, \beta_p, 0, \ldots) \in I_0^\infty$, because

$$(T_s^\alpha h_\beta^\alpha, h_\gamma^\alpha)_{L_2(\mathbb{R}^\infty, dg_\alpha)} = \prod_{k=1}^{\max\{\nu(s),\nu(\beta),\nu(\gamma)\}} (T_{s_k}^{\alpha_k} h_{\beta_k}^{\alpha_k}, h_{\gamma_k}^{\alpha_k})_{L_2(\mathbb{R}^1, dg_{\alpha_k})},$$

$$(T_{s_k}^{\alpha_k} h_{\beta_k}^{\alpha_k}, h_{\gamma_k}^{\alpha_k})_{L_2(\mathbb{R}^1, dg_{\alpha_k})} = (T_{s_k/\sqrt{\alpha_k}} h_{\beta_k}, h_{\gamma_k})_{L_2(\mathbb{R}^1, dg)}$$

and because the operator $T_s : \mathbf{A}(\mathbb{R}^1) \to \mathbf{A}(\mathbb{R}^1)$ is continuous in the one-dimensional case (Lemma 2), it follows that the operator T_s^α is continuous.

4. We will prove that D is contained in the domains of the generators of all one-parameter subgroups and that D is invariant. Consider the generators A_k, $k = 1,2,\ldots$ which are multiplication operators x_k in $L_2(\mathbb{R}^\infty, d\mu)$:

$$\mathbf{D}(A_k) = \{f \in L_2(\mathbb{R}^\infty, d\mu) \mid \int_{\mathbb{R}^\infty} |x_k f(x)|^2 \, d\mu(x) < \infty\},$$

$$(A_k f)(x) = x_k f(x), \; f \in \mathbf{D}(A_k).$$

Other generators are linear combinations of A_k.

Using (14) we have similar to the one-dimensional case $D \subset \mathbf{D}(A_k)$, $A_k(D) \subset D$.

5. The proof that D consists of the entire vectors for the generators of all one-parameter subgroups of the group $\mathbb{R}_0^\infty$ is the same as in the one-dimensional case.

6. The proof that the function $\mathbb{R}_0^\infty \ni s \mapsto U_s f \in D$ is continuous for all $f, h \in D$ is done similarly to the one-dimensional case using relations similar to (12). □

Remark 5. If the initial representation is not cyclic, then there exists a countable sequence $(\mu_n(\cdot))_{n=1}^\infty$ of measures on $(\mathbb{R}^\infty, \mathbf{B}(\mathbb{R}^\infty))$, such that the space H can be realized as $\oplus \sum_{n=1}^{\infty} L_2(\mathbb{R}^\infty, d\mu_n)$ and to every operator U_t, on every space $L_2(\mathbb{R}^\infty, d\mu_n)$ there corresponds the multiplication operator $e^{i(t,x)}$ (§4). In every space we construct the Gårding domain D_n satisfying the properties 1) - 6), and define $D = \oplus \sum_{n=1}^{\infty} D_n$. □

As inductive limits of commutative locally compact Lie groups $\mathbb{R}^n$, $G = j - \lim\limits_{\rightarrow} \mathbb{R}^n$, are isomorphic to the group $\mathbb{R}_0^\infty$ the given procedure can be applied to construct the Gårding domain for unitary strongly continuous representations of the groups $G = n - \lim\limits_{\rightarrow} \mathbb{R}^n$.

Remark 6. Consider the group $\mathbb{R}_0^{[0,1]}$ of functions $f : [0,1] \ni x \mapsto f(x) \in \mathbb{R}^1$ such that $f(x) = 0$ for all but a finite number of $x \in [0,1]$. Endow it with the inductive limit topology for finite dimensional spaces. Consider any strongly continuous unitary representation $\mathbb{R}_0^{[0,1]} \ni f \mapsto U_f = \exp(i \sum_{x \in [0,1]} f(x) A_x) \in \mathbf{U}(H)$, where $H = L_2([0,1], dx)$ and A_{x_0} is a multiplication operator $[0,1] \ni x \mapsto \dfrac{1}{x - x_0} \in \mathbb{R}^1$ in $L_2([0,1], dx)$.

For this unitary representation of the group $\mathbb{R}_0^{[0,1]}$ the Gårding domain does not exist. □

2.6. Unitary representations of $\mathbb{R}_{\pi(2)}^\infty$.

The group $\mathbb{R}_{\pi(2)}^\infty$ is a group of real step functions on $[0,1]$ with discontinuities in the binary rational points, $\mathbb{R}_{\pi(2)}^\infty = j - \lim\limits_{\rightarrow} \mathbb{R}^{2^n}$ with the imbeddings $j_{n+1}^n : \mathbb{R}^{2^n} \hookrightarrow \mathbb{R}^{2^{n+1}}$ defined as follows: $j_{n+1}^n(t_1, t_2, \ldots, t_{2^n}) = (t_1, t_1, t_2, t_2, \ldots, t_{2^n}, t_{2^n}) \in \mathbb{R}^{2^{n+1}}$.

The theory of unitary representations of the commutative group $\mathbb{R}_{\pi(2)}^\infty$ of binary currents on $[0,1]$ could be developed as in §3. But we will use the infinitesimal method: turn to the self-adjoint representations of the basis in the Lie algebra $\mathbf{g}_{\pi(2)}^\infty$ of the group

$\mathbb{R}^\infty_{\pi(2)}$ i.e. use the spectral theory of a countable collection of commuting self-adjoint operators $(A_{i_1 \cdots i_n \cdots})_{i_1 \cdots i_n \cdots \in \mathbb{Z}^\infty_{2,0}}$ (see §1.8, Chapter 1).

A countable collection $(A_{i_1 \cdots i_n \cdots})_{i_1 \cdots i_n \cdots \in \mathbb{Z}^\infty_{2,0}}$ of CSO defines a unitary representation $\mathbb{R}^\infty_{\pi(2)} \ni j_n(t_1, \ldots, t_{2^n}) \mapsto \exp(i \sum_{k=1}^{2^n} t_k A_k^{(n)})$, where the operators $A_k^{(n)}$ can be linearly expressed by $(A_{i_1 \cdots i_n \cdots})_{i_1 \cdots i_n \cdots \in \mathbb{Z}^\infty_{2,0}}$ in the same way as the function

$$U_k(x) = \begin{cases} 1, & \dfrac{k-1}{2^n} < x \leq \dfrac{k}{2^n}, \\ 0, & x \notin \left[\dfrac{k-1}{2^n}, \dfrac{k}{2^n}\right] \end{cases}$$

is linearly expressed by Walsh functions.

Conversely, by the Stone theorem any unitary representation of the group $\mathbb{R}^{2^n} \ni (t_1, \ldots, t_{2^n}) \mapsto U^{(n)}_{(t_1, \ldots, t_{2^n})}$ has the form $U^{(n)}_{(t_1, \ldots, t_{2^n})} = \exp(i \sum_{k=1}^{2^n} t_k A_k^{(n)})$ where $(A_k^{(n)})_{k=1}^{2^n}$ are CSO.

Setting

$$A_{i_1 \cdots i_n \cdots} = \sum_{k=1}^{2^n} u_{i_1 \cdots i_n \cdots}(\tfrac{k}{2n}) A_k^{(n)}$$

where $u_{i_1 \cdots i_n \cdots}$ are Walsh functions with last nonzero index i_n, we construct a system of CSO $(A_{i_1 \cdots i_n \cdots})_{i_1 \cdots i_n \cdots \in \mathbb{Z}^\infty_{2,0}}$.

Because the spectrum of the collection $(A_{i_1 \cdots i_n \cdots})_{i_1 \cdots i_n \cdots \in \mathbb{Z}^\infty_{2,0}}$ of CSO is a subset of $\mathbb{R}^{\mathbb{Z}^\infty_{2,0}}$ we have the proposition.

PROPOSITION 8.

$$\hat{\mathbb{R}}^\infty_{\pi(2)} = \mathbb{R}^{\mathbb{Z}^\infty_{2,0}}.$$

Comments to Chapter 2.

1. General known examples of inductive limits of locally-compact Lie groups and their Lie algebras are considered.

2. Character theory for inductive limits of commutative locally compact groups has been considered in N.Ya. Vilenkin [1], S. Kaplan [1] and others (also see references in S. Morris [1]).

3. The proof of Theorem 5 essentially coincides with the proof of Theorem 1 (Kolmogorov theorem for the projection-valued measures).

4. The exposition is parallel to §1.4 and 1.5 of Chapter 1.

5. The construction and the study of the Gårding domain for unitary representations of $\mathbb{R}_0^\infty$ follows the article A.V. Kosyak and Yu.S. Samoĭlenko [3]. For the construction of the spaces of functions of infinitely many variables see Yu.G. Kondrat'ev and Yu.S. Samoĭlenko [2] (see also Yu.M. Berezanskiĭ [9]).

6. Exposition follows §1.8 of Chapter 1. For Walsh functions, see, for example, S. Kaczmarz, H. Steinhaus [1].

Chapter 3.
DIFFERENTIAL OPERATORS WITH CONSTANT COEFFICIENTS IN SPACES OF FUNCTIONS OF INFINITELY MANY VARIABLES

In the space $L_2(\mathbb{R}^n, dx_1 \cdots dx_n)$ a differential operator with constant coefficients is defined to be a polynomial of the commuting family of self-adjoint differential operators $(D_k)_{k=1}^n$ (D_k is the minimal operator in $L_2(\mathbb{R}^n, dx_1 \cdots dx_n)$ generated by the expression $-i \dfrac{\partial}{\partial x_k}$). The D_k are generators of the unitary representation of $\mathbb{R}^n$

$$\mathbb{R}^n \ni (t_1, \ldots, t_n) \mapsto (V_{(t_1,\ldots,t_n)} f)(x_1, \ldots, x_n) = f(x_1+t_1, \ldots, x_n+t_n) \tag{1}$$

by the shifts in $L_2(\mathbb{R}^n, dx_1 \cdots dx_n) \ni f(\cdot)$. If we change the Lebesgue measure $dx_1 \cdots dx_n$ in $\mathbb{R}^n$ to any other quasi-invariant measure $d\omega(x_1, \ldots, x_n)$ in $\mathbb{R}^n$ and correspondingly representation (1) to the unitary representation

$$\mathbb{R}^n \ni (t_1, \ldots, t_n) \mapsto (V_{(t_1,\ldots,t_n)} \phi)(x_1, \ldots, x_n) =$$

$$= \sqrt{\frac{d\omega(x_1+t_1, \ldots, x_n+t_n)}{d\omega(x_1, \ldots, x_n)}}\ \phi(x_1+t_1, \ldots, x_n+t_n)$$

in the space $L_2(\mathbb{R}^n, d\omega(x_1, \ldots, x_n)) \ni \phi(\cdot)$, then this will lead to a unitarily equivalent class of the generators $(D_k')_{k=1}^\infty$ (because in the finite dimensional space $\mathbb{R}^n$ any quasi-invariant measure is equivalent to the Lebesgue measure). Consequently, we get a unitarily equivalent class of differential operators with constant coefficients.

Using the Fourier transform, constructed for the spectral decomposition of the collection $(D_k)_{k=1}^\infty$ we can reduce the study of the differential operators $P(D_1, \ldots, D_n)$ to the study of operators of multiplication by the polynomials $P(\lambda_1, \ldots, \lambda_n)$.

In this chapter we study the infinite dimensional analog of the class of differential operators with constant coefficients, viz. polynomials of a countably family $(D_k)_{k=1}^\infty$ of CSO of differentiation. The operators $(D_k)_{k=1}^\infty$ are introduced as generators of a unitary representation in $L_2(\mathbb{R}^\infty, d\omega(x))$ where ω is a measure, quasi-invariant with respect to the additive group of shifts by vectors from $\mathbb{R}_0^\infty = \lim\limits_{\to} \mathbb{R}^n$. So, as mentioned in the Preface, the material considered in this chapter may serve as a particular but important example to some constructions introduced in Chapter 1 and Chapter 2.

Note that the existence in $(\mathbb{R}^\infty, \mathrm{B}(\mathbb{R}^\infty))$ of nonequivalent $\mathbb{R}_0^\infty$-quasi-invariant measures implies the existence of unitarily nonequivalent classes of differential operators with constant coeffficients operating on functions of countably many variables. There appears a whole class of infinite dimensional Fourier transforms that are used to study the corresponding classes of differential operators with constant coefficients acting in the space of functions of countably many variables.

3.1. $\mathbb{R}_0^\infty$-quasi-invariant measures on $(\mathbb{R}^\infty, \mathbf{B}(\mathbb{R}^\infty))$.

Let X be a set, $\mathbf{B}(X)$ *a* σ-algebra of subsets of X, $\mu(\cdot)$ a σ-finite measure on the measurable space $(X, \mathbf{B}(X))$.

Now , let $G \ni g$ be a group of measurable one-to-one mappings of X into X. Define a measure $\mu_g(\cdot)$ by setting

$$\mu_g(\Delta) = \mu(g\,\Delta) \quad (\Delta \in \mathbf{B}(X)).$$

DEFINITION 1. We say that

a) $\mu(\cdot)$ is G-invariant if $\mu_g(\cdot) = \mu(\cdot)$ $\forall g \in G$;

b) $\mu(\cdot)$ is G-quasi-invariant if the measures $\mu_g(\cdot)$ are equivalent (absolutely continuous with respect to each other) $\forall g \in G$;

c) $\mu(\cdot)$ is G-ergodic if $\mu(\cdot)$ is G-quasi-invariant and any G-quasi-invariant measure $\mu'(\cdot)$ $(\neq 0)$ which is absolutely continuous with respect to $\mu(\cdot)$ is equivalent to $\mu(\cdot)$.

On a linear space X the group X operates as translations on the linear space.

PROPOSITION 1. If a σ-algebra $\mathbf{B}(X)$ satisfies the following condition: $(x,y) \mapsto x+y$ is a $\mathbf{B}(X)$ measurable mapping of $(X \times X, \mathbf{B}(X) \times \mathbf{B}(X)) \to (X, \mathbf{B}(X))$ (i.e. if $\forall \Delta \in \mathbf{B}(X)$ $\{(x,y) \mid x+y \in \Delta\} \in \mathbf{B}(X) \times \mathbf{B}(X)$), then any two X-quasi-invariant measures $\mu_1(\cdot)$ and $\mu_2(\cdot)$ are equivalent.

Proof. We can assume that the measures $\mu_1(\cdot)$ and $\mu_2(\cdot)$ are finite. Calculate the double integral

$$J = \int \chi_\Delta(x+y)\, d\mu_1(x)\, d\mu_2(y).$$

From the Fubini's theorem we have

$$J = \int \mu_1(\Delta - y)\, d\mu_2(y) = \int \mu_2(\Delta - x)\, d\mu_1(x)$$

so assuming that $\mu_1(\Delta) = 0$ because $\mu_1(\cdot)$ is quasi-invariant we get that $J = 0$, and consequently $\mu_2(\Delta) = 0$. ▯

On the measurable linear topological space $(\mathbb{R}^n, \mathbf{B}(\mathbb{R}^n))$ there exists a $\mathbb{R}^n$-invariant σ-finite Lebesgue measure. Consequently, it follows from the Proposition 1 that any σ-finite $\mathbb{R}^n$-quasi-invariant measure $\mu(\cdot)$ on $(\mathbb{R}^n, \mathbf{B}(\mathbb{R}^n))$ is equivalent to a Lebesgue measure, i.e.

$$d\mu(x_1, \ldots, x_n) = p(x_1, \ldots, x_n)\, dx_1 \cdots dx_n \quad ((x_1, \ldots, x_n) \in \mathbb{R}^n)$$

where $p(x_1, \ldots, x_n) > 0$ for almost all $(x_1, \ldots, x_n)$ with respect to the Lebesgue measure $dx_1 \cdots dx_n$.

THEOREM 1. (V.N. Sudakov). There does not exist a σ-infinite Borel X-quasi-invariant measure on an infinite dimensional locally convex space X.

The situation will be different if instead of being X-quasi-invariant, we require the measure to be quasi-invariant with respect to a dense linear set of shifts.

Consider the following basic situtation: we have a measurable linear space $(\mathbb{R}^\infty, \mathbf{B}(\mathbb{R}^\infty))$, where $\mathbb{R}^\infty = \mathbb{R}^1 \times \mathbb{R}^1 \times \cdots$ and $\mathbf{B}(\mathbb{R}^\infty)$ is the σ-algebra generated by the cylindrical subsets of $\mathbb{R}^\infty$ with the Borel bases.

Let then S_t be the shift $\mathbb{R}^\infty \ni x \mapsto x + t \in \mathbb{R}^\infty$ by a vector $t \in \mathbb{R}_0^\infty = \bigcup_{n=1}^{\infty} \mathbb{R}^n \times (0,0,...)$ i.e. the space of all finite real sequences. This is a one-to-one measurable mapping of the space $(\mathbb{R}^\infty, \mathbf{B}(\mathbb{R}^\infty))$.

We recall some of the properties of $\mathbb{R}_0^\infty$-quasi-invariant measures $\omega(\cdot)$ on $(\mathbb{R}^\infty, \mathbf{B}(\mathbb{R}^\infty))$ viz. measures $\omega(\cdot)$ on $(\mathbb{R}^\infty, \mathbf{B}(\mathbb{R}^\infty))$ with the property $\forall t \in \mathbb{R}_0^\infty$ $\omega_t(\cdot)$ and $\omega(\cdot)$ are equivalent.

Denote by $\omega_n(\cdot)$ $(n=1,2,...)$ the restriction of the measure $\omega(\cdot)$ to the σ-algebra $\mathbf{B}_n(\mathbb{R}^\infty) = \mathbf{B}(\mathbb{R}^n) \times \mathbb{R}^1 \times \mathbb{R}^1 \times \cdots$. It can be considered as a measure on $(\mathbb{R}^n, \mathbf{B}(\mathbb{R}^n))$.

PROPOSITION 2. The measure $\omega_n(\cdot)$ on $(\mathbb{R}^n, \mathbf{B}(\mathbb{R}^n))$ is $\mathbb{R}^n$-quasi-invariant $(n = 1,2,..)$, i.e. it is equivalent to the Lebesgue measure $dx_1 \cdots dx_n$.

The converse does not hold, that is, if the measures $\omega_n(\cdot)$ are quasi-invariant for all $n = 1,2,...$ it does not necessarily mean that the measure $\omega(\cdot)$ is $\mathbb{R}_0^\infty$-quasi-invariant. Below (see Example 1) we will give an example of a Gaussian measure on $\mathbb{R}^\infty$ which is not $\mathbb{R}_0^\infty$ quasi-invariant (its projections on $\mathbb{R}^n$ are, of course, quasi-invariant).

Now denote by $\omega^n(\cdot)$ the restrictions of the measure $\omega(\cdot)$ to the σ-algebra $\mathbf{B}^n(\mathbb{R}^\infty)$ generated by the cylindrical sets of the form $\mathbb{R}^n \times S \times \mathbb{R}^\infty$ $(S \in \mathbf{B}(\mathbb{R}^p), p = 1,2,...)$.

PROPOSITION 3. For any $n = 1,2,...$ the measures $\omega(\cdot)$ and $\omega_n(\cdot) \otimes \omega^n(\cdot)$ are equivalent.

Denote by $\rho_t(x) = \dfrac{d\omega_t}{d\omega}(x)$ the Radon-Nikodym derivative of the shifted measure $\omega_t(\cdot)$ with respect to its equivalent measure $\omega(\cdot)$ and let $p_n(x_1, \ldots, x_n) = \dfrac{d\omega_n}{dx_1 \cdots dx_n}(x_1, \ldots, x_n)$ $(p_n(x_1, \ldots, x_n) > 0 \pmod{dx_1 \cdots dx_n})$.

Then

$$\frac{d\omega_{n,t}}{d\omega_n}(x_1, \ldots, x_n) = \frac{p_n(x_1 + t_1, \ldots, x_n + t_n)}{p_n(x_1, \ldots, x_n)}. \tag{2}$$

PROPOSITION 4. For all $t \in \mathbb{R}_0^\infty$ the function sequence $\left\{\dfrac{p_n(x_1 + t_1, \ldots, x_n + t_n)}{p_n(x_1, \ldots, x_n)}\right\}_{n=1}^{\infty}$ converges to $\rho_t(x)$ with respect to the norm in $L_1(\mathbb{R}^\infty, d\omega(x))$.

Note that the $\mathbb{R}_0^\infty$-quasi-invariance of the measure $\omega(\cdot)$ on $(\mathbb{R}^\infty, \mathbf{B}(\mathbb{R}^\infty))$ actually means that this measure is quasi-invariant with respect to a larger set of shifts. We have the following

PROPOSITION 5. For any $\mathbb{R}_0^\infty$-quasi-invariant measure $\omega(\cdot)$ on $(\mathbb{R}^\infty, \mathrm{B}(\mathbb{R}^\infty))$ there exists a set of weights $(p_k)_{k=1}^\infty$ such that $\omega(\cdot)$ is $l_2(p_k)$-quasi-invariant.

Nevertheless we have

PROPOSITION 6. The $\omega(\cdot)$ measure of the set T_ω of all admissible shifts, i.e. of the shifts $t \in \mathbb{R}^\infty$ such that the measure $\omega_t(\cdot)$ is equivalent to $\omega(\cdot)$ is equal to zero.

Consider the question of the product measures and the Gaussian measures on $(\mathbb{R}^\infty, \mathrm{B}(\mathbb{R}^\infty))$ being $\mathbb{R}_0^\infty$-quasi-invariant.

PROPOSITION 7. For the product-measures $d\omega(x) = \bigotimes_{k=1}^\infty d\omega_k(x_k)$ to be $\mathbb{R}_0^\infty$-quasi-invariant it is necessary and sufficient that the factors $d\omega_k(x_k)$ be $\mathbb{R}^1$-quasi-invariant, i.e. $d\omega(x_1, \ldots, x_n, \ldots) = \bigotimes_{k=1}^\infty p_k(x_k)\, dx_k \quad (p_k(x_k) > 0 \ (\mathrm{mod}\, dx_k))$ and $\int_{\mathbb{R}^1} p_k(x_k)\, dx_k = 1, k = 1,2,\ldots)$.

For the stationary product-measures, i.e. for the measures of the form $\bigotimes_{k=1}^\infty p(x_k)\, dx_k$ the set of admissible shifts $T_{\bigotimes_{k=1}^\infty p(x_k)dx_k} \subseteq l_2$.

For the equality $T_{\bigotimes_{k=1}^\infty p(x_k)dx_k} = l_2$ it is necessary and sufficient that

$$\int_{\mathbb{R}^1} \left| \frac{d}{dx_1} \sqrt{p(x_1)} \right|^2 dx_1 = \int_{\mathbb{R}^1} \lambda_1^2 \left| \widetilde{\sqrt{p(\lambda_1)}} \right|^2 d\lambda_1 < \infty$$

(here $\tilde{\cdot}$ is the Fourier transform of a function on $\mathbb{R}^1$).

For the Gaussian measures $g_B(\cdot)$ on $(\mathbb{R}^\infty, \mathrm{B}(\mathbb{R}^\infty))$ the property of being $\mathbb{R}_0^\infty$-quasi-invariant depends on B which is a real symmetric positive definite infinite matrix such that the characteristic function satisfies

$$k(t) = \int_{\mathbb{R}^\infty} e^{i(t,x)}\, dg_B(x) = e^{-\frac{1}{2}(Bt,t)} \quad (t \in \mathbb{R}_0^\infty).$$

LEMMA 1. The matrix B can be represented in the form $B = A^* A$, where A is an upper triangular matrix with nonzero diagonal elements and where A^* is its transpose.

Proof. Suppose that the representation $B = A^* A$ holds:

$$\begin{pmatrix} b_{11} & b_{12} & \cdots & b_{1k} & \cdots \\ b_{12} & b_{22} & \cdots & b_{2k} & \cdots \\ \cdots & \cdots & \cdots & \cdots & \cdots \\ b_{1k} & b_{2k} & \cdots & b_{kk} & \cdots \\ \cdots & \cdots & \cdots & \cdots & \cdots \end{pmatrix} = \begin{pmatrix} a_{11} & & & \\ a_{12} & a_{22} & & 0 \\ \cdots & \cdots & & \\ a_{1k} & a_{2k} & \cdots & a_{kk} \\ \cdots & \cdots & \cdots & \cdots \end{pmatrix} \times$$

$$\times \begin{bmatrix} a_{11} & a_{12} & \cdots & a_{1k} & \cdots \\ & a_{22} & \cdots & a_{2k} & \cdots \\ & & \cdots & \cdots & \cdots \\ 0 & & & a_{kk} & \cdots \\ & & & & \cdots \end{bmatrix}.$$

Multiplying the matrices we see that any element of the matrix A can be expressed by elements with smaller indices:

$$a_{rk} = \frac{b_{rk} - (a_{1r}\, a_{1k} + \cdots + a_{r-1r}\, a_{r-1k})}{a_{rr}} \quad r \le k \quad (a_{11} > 0).$$

So the elements of the matrix A are uniquely defined.

We show that $a_{rr} \neq 0$. Let $B_r = (b_{ij})_{i,j=1}^{r}$. Obviously, $B_r = A_r^* A_r$. Then $|B_r| = a_{11}^2 \cdots a_{rr}^2 \neq 0$ because B is positive definite. □

Next we show that the matrices $(A^*)^{-1}$ and A^{-1} exist and have correspondingly upper triangular and lower triangular forms. To do that, calculate the left inverse of the matrix A^* that has the lower triangular form: $C A^* = I$, i.e.

$$\begin{bmatrix} c_{11} & & & 0 & \\ c_{12} & c_{22} & & & \\ \cdots & \cdots & \cdots & & \\ c_{1k} & c_{2k} & \cdots & c_{kk} & \\ \cdots & \cdots & \cdots & \cdots & \cdots \end{bmatrix} \begin{bmatrix} a_{11} & & & 0 & \\ a_{12} & a_{22} & & & \\ \cdots & \cdots & \cdots & & \\ a_{1k} & a_{2k} & \cdots & a_{kk} & \cdots \\ \cdots & \cdots & \cdots & \cdots & \cdots \end{bmatrix} =$$

$$= \begin{bmatrix} 1 & & 0 & \\ & 1 & & \\ & & \ddots & \\ & & 1 & \\ 0 & & & \ddots \end{bmatrix}.$$

Multiplying the matrices, we get the system of equations:

$$c_{11}\, a_{11} = 1\,,\ c_{12}\, a_{11} + c_{22}\, a_{12} = 0, \ldots, c_{1n}\, a_{11} + \cdots + c_{nn}\, a_{1n} = 0\,,$$

$$c_{22}\, a_{22} = 1\,,\ c_{23}\, a_{22} + c_{33}\, a_{23} = 0, \ldots, c_{2n+1}\, a_{22} + \cdots + c_{n+1n+1}\, a_{2n+1} = 0\,,$$

....

$$c_{kk}\, a_{kk} = 1\,,\ c_{kk+1}\, a_{kk} + c_{k+1k+1}\, a_{kk+1} = 0, \ldots, c_{kn+k-1}\, a_{kk} + \cdots +$$

$$+ c_{n+k-1n+k-1}\, a_{kn+k-1} = 0\,;$$

Solving this system step by step we get all the elements of the matrix C.

It's not difficult to check that the matrix C is the right inverse of A^*, i.e. $C = (A^*)^{-1}$. Similarly we calculate the matrix A^{-1}.

The matrix A generates an operator from $\mathbb{R}_0^\infty$ into $\mathbb{R}_0^\infty$; the matrix A^* generates the adjoint operator from $\mathbb{R}^\infty$ into $\mathbb{R}^\infty$ and the two are nondegenerate and defined on the whole space. Their inverses generated by A^{-1} and $(A^*)^{-1}$ are also defined on the whole space.

LEMMA 2. The transformation $y = A^* x$ of $\mathbb{R}^\infty$ transforms the measure $g_B(\cdot)$ into the canonical measure

$$dg(x) = \bigotimes_{k=1}^{\infty} \frac{1}{\sqrt{2\pi}} e^{-x_k^2/2}\, dx_k = dg_I(x).$$

Proof. Calculate the characteristic function $k(t)$ of the measure $g_B(A^*\cdot)$

$$k(t) = \int_{\mathbb{R}^\infty} e^{i(t,x)}\, dg_B(A^* x) = \int_{\mathbb{R}^\infty} e^{i(t,(A^*)^{-1}x)}\, dg_B(x) =$$

$$= \int_{\mathbb{R}^\infty} e^{i(A^{-1}t,x)}\, dg_B(x) = e^{-\frac{1}{2}(BA^{-1}t,A^{-1}t)} = e^{-\frac{1}{2}(t,t)}.$$

Extending $k(t)$ on all of l_2 using the continuity, we get a measure canonical in $\mathbb{R}^\infty$. □

It is known that for the product measure $g(\cdot)$ the space l_2 is the set of all admissible shifts (i.e $g(\cdot)$ is quasi-invariant with respect to the elements from and only from l_2). Using the inverse change of variables we get

PROPOSITION 8. The measure $g_B(\cdot)$ is quasi-invariant with respect to the shift by the elements of the set $A^*(l_2)$ and

$$T_{g_B}(\cdot) = A^*(l_2).$$

Requiring that the condition $\mathbb{R}_0^\infty \subset A^*(l_2)$ or $(A^*)^{-1}\, \mathbb{R}_0^\infty \subset l_2$ holds and applying $(A^*)^{-1}$ to the basis vectors of $\mathbb{R}_0^\infty$ we get the following proposition.

PROPOSITION 9. The measure $g_B(\cdot)$ is quasi-invariant with respect to finite shifts if and only if the columns of the matrix $(A^*)^{-1}$ are square summable.

Example 1. A Gaussian measure on $(\mathbb{R}^\infty, \mathrm{B}(\mathbb{R}^\infty))$ that is not $\mathbb{R}_0^\infty$-quasi-invariant. The matrix

$$B = \begin{pmatrix} 1 & -\frac{1}{\sqrt{2}} & -\frac{1}{\sqrt{3}} & \cdots & -\frac{1}{\sqrt{n}} & \cdots \\ -\frac{1}{\sqrt{2}} & 1+\frac{1}{2} & \frac{1}{\sqrt{2\cdot 3}} & \cdots & \frac{1}{\sqrt{2\cdot n}} & \cdots \\ -\frac{1}{\sqrt{3}} & \frac{1}{\sqrt{3\cdot 2}} & 1+\frac{1}{3} & \cdots & \frac{1}{\sqrt{3\cdot n}} & \cdots \\ \cdots & \cdots & \cdots & \cdots & \cdots & \cdots \\ -\frac{1}{\sqrt{n}} & \frac{1}{\sqrt{n\cdot 2}} & \frac{1}{\sqrt{n\cdot 3}} & \cdots & 1+\frac{1}{n} & \cdots \\ \cdots & \cdots & \cdots & \cdots & \cdots & \cdots \end{pmatrix}$$

is positive definite.

The Gaussian measure $g_B(\cdot)$ on $\mathbb{R}^\infty$ is not $\mathbb{R}_0^\infty$-quasi-invariant because

$$(A^*)^{-1} = \begin{bmatrix} \frac{1}{\sqrt{2}} & 1 & 0 & & \\ \frac{1}{\sqrt{3}} & 0 & 1 & & \\ \cdots & \cdots & \cdots & & \\ \frac{1}{\sqrt{n}} & 0 & 0 & 1 & \\ \cdots & \cdots & \cdots & \cdots & \cdots \end{bmatrix}.$$

□

We look also at some properties of G-ergodic measures on $(X, \mathbf{B}(X))$ and at the question how to decompose G-quasi-invariant measures into G-ergodic components.

PROPOSITION 10. Let $\mu(\cdot)$ be a G-quasi-invariant measure on X and let the space $L_2(X, d\mu(x))$ be separable. The following are equivalent:

1) (definition of ergodicity of $\mu(\cdot)$) for every G-quasi-invariant measure $\mu'(\cdot)$ the relation $\mu'(\cdot) \ll \mu(\cdot)$ implies either $\mu'(\cdot) = 0$ or $\mu'(\cdot) \sim \mu(\cdot)$;

2) for any measurable set Δ the equality $\Delta = g\Delta \,(\mathrm{mod}\,\mu(\cdot))$ $\forall g \in G$ implies $\Delta = \varnothing$ or $\Delta = X$;

3) for any measurable function $\phi(\cdot)$ the equality $\phi(gx) = \phi(x)$ $(\mathrm{mod}\,\mu(\cdot))$ $\forall g \in G$ implies $\phi(x) = \mathrm{const}(\mathrm{mod}\,\mu(\cdot))$.

The following proposition follows from the definition of G-ergodic measures.

PROPOSITION 11. If $\mu(\cdot)$ and $\nu(\cdot)$ are two G-ergodic measures, then either $\mu(\cdot) \sim \nu(\cdot)$ or $\mu(\cdot) \perp \nu(\cdot)$.

Now we come back to the situation being basic for us and consider $\mathbb{R}_0^\infty$-quasi-invariant measures on $(\mathbb{R}^\infty, \mathbf{B}(\mathbb{R}^\infty))$. We consider the σ-algebra $\mathbf{B}_{\mu(\cdot)}^\infty(\mathbb{R}^\infty) = \bigcap_{n=1}^{\infty} \mathbf{B}_{\mu(\cdot)}^n(\mathbb{R}^\infty) \cap \mathbf{B}(\mathbb{R}^\infty)$ (here $\mathbf{B}_{\mu(\cdot)}^n(\mathbb{R}^\infty)$ is the complement of $\mathbf{B}^n(\mathbb{R}^n)$ with respect to the measure $\mu(\cdot)$ (see Chapter 1, §1.7).

PROPOSITION 12. For a $\mathbb{R}_0^\infty$-quasi-invariant probability measure $\mu(\cdot)$ to be $\mathbb{R}_0^\infty$-ergodic it is necessary and sufficient that the measure $\mu(\cdot)$ takes only the values 0 and 1 on $\mathbf{B}_{\mu(\cdot)}^\infty$.

Following A.V. Skorokhod [3] we can describe all $\mathbb{R}_0^\infty$-ergodic probability measures on $(\mathbb{R}^\infty, \mathbf{B}(\mathbb{R}^\infty))$.

DEFINITION 2. A probability measure on $(\mathbb{R}^\infty, \mathbf{B}(\mathbb{R}^\infty))$ is called locally dependent if for any $n \in \mathbb{N}$ there can be found $m > n$ such that for any $\Delta_1 \in \mathbf{B}_n(\mathbb{R}^\infty)$ and $\Delta_2 \in \mathbf{B}^m(\mathbb{R}^\infty)$

$$\mu(\Delta_1 \cap \Delta_2) = \mu(\Delta_1)\,\mu(\Delta_2).$$

THEOREM 2. (A.V. Skorokhod) A measure is $\mathbb{R}_0^\infty$-ergodic if and only if it is equivalent to a locally dependent measure.

PROPOSITION 13. All $\mathbb{R}_0^\infty$-quasi-invariant product measures $\mu(\cdot) = \bigotimes_{k=1}^{\infty} \mu_k(\cdot)$ are $\mathbb{R}_0^\infty$-ergodic. For a $\mathbb{R}_0^\infty$-quasi-invariant Gaussian measure $g_B(\cdot)$ to be $\mathbb{R}_0^\infty$-ergodic, it is necessary and sufficient that the set $(A^*)^{-1}(\mathbb{R}_0^\infty)$ be dense in l_2 (see Proposition 9).

An arbitrary $\mathbb{R}_0^\infty$-quasi-invariant measure can be decomposed into $\mathbb{R}_0^\infty$-quasi-invariant ergodic factors.

To finish this section we give a description, following Y. Umemura [1], of $O(\infty)$ (S_∞)-invariant $\mathbb{R}_0^\infty$-ergodic measures on $(\mathbb{R}^\infty, \mathrm{B}(\mathbb{R}^\infty))$ with respect to which any $O(\infty)$ (S_∞)-invariant measure can be decomposed.

PROPOSITION 14. Let $\mu(\cdot)$ be a S_∞ invariant $\mathbb{R}_0^\infty$-ergodic measure. Then there exists a Borel measure $m(\cdot)$ on $\mathbb{R}^1$ such that $\mu(\cdot) = \bigotimes_{k=1}^{\infty} m(\cdot)$.

Let now $\omega(\cdot)$ be a $O(\infty)$-invariant $\mathbb{R}_0^\infty$-ergodic measure. Then $\exists\, \alpha > 0$ such that

$$d\omega(x) = dg_{\alpha I}(x) = \bigotimes_{k=1}^{\infty} \sqrt{\frac{\alpha}{\pi}} \exp(-\alpha x \frac{2}{k})\, dx_k.$$

3.2. The countable family $(D_k)_{k=1}^\infty$ of CSO.

Let $\omega(\cdot)$ be a probability measure on $(\mathbb{R}^\infty, \mathrm{B}(\mathbb{R}^\infty))$ quasi-invariant with respect to all shifts by a vector of $\mathbb{R}_0^\infty$. Consider the unitary representation in the space $L_2(\mathbb{R}^\infty, d\omega(x))$ of the group $\mathbb{R}_0^\infty$ by the operators

$$(U_t f)(x) = \sqrt{\rho_t(x)}\, f(x+t) \tag{3}$$

where $\rho_t(\cdot) = \dfrac{d\omega_t}{d\omega}(\cdot)$.

PROPOSITION 15. The unitary representation $\mathbb{R}_0^\infty \ni t \mapsto U_t$ is strongly continuous.

Proof. Show that for any n the restriction to $\mathbb{R}^n \times (0,0,\dots)$ of the representation is strongly continuous. It follows from Proposition 3 that the measures $\omega(\cdot)$ and $\omega_n(\cdot) \otimes \omega^n(\cdot)$ are equivalent. Working with the equivalent representation V_t in $L_2(\mathbb{R}^\infty, \omega_n(\cdot) \otimes \omega^n(\cdot))$, we use the fact that its restriction to $\mathbb{R}^n \times (0,0,\dots)$ *sucth* $(t_1, \dots, t_n, 0,\dots)$ has the form $(t_1, \dots, t_n, 0,\dots) \mapsto V(t_1, \dots, t_n) \otimes I$, where $V(t_1, \dots, t_n)$ are the unitary operators in $L_2(\mathbb{R}^n, d\omega_n(x_1, \dots, x_n)) \ni f(x_1, \dots, x_n)$ given by

$$(V(t_1,\ldots,t_n)f)(x_1,\ldots,x_n)=\sqrt{\frac{d\omega_{n,t}}{d\omega_n}}\,f(x_1+t_1,\ldots,x_n+t_n).$$

But because by Proposition 1 the measure $\omega_n(\cdot)$ on $(\mathbb{R}^n, \mathrm{B}(\mathbb{R}^n))$ is equivalent to the Lebesgue measure, it remains to show that the unitary shift operators by vectors

$$(t_1,\ldots,t_n):(W(t_1,\ldots,t_n)\phi)(x_1,\ldots,x_n)=\phi(x_1+t_1,\ldots,x_n+t_n)$$

on $L_2(\mathbb{R}^n, dx_1\cdots dx_n)$ are strongly continuous. But the latter is equivalent to continuity in the mean of the functions $\phi(\cdot)\in L_2(\mathbb{R}^n, dx_1\cdots dx_n)$. □

Now, denote the generator of the strongly continuous one-parameter group of unitary operators $U(0,\ldots,0,t_n,0,\ldots)$ by D_n. The family $(D_n)_{n=1}^{\infty}$ is a family of commuting self-adjoint operators on $L_2(\mathbb{R}^{\infty}, d\omega(x))$.

In what follows we will assume that the joint spectrum of the family $(D_k)_{k=1}^{\infty}$ of CSO is simple (that is the same as assuming that the unitary representation $\mathbb{R}_0^{\infty}\ni t\mapsto U_t$ in $L_2(\mathbb{R}^{\infty}, d\omega(x))$ is cyclic).

PROPOSITION 16. If $d\omega(x)=\bigotimes_{n=1}^{\infty} p_n(x_n)\,dx_n$ is a product measure, then the representation (3) is cyclic. The vector $\Omega(x)\equiv 1$ is a cyclic vector of the representation if and only if $\forall n\in\mathbb{N}$

$$q_n(\lambda_n)=\widetilde{\sqrt{p_n}}\,(\lambda_n)\neq 0 \pmod{d\lambda_n} \tag{4}$$

($\widetilde{\ }$ denotes the Fourier transform of a function of $L_2(\mathbb{R}^1, dx_n)$).

Proof. Because the vector $\Omega(x)=1=1\otimes 1\otimes\cdots$ is a product-vector in $\bigotimes_{n=1;1}^{\infty} L_2(\mathbb{R}^1, p_n(x_n)\,dx_n)$, for it to be a cyclic vector of the CSO $(D_k)_{k=1}^{\infty}$ it is necessary and sufficient that for every $L_2(\mathbb{R}^1, p_n(x_n)\,dx_n)$ the vector $\Omega_n(x_n)=1$ is a cyclic vector of the operator D_n i.e. that condition (4) is satisfied. □

For the $\mathbb{R}_0^{\infty}$-quasi-invariant Gaussian measures $g_B(\cdot)$ on $(\mathbb{R}^{\infty}, \mathrm{B}(\mathbb{R}^{\infty}))$, the unitary strongly continuous representation $\mathbb{R}_0^{\infty}\ni t\mapsto (U_t^B f)(x)=\sqrt{\rho_t(x)}\,f(x+t)$, where $\rho_t(\cdot)=\dfrac{dg_{B,t}}{dg_B}(\cdot)$, is not cyclic for all admissable operators B.

PROPOSITION 17. The unitary representation

$\mathbb{R}_0^{\infty}\ni t\mapsto (U_t f)(x)=\sqrt{\dfrac{dg_t(x)}{dg(x)}}\,f(x+t)$ in $L_2(\mathbb{R}^{\infty}, dg(x))$ where $g(\cdot)$ is the canonical Gaussian measure, is cyclic with cyclic vector $\Omega(x)=1$.

The proof will be given later for any dense linear set of shifts (see Lemma 4) in l_2.

LEMMA 3. The representation U_t^B is unitarily equivalent to the representation $U_{(A^*)^{-1}t}$.

Proof. Define the operator V as follows: $(Vf)(x)=f((A^*)^{-1}x)$. From the equality

$$\int_{\mathbb{R}^\infty} f(x)\, dg(x) = \int_{\mathbb{R}^\infty} f((A^*)^{-1} x)\, dg_B(x)$$

it follows that V is a unitary operator from $L_2(\mathbb{R}^\infty, dg(x))$, onto $L_2(\mathbb{R}^\infty, dg_B(x))$.

Let $y = (A^*)^{-1} x$. Then

$$(U_t^B f)(x) = \sqrt{\frac{dg_{B,t}(x)}{dg_B(x)}}\, f(x+t) =$$

$$= \sqrt{\frac{dg_{(A^*)^{-1}t}(y)}{dg(y)}}\, f(A^* y + t) = (U_{(A^*)^{-1}t} V^{-1} f)(y) =$$

$$= (U_{(A^*)^{-1}t} V^{-1} f)((A^*)^{-1} x) = (V U_{(A^*)^{-1}t} V^{-1} f)(x).$$

▯

Thus, instead of studying the spectral properties of the representation U_t^B, we can study the representation $U_{(A^*)^{-1}t}$.

LEMMA 4. Let $\mathbf{R}$ be a dense set in l_2. Then the function $\Omega(x) \equiv \mathbf{1}$ is a cyclic vector for the group $(U_\tau)_{\tau\in\mathbf{R}}$.

Proof. The mapping $l_2 \ni \tau \mapsto (U_\tau 1)(x) \in L_2(\mathbb{R}^\infty, dg(x))$ is continuous. So, because $\mathbf{R}$ is dense in l_2, it follows that

$$\underset{\tau\in\mathbf{R}}{\mathrm{CLS}}\,\{U_\tau \mathbf{1}\} = \underset{\tau\in l_2}{\mathrm{CLS}}\,\{U_\tau \mathbf{1}\}.$$

The set $\underset{\tau\in l_2}{\mathrm{CLS}}\,\{U_\tau\}$ contains the derivatives of all order with respect to all the coordinates of the vector τ of the function $(U_\tau 1)(x) = \exp(-\frac{1}{2}(\tau,x) - \frac{1}{4}\sum_{k=1}^{\infty}\tau_k^2)$. So $\underset{\tau\in l_2}{\mathrm{CLS}}\,\{U_t \mathbf{1}\}$ contains all cylindrical polynomials. But they form a dense set in $L_2(\mathbb{R}^\infty, dg(x))$ i.e.

$$\underset{t\in l_2}{\mathrm{CLS}}\,\{U_t \mathbf{1}\} = L_2(\mathbb{R}^\infty, dg(x)).$$

▯

THEOREM 3. If the columns of the matrix $(A^*)^{-1}$ are total in l_2, the representation U_t^B has a simple spectrum. Otherwise, the spectrum of U_t^B has countable multiplicity. The function $\Omega(x) \equiv \mathbf{1}$ is a vector of maximal spectral type.

Proof. Instead of the representation U_t^B we will consider its unitarily equivalent representation $U_{(A^*)^{-1}t}$. We use the notation $\mathbf{R}_1 = (A^*)^{-1}(\mathbb{R}_0^\infty)$.

If the columns of the matrix $(A^*)^{-1}$ are total in l_2, then $\overline{\mathbf{R}}_1 = l_2$ and the group $(U_\tau)_{\tau\in\mathbf{R}_1}$ has the cyclic vector 1, i.e. it has a simple spectrum.

Let $\overline{\mathbf{R}}_1 \neq l_2$. Then l_2 can be decomposed into the orthogonal sum of two subspaces: $l_2 = \overline{\mathbf{R}}_1 \oplus \mathbf{R}_2$ and the measure $g(\cdot)$ is decomposed into the product of the two measures: $g(\cdot) = g_1(\cdot) \otimes g_2(\cdot)$ which are canonical respectively on $\overline{\mathbf{R}}_1$ and $\mathbf{R}_2$.

The measure $g(\cdot)$ has the support on an extension of l_2: $\mathbf{H} = \mathbf{H}_1 \oplus \mathbf{H}_2$, where $\mathbf{H}_1$ and $\mathbf{H}_2$ are the corresponding extensions of $\overline{\mathbf{R}}_1$ and $\mathbf{R}_2$, on which the supports of the measures $g_1(\cdot)$ and $g_2(\cdot)$ lie. Then

$$L_2(\mathbb{R}^\infty, dg(x)) = L_2(\mathbf{H}, dg(x)) = L_2(\mathbf{H}_1, dg_1(x_1)) \otimes L_2(\mathbf{H}_2, dg_2(x_2)).$$

Choose in $L_2(\mathbf{H}_2, dg_2(x_2))$ an orthonormal basis $(\phi_j(x_2))_{j=1}^\infty$. Then $L_2(\mathbf{H}, dg(x))$ decomposes into the orthogonal sum of the isomorphic subspaces $H_j = L_2(\mathbf{H}_1, dg_1(x_1)) \otimes \phi_j(x_2)$. The representation $U_{(A^*)^{-1}t}$ decomposes into the product of two representations:

$$U_{(A^*)^{-1}t} = \overline{U}_{(A^*)^{-1}t} \otimes I\,,$$

where $\overline{U}_{(A^*)^{-1}t}$ is the operator analog of $U_{(A^*)^{-1}t}$ but which act on the space $L_2(\mathbf{H}_1, dg_1(x_1))$. The restrictions of $U_{(A^*)^{-1}t}$ on each of the subspaces H_j are unitarily equivalent; each has a cyclic vector $1 \otimes \phi_j(x_2)$. It is evident, that every vector $1 \otimes \phi_j(x_2)$ is a vector of maximal spectral type.

Thus the spectrum of the representation $U_{(A^*)^{-1}t}$ has countable multiplicity.

Choosing $\phi_1(x_2) = 1$ we see that 1 is a vector of maximal spectral type. $\square$

Example 2. We give an example of a representation U_t^B which has a spectrum with countable multiplicity. Let

$$A^* = \begin{bmatrix} 1 & & & & 0 \\ 2 & 1 & & & \\ 4 & 2 & 1 & & \\ 8 & 4 & 2 & 1 & \\ \cdots & \cdots & \cdots & \cdots & \cdots \end{bmatrix},$$

$B = A^* A$. The inverse matrix $(A^*)^{-1}$ has the form:

$$(A^*)^{-1} = \begin{bmatrix} 1 & & & & 0 \\ -2 & 1 & & & \\ 0 & -2 & 1 & & \\ 0 & 0 & -2 & 1 & \\ \cdots & \cdots & \cdots & \cdots & \cdots \end{bmatrix},$$

The columns of the matrix $(A^*)^{-1}$ are not total in l_2, they have an orthogonal vector $\begin{bmatrix} 1 \\ 1/2 \\ 1/4 \\ \cdots \end{bmatrix}$. Thus, by Theorem 1, the spectrum of the representation U_t^B has countable multiplicity. $\square$

3.3. The Fourier transforms.

Consider the family $(D_k)_{k=1}^{\infty}$ of CSO on $L_2(I\!R^{\infty}, d\omega(x))$. Let $E(\cdot)$ be the resolution of the identity of the family $(D_k)_{k=1}^{\infty}$, and let Ω be a vector of maximal spectral type, $\mu(\cdot) = (E(\cdot)\,\Omega, \Omega)$ the spectral measure of the family corresponding to the vector Ω. We construct the Fourier transform related to the spectral decomposition for the family $(D_k)_{k=1}^{\infty}$, assuming for simplicity that the representation (3) is cyclic.

So, let Ω be a cyclic vector of the representation (3). Consequently it is also a cyclic vector for the resolution of the identity $E(\cdot)$ (see Chapter 2). The Fourier transform corresponding to the vector Ω is a unitary operator $F : L_2(I\!R^{\infty}, d\omega(x)) \to L_2(I\!R^{\infty}, d\mu(\lambda))$ such that $\forall \Delta \in \mathbf{B}(I\!R^{\infty})\ FE(\Delta)\,\Omega = h_{\Delta}(\cdot)$ ($h_{\Delta}(\cdot)$ is the indicator of the set Δ). The operators $F D_n F^{-1}$ acting in $L_2(I\!R^{\infty}, d\mu(\lambda))$ are operators of multiplication by $\lambda_n (n \in I\!N)$. The joint resolution of the identity $\tilde{E}(\Delta) = FE(\Delta)\,F^{-1}$ of the family $(\lambda_n)_{n=1}^{\infty}$ coincides with the operators of multiplication by $h_{\Delta}(\cdot)$, and the spectral measure is $\mu(\cdot)$.

It is clear that by choosing another cyclic vector Ω' we get a measure $\mu'(\cdot)$ equivalent to $\mu(\cdot)$ and the Fourier transform $F' : L_2(I\!R^{\infty}, d\omega(x)) \to L_2(I\!R^{\infty}, d\mu'(\lambda))$ unitarily equivalent to F.

Let us look at the calculation of the measure $\mu(\cdot)$.

THEOREM 4. If $d\omega(x) = \bigotimes_{n=1}^{\infty} p_n(x_n)\,dx_n$ is a product-measure and

$$q_n(\lambda_n) = \widetilde{\sqrt{p_n}}\ (\lambda_n) \neq 0 \ \ (\mathrm{mod}\ d\lambda_n) \ \ (n = 1,2,...) \tag{5}$$

($\tilde{u}(\lambda_n)$ is the Fourier transform of the function $u(x_n) \in L_2(I\!R^1, dx_n)$) then the measure $\mu(\cdot)$ corresponding to the vector $\Omega(x) = 1$ equals the product measure $\bigotimes_{n=1}^{\infty} |q_n(\lambda_n)|^2\,d\lambda_n$.

Proof. The restriction of the representation (3) to $L_2(I\!R^1, p_n(x_n)\,dx_n)$ has the form

$$(U_{(0,\ldots,0,t_n,0,\ldots)}\,u)\,(x_n) = \sqrt{\frac{p_n(x_n+t_n)}{p_n(x_n)}}\ u(x_n+t_n). \tag{6}$$

For the space $L_2(I\!R^1, dx_n)$ we find that the image of the operators $U_{(0,\ldots,0,t_n,0,\ldots)}$ has the form $(W_{t_n}\,v)\,(x_n) = v(x_n+t_n)$ where $v(x_n) \in L_2(I\!R^1, dx_n)$. We know that the condition (5) is necessary and sufficient for the function $q_n(\lambda)$ to be cyclic vector for the representation $I\!R^1 \ni t_n \mapsto W_{t_n}$ on $L_2(I\!R^1, dx_n)$ which is unitarily equivalent to (6). It follows from this that the vector $\Omega(x) = 1$ is cyclic.

Calculate the spectral measure $\mu(\cdot)$. For $\Delta = \prod_{n=1}^{\infty} \Delta_n$ we have

$$(\bigotimes_{n=1}^{\infty} E_n(\Delta_n)\mathbf{1}, \mathbf{1}) = \prod_{n=1}^{\infty} (E_n(\Delta_n)\,1, 1) =$$

$$= \prod_{n=1}^{\infty} \int_{\Delta_n} |q_n(\lambda_n)|^2 \, d\lambda_n = \int_{\Delta} \bigotimes_{n=1}^{\infty} |q_n(\lambda_n)|^2 \, d\lambda_n , \tag{7}$$

where $E_n(\cdot)$ is the decomposition of the identity for the operator D_n.

Now consider the case that the condition (5) does not hold. It means that for some n the Fourier transform of the function $\sqrt{p_n(\cdot)}$ is zero on a set of positive measure. For every such n choose a function $\omega_n(\cdot) \in L_2(\mathbb{R}^1, dx_n)$ such that $\|\omega_n(\cdot)\| = 1$, $(F\,\omega_n)\,(\lambda_n) \neq 0 \pmod{d\lambda_n}$ and such that the series

$$\sum_{n=1}^{\infty} |1 - (\sqrt{p_n}\,, \omega_n)| < \infty \tag{8}$$

converges. This can be done by choosing ω_n the sum of N_n terms in the expansion of the function $\sqrt{p_n(\cdot)} \in L_2(\mathbb{R}^1, dx_n)$ with respect to the Hermite functions $(h_k(x_n))_{k=1}^{\infty}$, because we have

$$(\sum_{k=0}^{N_n} c_k \tilde{h}_k(\cdot))\,(\lambda_n) = \sum_{k=0}^{N_n} (-i)^k\, c_k\, h_k(\lambda_n) \neq 0$$

almost everywhere with respect to the Lebesgue measure on the line.

As before we can show that the function $\omega_n(\cdot)/\sqrt{p_n(\cdot)}$ is a cyclic vector for the representation U_{t_n} which is unitarily equivalent to the representation (6). Put $\Omega'(x) = \prod_{n=1}^{\infty} \omega_n(x_n)/\sqrt{p_n(x_n)}$. The condition (8) guarantees that the vector $\Omega'(x)$ belongs to the space $L_2(\mathbb{R}^2(\mathbb{R}^{\infty}, d\omega(x))$. This vector is cyclic for the representation (3). By a direct calculation we see that the spectral measure corresponding to $\Omega'(x)$ equals $\bigotimes_{n=1}^{\infty} |\tilde{\omega}_n(\lambda_n)|^2 \, d\lambda_n$. □

Note that if the measure $\omega(\cdot)$ is exchanged, there appear unitarily nonequivalent families of differential operators.

Example 3. Let $\omega_1(\cdot)$ and $\omega_2(\cdot)$ be orthogonal Gaussian product-measures. From the known equality

$$(\exp\{-\alpha x_k^2\})^{\tilde{}}\,(\lambda_k) = \sqrt{\frac{\pi}{a}}\, \exp\left\{-\frac{\lambda_k^2}{4\alpha}\right\} \qquad (k = 1, 2, \ldots)$$

and Theorem 4 it follows that the spectral measures $\mu_1(\cdot)$ and $\mu_2(\cdot)$ are also orthogonal Gaussian product-measures. □

Let us compute the unitary Fourier operator $F : L_2(\mathbb{R}^{\infty}, d\omega(x)) \to L_2(\mathbb{R}^{\infty}, d\mu(\lambda))$ under the assumption that condition (5) in Theorem 4 is satisfied. So the vector $\Omega(x) \equiv \mathbf{1}$ is cyclic for the representation (3) of $\mathbb{R}_0^{\infty}$ in $L_2(\mathbb{R}^{\infty}, \bigotimes_{n=1}^{\infty} p_n(x_n)\, dx_n)$.

THEOREM 5. The unitary Fourier transform

$$F : L_2(\mathbb{R}^\infty, \bigotimes_{k=1}^{\infty} p_k(x_k)\,dx_k) \to L_2(\mathbb{R}^\infty, \bigotimes_{k=1}^{\infty} |q_k(\lambda_k)|^2\,d\lambda_k)$$

defined on the cylindrical functions $f(x_1, \ldots, x_n) \in L_2(\mathbb{R}^\infty, \bigotimes_{k=1}^{\infty} p_k(x_k)\,dx_k)$ is given by the formula

$$(F f)(\lambda_1, \ldots, \lambda_n) = \widetilde{(f(x_1, \ldots, x_n)\sqrt{p_1(x_1)\cdots p_n(x_n)})}(\lambda_1, \ldots, \lambda_n)\,/$$
$$/\,q_1(\lambda_1)\cdots q_n(\lambda_n), \tag{9}$$

where $\tilde{\cdot}$ is the n-dimensional Fourier operator on $L_2(\mathbb{R}^n, dx_1\cdots dx_n)$.

One proves this by a direct calculation of the joint resolution of the identity of the CSO $(D_k)_{k=1}^{\infty}$ on $L_2(\mathbb{R}^\infty, \bigotimes_{k=1}^{\infty} p_k(x_k)\,dx_k)$. □

We also give a calculation of the spectral measure of the operators $(D_k)_{k=1}^{\infty}$ on $L_2(\mathbb{R}^\infty, dg_B(x))$ where $g_B(x)$ is a $\mathbb{R}_0^\infty$-quasi-invariant Gaussian measure.

THEOREM 6. The spectral measure of the representation U_t^B can be chosen to be a Gaussian measure on $\mathbb{R}^\infty$ with characteristic functional of the form

$$\chi(t) = e^{-\frac{1}{8}(B^{-1}t,t)} \qquad (t \in \mathbb{R}_0^\infty)$$

where $B^{-1} = A^{-1}(A^*)^{-1}$.

Proof. Because the vector $\Omega(x) \equiv \mathbf{1}$ is a vector of maximal spectral type of the representation U_t^B, the measure $\mu(\Delta) = (E(\Delta)\,\mathbf{1}, \mathbf{1})$ is spectral. Its characteristic functional

$$\chi(t) = \int_{\mathbb{R}^\infty} e^{i(t,\lambda)}\,d(E(\lambda)\,\mathbf{1}, \mathbf{1}) = (U_t^B\,\mathbf{1}, \mathbf{1}).$$

So

$$\chi(t) = \int_{\mathbb{R}^\infty} (U_t^B\,1)(x)\,dg_B(x) = \int_{\mathbb{R}^\infty} (U_{(A^*)^{-1}t}\,1)(x)\,dg(x) =$$

$$= \int_{\mathbb{R}^\infty} \exp(-\frac{1}{4}\|(A^*)^{-1}t\|^2 - \frac{1}{2}((A^*)^{-1}t,x))\,dg(x) = \exp(-\frac{1}{8}\|(A^*)^{-1}t\|^2).$$

Denote $B^{-1} = A^{-1}(A^*)^{-1}$. The matrix B^{-1} exists because the columns of $(A^*)^{-1}$ and the rows of A^{-1} are vectors in l_2, and B^{-1} generates a continuous operator from $\mathbb{R}_0^\infty$ into $\mathbb{R}^\infty$. Then

$$\chi(t) = e^{-\frac{1}{8}(B^{-1}t,t)}.$$

□

Define a unitary transformation F^B which, in the case of a simple spectrum of the representation, transforms the action of the operator U_t^B into multiplication by the function $e^{i(t,\lambda)}$. And so the CSO $(D_k)_{k=1}^{\infty}$ will be transformed into the CSO $(\lambda_k)_{k=1}^{\infty}$.

LEMMA 5. Let $f(\cdot)$ be a cylindrical function, $f(\cdot) \in L_2(\mathbb{R}^\infty, dg_I(x))$. Then

$$(F^I f)(\lambda) = e^{\|\lambda\|^2} \int_{\mathbb{R}^\infty} f(x)\, e^{-i(\lambda,x)}\, dg_{2I}(x)$$

and $(F^I f)(\cdot)$ is a cylindrical function of $L_2(\mathbb{R}^\infty, dg_{\frac{1}{4}I}(\lambda))$.

Proof. We show that $(F^I f)(\lambda)$ is a cylindrical function. Let $f(x) = f(p_n x)$, where p_n is a projection on the finite-dimensional space $\mathbb{R}^n$.

Using the equality $g_I(\cdot) = g_I^{p_n}(\cdot) \otimes g_I^{I-p_n}(\cdot)$, where $g_I^{p_n}(\cdot)$ is the restriction of the measure $g_I(\cdot)$ to $\mathbf{B}_n(\mathbb{R}^\infty)$, we get

$$(F^I f)(\lambda) = e^{\|\lambda\|^2} \int f(p_n x)\, e^{-i(\lambda,p_n x)}\, e^{-i(\lambda, I-p_n x)}\, dg_{2I}(x) =$$

$$= e^{\|\lambda\|^2} \int_{\mathbb{R}^n} f(x_1,\ldots,x_n)\, e^{-i(\lambda,x)}\, dg_{2I}^{p_n}(x) \times$$

$$\times \int_{\mathbb{R}^\infty} e^{-i(\lambda,x)}\, dg_{2I}^{I-p_n}(x) = (F^I f)(p_n \lambda).$$

Because $dg_I(x) = \bigotimes_{k=1}^{\infty} dg_I(x_k)$, from now on we can work in the one-dimensional case.

Let's prove that F^I is a unitary transformation from $L_2(\mathbb{R}^\infty, dg_I(x))$ into $L_2(\mathbb{R}^\infty, dg_{\frac{1}{4}I}(\lambda))$ ($g_{\frac{1}{4}I}(\cdot)$ is the spectral measure of the representation U_t^I; in the one-dimensional case $dg_{\frac{1}{4}I}(\lambda_k) = \sqrt{\frac{2}{\pi}}\, e^{-2\lambda_k^2}\, d\lambda_k$):

$$\|f\|^2_{L_2(\mathbb{R}^1, dg_I(x_1))} = \frac{1}{\sqrt{2\pi}} \int_{-\infty}^{\infty} |f(x_1)|^2\, e^{-\frac{x_1^2}{2}}\, dx_1 =$$

$$= \frac{1}{\sqrt{2\pi}} \int_{-\infty}^{\infty} |f(x_1)\, e^{-\frac{x_1^2}{4}}|^2\, dx_1 = \frac{1}{\sqrt{2\pi}} \int_{-\infty}^{\infty} \Bigl| \frac{1}{\frac{1}{\sqrt{2\pi}}} \times$$

$$\times \int_{-\infty} f(x_1)\, e^{-\frac{x_1^2}{4}}\, e^{i\lambda_1 x_1}\, dx_1 \Bigr|^2\, d\lambda_1 =$$

$$= \sqrt{\frac{2}{\pi}} \int_{-\infty}^{\infty} \Bigl| \frac{1}{2\sqrt{\pi}}\, e^{\lambda_1^2} \int_{-\infty}^{\infty} f(x_1)\, e^{-\frac{x_1^2}{4}}\, e^{ix_1\lambda_1}\, dx_1 \Bigr|^2\, e^{-2\lambda_1^2}\, d\lambda_1 =$$

$$= \sqrt{\frac{2}{\pi}} \int_{-\infty}^{\infty} |(F^I f)(\lambda_1)|^2\, e^{-2\lambda_1^2}\, d\lambda_1 = \|F^I f\|^2_{L_2(\mathbb{R}^1, dg_{\frac{1}{4}I}(\lambda_1))}.$$

Now we show that the transformation F^I transforms the operator U_t^I into multiplication by an exponent:

$$(F^I U_t^I f)(\lambda_1) = \frac{1}{2\sqrt{\pi}} e^{\lambda_1^2} \int_{-\infty}^{\infty} f(x_1+t_1)\, e^{-\frac{t_1^2}{4} - \frac{t_1 x_1}{2}}\, e^{-i\lambda_1 x_1}\, e^{-x_1^2/4}\, dx_1 =$$

$$\frac{1}{2\sqrt{\pi}} e^{\lambda_1^2} \int_{-\infty}^{\infty} f(x_1+t_1)\, e^{-i\lambda_1(x_1+t_1)}\, e^{-\frac{1}{4}(x_1+t_1)^2}\, d(x_1+t_1)\, e^{i\lambda_1 t_1} =$$

$$= e^{it_1\lambda_1} (F^I f)(\lambda_1).$$

Changing the variables we get the following theorem.

THEOREM 7. Let $f(\cdot)$ be a cylindrical function from $L_2(\mathbb{R}^\infty, dg_B(x))$. Then $(F^B f)(\lambda) = e^{(B\lambda,\lambda)} \int_{\mathbb{R}^\infty} f(x)\, e^{-i(\lambda,x)}\, dg_{2B}(x)$ and $(F^B f)(\lambda)$ is also a cylindrical function from $L_2(\mathbb{R}^\infty, dg_{\frac{1}{4}B^{-1}}(\lambda))$.

Note that for the representation $U_t^{\frac{1}{2}I}$ the transformation $F^{\frac{1}{2}I} : L_2(\mathbb{R}^\infty, dg_{\frac{1}{2}I}(x)) \to L_2(\mathbb{R}^\infty, dg_{\frac{1}{2}I}(\lambda))$ coincides with the usual Fourier-Wiener transformation, (see R.H. Cameron and W.T. Martin [1], R.V. Guseĭnov [1,2]).

3.4. Measurable polynomials of $(D_k)_{k=1}^\infty$. Differential operators with constant coefficients.

Let $(D_n)_{n=1}^\infty$ be a family of commutative differential operators on $L_2(\mathbb{R}^\infty, d\omega(x))$ corresponding to the $\mathbb{R}_0^\infty$-quasi-nvariant measure $\omega(\cdot)$, and let $\mu(\cdot)$ be its spectral measure. We recall the definition of $\mu(\cdot)$-measurable polynomial of infinitely many variables. Let $\mathrm{P}_0(\mathbb{R}^\infty)$ be the space of cylindrical (i.e. dependent only on finitely many variables) polynomials on $\mathbb{R}^\infty$, $\mathrm{P}_{0,n}(\mathbb{R}^\infty)$ be the space of cylindrical polynomials of degree not greater than n. The polynomial $\pi(\lambda) = \lim_{k\to\infty} P_k(\lambda) \pmod{\mu(\cdot)}$ where $P_k(\cdot) \in \mathrm{P}_{0,n}(\mathbb{R}^\infty)$ $k = (1,2,...)$ is called a $\mu(\cdot)$-measurable polynomial of the degree not greater than n. Denote the totality of $\mu(\cdot)$-measurable polynomials of the degree not greater than n by $\Pi_n(\mu)$. Then $\Pi(\mu) = \bigcup_{n=0}^{\infty} \Pi_n(\mu)$ is set of all $\mu(\cdot)$-measurable polynomials. In short, a measurable polynomial with respect to the measure $\mu(\cdot)$ is a $\mathbf{B}(\mathbb{R}^\infty)$-measurable function which can be represented as the limit of a sequence of cylindrical polynomials of degree not greater than some $n \in \mathbb{N}$ with convergence almost everywhere with respect to the measure $\mu(\cdot)$. The degree of a $\mu(\cdot)$-measurable polynomial $\pi(\cdot)$ is the smallest n for which $\pi(\cdot) \in \Pi_n(\mu)$.

DEFINITION 3. A differential operator on $L_2(\mathbb{R}^\infty, d\omega(x))$ with constant coefficients is a $\mu(\cdot)$-measurable polynomial $P(\cdot)$ of the family $(D_n)_{n=1}^\infty$, i.e.

$$P(D_1, \ldots, D_n, \ldots) = \int_{\mathbb{R}^\infty} P(\lambda_1, \ldots, \lambda_n, \ldots)\, dE(\lambda_1, \ldots, \lambda_n, \ldots). \tag{10}$$

Let us give an example of the dependence on the choice of $\omega(\cdot)$ of the class $\Pi(\mu)$ of polynomials, measurable with respect to the measure $\mu(\cdot)$.

Example 4. Let $d\omega(x) = \bigotimes_{n=1}^{\infty} (\pi n^2)^{-1/2} e^{-x_n^2/n^2} dx_n$. The spectral measure $\mu(\cdot)$ corresponding to the family $(D_n)_{n=1}^{\infty}$ and to the vacuum $\Omega \equiv 1$ equals $\bigotimes_{n=1}^{\infty} (\pi n^{-2})^{-1/2} e^{-n^2\lambda_n^2} d\lambda_n = d\mu(\lambda)$. The polynomial $\sum_{n=1}^{\infty} \lambda_n^2$ belongs to $\Pi(\mu)$ and so one can define the Laplacian $\Delta = D_1^2 + \cdots + D_n^2 + \cdots$ on $L_2(\mathbb{R}^\infty, d\omega(x))$. However, in the space $L_2(\mathbb{R}^\infty, dg(x))$ which is constructed by using the Gaussian measure $dg_{\frac{1}{2}I}(x) = \bigotimes_{k=1}^{\infty} \frac{1}{\sqrt{\pi}} e^{-x_k^2} dx_k$, the spectral measure of the family $(D_k)_{k=1}^{\infty}$ of differentiations again equals the measure $dg_{\frac{1}{2}I}(\cdot)$; the polynomial $\sum_{n=1}^{\infty} \lambda_n^2$ does not belong to $\Pi(g_{\frac{1}{2}I}(\cdot))$ and the corresponding operator $\Delta' = D_1^2 + \cdots + D_n^2 + \cdots$ is not well defined. □

All the results of Chapter 1, §1.7 hold for the operators of type (10).

We will also prove a theorem on absolute continuity with respect to the Lebesgue measure on the line of the spectral measure of the operator $P(D_1, \ldots, D_n, \ldots)$.

THEOREM 8. Let $\mu(\cdot)$ be a spectral measure of the family $(D_n)_{n=1}^{\infty}$, $\mathbf{H}$ be a Hilbert space of full $\mu(\cdot)$ measure. If $P(\lambda)$ $(\lambda \in \mathbf{H})$ is a continuous polynomial, then the spectrum of the operator $P(D_1, \ldots, D_n, \ldots)$ equals $\overline{\mathbf{R}(P)}$. If in addition $P(\lambda) \neq \text{const}$, then the spectral measure of the operator $P(D_1, \ldots, D_n, \ldots)$ is absolutely continuous with respect to the Lebesgue measure on $\overline{\mathbf{R}(P)}$.

Proof. From the definition of the operator $P(D_1, \ldots, D_n, \ldots)$ as a function of the family $(D_n)_{n=1}^{\infty}$ it follows that the spectral measure of this operator is equivalent to the $\mu(P^{-1})$-image of the measure $\mu(\cdot)$ under the mapping $P : \mathbf{H} \to \mathbb{C}^1$. We show that $d\mu(P^{-1}) \ll dz(\cdot)$. We know that for any $n \in \mathbb{N}$, $\mu(\cdot) \sim \mu_n(\cdot) \otimes \mu^n(\cdot)$, where $\mu_n(\cdot), \mu^n(\cdot)$ are the corresponding projections of $\mu(\cdot)$ on $\mathbb{R}^n$ and $\mathbf{H} \ominus \mathbb{R}^n$. Thus it will be sufficient to show that for some n $(d\mu_n(\cdot) \otimes d\mu^n(\cdot))(P^{-1}) \ll dz$.

Let $\Delta \in \mathbf{B}(\mathbb{C}^1)$ have Lebesgue measure zero. Then we have to show that for some n $(\mu_n(\cdot) \otimes \mu^n(\cdot))(P^{-1}(\cdot)) = 0$. By Fubini's theorem, this is equivalent to $\mu_n(P^{-1}(\Delta)(\cdot, \lambda_{n+1}, \ldots)) = \mu_n(\{(\lambda_1, \ldots, \lambda_n) \in \mathbb{R}^n : P(\lambda_1, \ldots, \lambda_n, \lambda_{n+1}, \ldots) \in \Delta\}) = 0$ for almost all $(\lambda_{n+1}, \lambda_{n+2}, \ldots) \in \mathbf{H} \ominus \mathbb{R}^n$ with respect to $\mu^n(\cdot)$.

For fixed $\lambda_{n+1}, \lambda_{n+2}, \ldots$, $P(\lambda_1, \ldots, \lambda_n, \lambda_{n+1}, \ldots)$ is a polynomial $P_n(\lambda_1, \ldots, \lambda_n)$ of n variables and if $P(\lambda) \neq \text{const}$ then there exists n such that $P_n(\lambda_1, \ldots, \lambda_n) \neq \text{const}$. The spectral measure of the operator $P_n(D_1, \ldots, D_n)$ is absolutely continuous with respect to the Lebesgue measure, and so the wanted result follows. □

Among the operators (10) there are operators that do not have finite-dimensional analogues. Such an operator is the one corresponding to the polynomial

$$P_L(\lambda) = \lim_{n\to\infty} n^{-1} \sum_{k=1}^{n} \lambda_k^2. \tag{11}$$

DEFINITION 4. *A* $\mathbf{B}(\mathbb{R}^\infty)$-measurable $\mu(\cdot)$-almost everywhere finite polynomial $P(\lambda_1, \ldots, \lambda_n, \ldots)$ on $\mathbb{R}^\infty$ is called essentially infinite-dimensional if for all $t \in \mathbb{R}_0^\infty$

$$P(\lambda_1 + t_1, \ldots, \lambda_n + t_n, \ldots) = P(\lambda_1, \ldots, \lambda_n, \ldots) \pmod{\mu(\cdot)}. \tag{12}$$

Here, the measure $\mu(\cdot)$ is supposed to be quasi-invariant and the polynomial $P(\lambda_1 + t_1, \ldots, \lambda_n + t_n, \ldots)$ is $\mathbf{B}(\mathbb{R}^\infty)$-measurable and finite almost everywhere with respect to the measure $\mu(\cdot)$.

The essentially infinite-dimensional polynomials are $\mathbf{B}_\mu^\infty(\cdot)$-measurable. If the measure $\mu(\cdot)$ is $\mathbb{R}_0^\infty$-ergodic then (12) is satisfied only by constants.

The properties of such a polynomial are closely related to the decomposition of the measure $\mu(\cdot)$ into $\mathbb{R}_0^\infty$-ergodic measures $\mu^\alpha(\cdot)$:

$$\mu(\cdot) = \int_{\mathbb{R}^1} \mu^\alpha(\cdot)\, d\nu(\alpha)$$

where $\mu^\alpha(\cdot) \perp \mu^{\alpha'}(\cdot)$ for $\alpha \neq \alpha'$. By the ergodicity theorem, $\lim_{n\to\infty} n^{-1} \sum_{k=1}^{n} \lambda_k^2$ either does not exist for $\mu^\alpha(\cdot)$-almost all $\lambda \in \mathbb{R}$ or exists and equals a constant $c_\alpha (\operatorname{mod} \mu^\alpha(\cdot))$. Then the operator of multiplication by $P_L(\lambda)$ in $L_2(\mathbb{R}^\infty, d\mu(\lambda)) = \oplus \int_{\mathbb{R}^1} L_2(\mathbb{R}^\infty, d\mu^\alpha(\lambda))\, d\nu(\alpha)$, has the form $\{c_\alpha I_\alpha\}$. The operator $P_L(D_1, \ldots, D_n, \ldots)$ has an analogous structure.

Example 5. Let

$$\omega(\Delta) = \int_0^\infty g_{\alpha I}(\Delta)\, d\rho(\alpha)$$

where $\Delta \in \mathbf{B}(\mathbb{R}^\infty)$, $dg_{\alpha I}(x) = \bigotimes_{n=1}^{\infty} (2\pi\alpha)^{-1/2} \exp(-x_n^2/2\alpha)\, dx_n$, $\rho(\cdot)$ is a probability measure on $[0, \infty)$. Because $g_{\alpha_1 I}(\cdot) \perp g_{\alpha_2 I}(\cdot)$ for $\alpha_1 \neq \alpha_2$ and $\forall \alpha \in [0, \infty)\ \exists B_\alpha \in \mathbf{B}(\mathbb{R}^\infty) : g_{\alpha' I}(B_\alpha) = \delta_{\alpha\alpha'}$ we have $L_2(\mathbb{R}^\infty, d\omega(x)) = \oplus \int_{[0,\infty)} L_2(\mathbb{R}^\infty, dg_{\alpha I}(x))\, d\rho(\alpha)$. In each $L_2(\mathbb{R}^\infty, dg_{\alpha I}(x))$ the representation (3) is cyclic, and $\Omega(x) \equiv \mathbf{1}$ is a vacuum. So it is cyclic in $L_2(\mathbb{R}^\infty, d\omega(x))$ and we can take the function identically equal to one to be the vacuum. Let us calculate the spectral measure corresponding to $\Omega(x) \equiv \mathbf{1}$. Using Theorem 4 we get

$$\mu(\Delta) = \int_{[0,+\infty)} g_{\frac{1}{\alpha} I}(\Delta)\, d\rho(\alpha).$$

The measure $\mu(\cdot)$ is not ergodic with respect to the action of the group $\mathbb{R}_0^\infty$, and so there exist non-trivial essentially infinite-dimensional polynomials in $\Pi(\mu)$. One of them is $P_L(\lambda) = \lim_{n\to\infty} n^{-1} \sum_{k=1}^{n} \lambda_k^2$. It is shown that $P_L(\lambda) = \dfrac{1}{\alpha} \pmod{g_{\frac{1}{\alpha} I}(\cdot)}$. So, in $L_2(\mathbb{R}^\infty, d\mu(\lambda))$ the operator of multiplication by the polynomial $P_L(\cdot)$ has the form $\{\alpha^{-1}\}$ and its spectrum coincides with the support of the measure $\rho(\cdot)$. The operator $P_L(D_1, \ldots, D_n)$ acts in the same way according to (11). Introducing an isomorphism between the spaces $\oplus \int_{[0,\infty)} L_2(\mathbb{R}^\infty, dg_{\alpha I}(x))$ and $L_2(\mathbb{R}^\infty, dg(x)) \otimes L_2([0,\infty), d\rho(\alpha))$ we find that the operator $P_L(D_1, \ldots, D_n, \ldots)$ is unitarily equivalent to the operator $I \otimes \alpha^{-1}$, where I are the identity operators on $L_2(\mathbb{R}^\infty, dg(x))$. $\square$

Comments to Chapter 3.

1. The properties of $\mathbb{R}_0^\infty$-quasi-invariant measures are given according to V.N. Sudakov [1,2] (Theorem 1), A.V. Skorokhod [3] (Proposition 12 and Theorem 2) and the articles of Japanese mathematicians Y. Umemura [1], H. Shimomura [1-5] and Y. Yamasaki [1,2,3]. The question of $\mathbb{R}_0^\infty$-quasi-invariance of the Gaussian measure on $\mathbb{R}^\infty$ is considered in A.Yu. Daletskiĭ [1]. Example 1 is given by the author.

2. In a more general setting, the differential CSO $(D_k)_{k=1}^\infty$ and their spectral properties have been studied in Yu.S. Samoĭlenko and G.F. Us [1-3].

3. The simplest situation of a Gaussian measure with a diagonal correlation operator introduced in this book as "Fourier transform", is the known Fourier-Wiener transformation (see R. Cameron, W. Martin [1], R.V. Guseinov [1,2]). The Fourier transforms in the function spaces constructed, using the product-measure on $\mathbb{R}^\infty$, are considered in Yu.S. Samoĭlenko and G.F. Us [1], and, using the Gaussian measure - in A.Yu. Daletskiĭ [1,2].

4. The considered class of the operators is introduced in Yu.S. Samoĭlenko, G.F. Us [1] using the analogy with differential operators with constant coefficients operating on functions of finitely many variables. Note that these operators can be essentially different from the operators called infinitely-dimensional differential operators with constant coefficients in some other articles (T. Hida [1], Yu.L. Daletskiĭ [2], M.I. Vishik [1] and others). Theorem 8 is due to I.A. Fil' [1]. For the examples of essentially infinite-dimensional differential operators with constant coefficients see Yu.S. Samoĭlenko and G.F. Us [1-3].

PART II
INDUCTIVE LIMITS OF FINITE-DIMENSIONAL LIE ALGEBRAS AND THEIR REPRESENTATIONS

Considering a finite collection $(A_k)_{k=1}^n$ of CSO we always assume that the resolutions of the identity of these operators commute. Equivalently we can consider a unitary representation of the Lie group $\mathbb{R}^n \ni (t_1, \ldots, t_n) \mapsto U(t_1, \ldots, t_n)$ corresponding to this collection of CSO. The collection of CSO $(A_k)_{k=1}^n$ forms a representation of the basis of the real Lie algebra (real commutative Lie group $\mathbb{R}^n$). It can be extended to a unitary representation of the group. The spectral characteristics of the unitary representations of $\mathbb{R}^n$ correspond to the spectral characteristics of the collections $(A_k)_{k=1}^n$. Countable collections of CSO lead to unitary representations of the group $\mathbb{R}_0^\infty$.

Let $\mathbf{g}$ be a real algebra, $(X_k)_{k=1}^d$ be a basis in $\mathbf{g}$ and $(c_{ij}^k)_{i,j,k=1}^d$ be its structure constants.

A skew-adjoint representation $\pi(\cdot)$ of a real Lie algebra $\mathbf{g}$ on a Hilbert space H is a homomorphism $\mathbf{g} \ni X \mapsto \pi(X)$ into the set of linear operators which are essentially skew-adjoint on a set Φ which is dense in H and invariant relative to the operators $(\pi(X))_{X \in \mathbf{g}}$.

Thus, generally speaking, the operators $\pi(X)$ are not assumed to be bounded. But

$$[\pi(X_i), \pi(X_j)]\, f = \sum_{k=1}^{d} c_{ij}^k\, \pi(X_k)\, f \tag{$*$}$$

in the domain $\Phi \ni f$ of their essential skew-adjointness.

If $G \ni g \mapsto U_g$ is a unitary representation of the Lie group G in H, then the Gårding domain $\mathbf{D}_\Gamma$ of the representation (see Chapter 2, §2.5) is a commom invariant domain of essential skew-adjointness of the generators of one-parameter subgroups of G, and the relation $(*)$ holds on $\mathbf{D}_\Gamma$.

So, for any strongly continuous unitary representation of the group $G \ni g \mapsto U_g$ there corresponds a skew-adjoint representation $\pi_U(\cdot)$ of its Lie algebra $\mathbf{g}$.

A skew-adjoint representation $\pi(\cdot)$ on H of a real Lie algebra $\mathbf{g}$ can be extended to a unitary representation of the group $G \ni g \mapsto U_g$ if $\pi(X) = \pi_U(X)$ for all $X \in \mathbf{g}$.

Not every representation of a Lie algebra can be extended to a group representation. Examples of representations of Lie algebras which can not be extended to a group representation are given, for example, in E. Nelson [1] (also see R. Powers [2] and M. Reed and B. Simon [1] and I. Segal [4]). The extension criteria for an algebra representation to a group representation are given in E. Nelson [1] and developed further by M.

Flato, J. Simon, H. Snellman and D. Sternheimer [1] (see also Chapter 11 of A. Barut and R. Rączka [1] and the bibliography therein).

When we study spectral questions for collections of skew-adjoint operators which establish a representation of a basis in a Lie algebra we will always assume that it is possible to extend this representation to a unitary representation of the corresponding Lie group. So, the transition to countable collections of skew-adjoint operators which establish a representation of a basis in an infinite-dimensional Lie algebra, namely the inductive limit of finite-dimensional Lie algebras, is closely connected with the theory of unitary representations of the inductive limits of the corresponding Lie groups.

The main content of this chapter is the study of some classes of unitary representations of groups $G = j-\varinjlim G_n$, where G_n are connected, simply connected Lie groups $(j=(j^n_{n+1}),\ j^n_{n+1} : G_n \hookrightarrow G_{n+1})$ as well as the study of these representations corresponding classes of skew-adjoint representations of the generators in the Lie algebra $\mathbf{g} = J-\varinjlim \mathbf{g}_n$ where $\mathbf{g}$ is the Lie algebra of the group G, $\mathbf{g}_n$ is the Lie algebra of the group $G_n (J=(J^n_{n+1}),$ $J^n_{n+1} : \mathbf{g}_n \hookrightarrow \mathbf{g}_{n+1})$.

Chapter 4.
CANONICAL COMMUTATION RELATIONS (CCR) OF SYSTEMS WITH COUNTABLE DEGREES OF FREEDOM

In this chapter we consider countable collections of self-adjoint operators $(P_k, Q_k)_{k=1}^{\infty}$ such that

$$(Q_k)_{k=1}^{\infty} - \text{CSO}, \quad (P_k)_{k=1}^{\infty} - \text{CSO}$$

$$[P_k, Q_l] = -i\,\delta_{kl}\, I \quad (k, l = 1, 2, \ldots). \tag{1}$$

The relations (1) cannot be satisfied by operators on a finite-dimensional Hilbert space H, or by Hilbert-Schmidt operators on a separable space H, because $Tr\,[P_k, Q_k] = 0$ and $Tr(i\,I) \neq 0$. Moreover we have the following:

PROPOSITION 1. There do not exist two simultaneously bounded operators P and Q on a separable H such that $[P, Q] = -i\,I$.

Proof. Suppose there exist two bounded self-adjoint operators P and Q on H such that $PQ - QP = -i\,I$. Without loss of generality, we can suppose that P is non-singular (otherwise we could consider the operator $P + \lambda I$ with $\lambda > \|P\|$ satisfying the same relation). Then from $PQ = P(QP)P^{-1}$ it follows that the spectra of PQ and QP coincide. From the commutation relations it follows that if the spectrum contains a point z, then it also contains all the points $z - i\nu$ ($\nu \in \mathbb{Z}$) and so it is not bounded. But this contradicts the assumption that P and Q are bounded. □

The operators of the collection $(Q_k, P_k)_{k=1}^{\infty}$ are unbounded, and below we introduce the relations (1) (the representations of canonical commutation relations (CCR) of systems with countable degrees of freedom) using the skew-adjoint Lie algebra $\mathfrak{g}_0^{\infty}$ representations which can be extended to unitary representations of the corresponding group G_0^{∞}.

First of all we recall the Stone-von-Neumann theorem on the uniqueness of an irreducible representation of G^n (such that it is related to a representation of the CCR with finite degrees of freedom). However, in the case of systems with countable degrees of freedom, there exist non-equivalent irreducible representations. The main subject of this chapter is the study of some particular classes of representations of the CCR with countable degrees of freedom. We consider classes of independent, locally dependent, Gaussian representations.

4.1. Canonical commutation relations of systems with finite degrees of freedom. The Stone-Von-Neumann uniqueness theorem.

Consider the Lie group $G = (\mathbb{R}^1, \mathbb{R}^1, S^1) \ni (t_1, s_1, \alpha_1)$ with the group operation

$$(t_1^{(1)}, s_1^{(1)}, \alpha_1^{(1)})\,(t_1^{(2)}, s_1^{(2)}, \alpha_1^{(2)}) =$$

$$= (t_1^{(1)} + t_1^{(2)}, s_1^{(1)} + s_1^{(2)}, e^{it_1^{(2)} s_1^{(1)}}\, \alpha_1^{(1)} \alpha_1^{(2)}).$$

The one-parameter subgroups $(\mathbb{R}^1, 0, 1) \ni (t_1, 0, 1)$, $(0, \mathbb{R}^1, 1) \ni (0, s_1, 1)$ and $(0, 0, s^1) \ni (0, 0, \alpha_1)$ generate the whole group $G : (t_1, s_1, \alpha) = (t_1, 0, 1) \cdot (0, s_1, 1)\,(0, 0, \alpha_1)$.

The real Lie algebra **g** of the group G is three dimensional, a basis in **g** is q, p, e, where q is a tangent vector the subgroup $(\mathbb{R}^1, 0, 1)$, p is a tangent vector to $(0, \mathbb{R}^1, 1)$, e is tangent to the subgroup $(0, 0, s^1)$, the commutational relations in the algebra are: $[q, e] = [p, e] = 0$, $[p, q] = e$.

Now consider the unitary representation of the group $G \ni g \mapsto U_g$ on H such that for the self-adjoint generators of the one-parameter subgroups we have: $U_{(t_1, 0, 1)} = U_{t_1} = e^{it_1 Q_1}$, $U_{(0, s_1, 1)} = V_{s_1} = e^{is_1 P_1}$ and $U_{(0, 0, e^{i\Phi_1})} = e^{i\Phi_1} I = e^{i\Phi_1 I}$.

One can choose a dense subspace D_Γ in H (see Chapter 2, §2.5) such that $D_\Gamma \subset H^\infty(P_1) \cap H^\infty(Q_1)$ is invariant under P_1 and Q_1 and

$$[P_1, Q_1] f = -i f \quad \forall f \in D_\Gamma.$$

Thus, for any unitary representation of the group $G \ni g \mapsto U_g$ on H such that $U_{(0, 0, \alpha_1)} = \alpha_1 I$, there correspond a pair of self-adjoint operators, P_1 and Q_1 such that $[P_1, Q_1] = -i I$ on a dense set in H which is invariant for P_1 and Q_1.

DEFINITION 1. The representation of the canonical commutation relations in the Weyl form of systems with one degree of freedom is taken to be the unitary representation of the group $G \ni (t_1, s_1, \alpha_1) \mapsto U_{(t_1, s_1, \alpha_1)}$ on H such that $U_{(0, 0, \alpha_1)} \mapsto \alpha_1 I$.

DEFINITION 2. The representation of the canonical commutation relations (CCR) in the Weyl form of systems with n degrees of freedom is taken to be the unitary representation of the group $G^n = G \times G \times \cdots \times G$

$$((t_1, \ldots, t_n), (s_1, \ldots, s_n), (\alpha_1, \ldots, \alpha_n)) \mapsto U_{((t_1, \ldots, t_n), (s_1, \ldots, s_n), (\alpha_1, \ldots, \alpha_n))}$$

such that $U_{((0, \ldots, 0), (0, \ldots, 0), (\alpha_1, \ldots, \alpha_n))} = \alpha_1 \cdots \alpha_n I$.

The self-adjoint generators

$$U_{((0, \ldots, 0, t_k, 0, \ldots, 0), (0, \ldots, 0), (1, \ldots, 1))} = U_{(0, \ldots, 0, t_k, 0, \ldots, 0)} = e^{it_k Q_k}$$

and

$$U_{((0, \ldots, 0), (0, \ldots, 0, s_k, 0, \ldots, 0), (1, \ldots, 1))} = V_{(0, \ldots, 0, s_k, 0, \ldots)} = e^{is_k P_k}, \quad k = 1, \ldots, n)$$

of the one-parameter subgroups satisfy the relations

$$[P_k, Q_j] f = -i\, \delta_{kj} f \quad (k, j = 1, \ldots, n)$$

on a dense set of vectors $f \in D_\Gamma \subset H^\infty(Q_1, \ldots, Q_n, P_1, \ldots, P_n) \subset H$, invariant with respect to $(P_k, Q_k)_{k=1}^n$.

First of all, consider the unitary representation of the group $\Phi^n \ni ((t_1, \ldots, t_n), (s_1, \ldots, s_n), (\alpha_1, \ldots, \alpha_n)) \mapsto U_{((t_1,\ldots,t_n),(s_1,\ldots,s_n),(\alpha_1,\ldots,\alpha_n))}$ on $L_2(\mathbb{R}^n, d\lambda_1 \cdots d\lambda_n)$:

$$(U_{(t_1,\ldots,t_n)} f)(\lambda_1, \ldots, \lambda_n) = \exp(i \sum_{k=1}^{n} t_k \lambda_k) f(\lambda_1, \ldots, \lambda_n),$$

$$(V_{(s_1,\ldots,s_n)} f)(\lambda_1, \ldots, \lambda_n) = f(\lambda_1 + s_1, \ldots, \lambda_n + s_n),$$

$$(U_{((0,\ldots,0),(0,\ldots,0),(\alpha_1,\ldots,\alpha_n))} f)(\lambda_1, \ldots, \lambda_n) = \alpha_1 \cdots \alpha_n f(\lambda_1, \ldots, \lambda_n). \quad (2)$$

The representation (2) of the CCR is irreducible, because any bounded operator which commutes with the operators

$$(U_{(t_1,\ldots,t_n)})_{(t_1,\ldots,t_n)\in \mathbb{R}^n} = (\exp(i \sum_{k=1}^{n} t_k \lambda_k))_{(t_1,\ldots,t_n)\in \mathbb{R}^n}$$

is a multiplication operator by a measurable essentially bounded function. The commutation of such an operator with the shifts yields that this function is constant almost everywhere, i.e the operator is a multiple of the identity. We show that the representation (2) is unique up to unitary equivalence an irreducible representation of a system with finite degrees of freedom.

THEOREM 1. (Stone-von Neumann). An irreducible representation of the CCR of a system with finite degrees of freedom is unique (up to unitary equivalence) and coincides with the representation (2). Any representation of the CCR of a system with finite degrees of freedom is the direct sum of the representations (2).

Proof. We give a proof for $n = 1$ and for a representation of the CCR such that the spectrum of the self-adjoint operator Q is simple. Choose a cyclic vector $\Omega \in H^\infty(Q,P)$. The representation of the group Φ is unitarily equivalent to the representation on $L_2(\mathbb{R}^1, d\rho(\lambda_1)) \ni f(\cdot)$ where the operators of the representation satisfy

$$(U_{t_1,0,1} f)(\lambda_1) = (U_{t_1} f)(\lambda_1) = e^{it_1\lambda_1} f(\lambda_1)$$

and where the function $\Omega(\lambda) \equiv 1$ corresponds to the cyclic vector Ω. In this representation, the operators $U_{(0,s_1,1)} = V_{s_1}$ act as

$$(V_{s_1} f)(\lambda_1) = \alpha_{s_1}(\lambda_1) f(\lambda_1 + s_1)$$

where $\alpha_{s_1}(\lambda_1)$ is a smooth function with respect to s_1 and λ_1 and $V_{s_1} 1$ is such that from $V_{s_1^{(1)}+s_1^{(2)}} 1 = V_{s_1^{(2)}} V_{s_1^{(2)}} 1$ it follows that

$$\alpha_{s_1^{(1)}+s_1^{(2)}}(\lambda_1) = \alpha_{s_1^{(1)}}(\lambda_1)\, \alpha_{s_1^{(2)}}(\lambda_1 + s_1^{(2)}).$$

Indeed, consider the function

$$T(\lambda_1) = \sum_{k=1}^{n} c_k e^{is_1^{(k)}\lambda_1} = \sum_{k=1}^{n} c_k U_{s_1^{(k)}} 1.$$

The commutation relations yield

$$V_{s_1} T(\lambda_1) = \sum_{k=1}^{n} c_k V_{s_1} U_{s_1^{(k)}} 1 = \sum_{k=1}^{n} c_k e^{is_1 s_1^{(k)}} U_{s_1^{(k)}} V_{s_1} 1 =$$

$$= \sum_{k=1}^{n} c_k e^{is_1 s_1^{(k)}} e^{is_1^{(k)}\lambda_1} \alpha_{s_1}(\lambda_1) =$$

$$= \alpha_{s_1}(\lambda_1) \sum_{k=1}^{n} c_k e^{is_1^{(k)}(\lambda_1+s_1)} = \alpha_{s_1}(\lambda_1)\, T(\lambda_1 + s_1).$$

But because the functions $T(\lambda_1)$ are dense in $L_2(\mathbb{R}^1, d\rho(\lambda_1))$, we have

$$(V_{s_1} f)(\lambda_1) = \alpha_{s_1}(\lambda_1) f(\lambda_1 + s_1).$$

for all the functions in $L_2(\mathbb{R}^1, d\rho(\lambda_1))$. Because

$$\alpha_{s_1+\lambda_1}(0) = \alpha_{\lambda_1}(0)\, \alpha_{s_1}(\lambda_1)$$

setting $k(\lambda_1) = \alpha_{\lambda_1}(0)$ we get: $\alpha_{s_1}(\lambda_1) = \dfrac{k(\lambda_1+s_1)}{k(\lambda_1)}$. So,

$$(V_{s_1} f)(\lambda_1) = \frac{k(\lambda_1 + s_1)}{k(\lambda_1)} f(\lambda_1 + s_1).$$

Now, because the V_{s_1} are unitary operators, for any $f(\cdot) \in L_2(\mathbb{R}^1, d\rho(\lambda_1))$ we must have:

$$\int_{\mathbb{R}^1} |f(\lambda_1)|^2 d\rho(\lambda_1) = \int_{\mathbb{R}^1} |\alpha_{s_1}(\lambda_1)|^2 |f(\lambda_1 + s_1)|^2 d\rho(\lambda_1) =$$

$$= \int_{\mathbb{R}^1} \frac{|k(\lambda_1 + s_1)|}{|k(\lambda_1)|^2} |f(\lambda_1 + s_1)|^2 d\rho(\lambda_1).$$

Consequently

$$d\rho(\lambda_1 + s_1) = |\alpha_{s_1}(\lambda_1)|^2 d\rho(\lambda_1) = \frac{|k(\lambda_1 + s_1)|^2}{|k(\lambda_1)|^2} d\rho(\lambda).$$

But this means that the measure $d\rho(\lambda_1)$ is quasi-invariant on $(\mathbb{R}^1, \mathbf{B}(\mathbb{R}^1))$ and $d\rho(\lambda_1) = \dfrac{1}{|k(\lambda_1)|^2} d\lambda_1$.

The unitary operator $U : L_2(\mathbb{R}^1, d\rho(\lambda_1)) \to L_2(\mathbb{R}^1, d\lambda_1)$ defined by

$$(Uf)(\lambda_1) = k(\lambda_1) f(\lambda_1) \quad (f \in L_2(\mathbb{R}^1, d\rho(\lambda_1)))$$

gives now a unitary equivalence of the considered representation and the representation

(2) on $L_2(\mathbb{R}^1, d\lambda_1)$.

One proves similarly an equivalence of an irreducible representation of CCR of a system with finite degree of freedom, $n = 2,3,\ldots$, and the representation (2) (under the assumption that the spectrum of the CSO $(Q_k)_{k=1}^n$ is simple).

If the spectrum of Q_1 is multiple, then as in Chapter 2, §2.5, we prove that a representation of the group Φ is unitarily equivalent to a representation on $\oplus \sum L_2(\mathbb{R}^1, d\lambda_1) = \int_{\mathbb{R}^1} H_{\lambda_1} d\lambda_1$ ($\dim H_{\lambda_1} = \text{const}$).

Here, the operators are:

$$(U_{t_1} \vec{f})(\lambda_1) = e^{it_1\lambda_1} \vec{f}(\lambda_1)$$

$$(V_{s_1} \vec{f})(\lambda_1) = C_{s_1}(\lambda_1) \vec{f}(\lambda_1 + s_1)$$

where $C_{s_1}(\lambda_1) : H_{\lambda_1+s_1} \to H_{\lambda_1}$ is a smooth operator function such that

1) $C_{s_1}(\lambda_1)$ is unitary;

2) $C_{s_1^{(1)}+s_1^{(2)}}(\lambda_1) = C_{s_1^{(1)}}(\lambda_1)\, C_{s_1^{(2)}}(\lambda_1 + s_1^{(1)})$.

Then for the unitary operator function $C_{\lambda_1}(0) : H_{\lambda_1} \to H_0$ it follows from the commutation relations that

$$C_{\lambda_1+s_1}(0) = C_{\lambda_1}(0)\, C_{s_1}(\lambda_1)$$

and so the unitary operator $C_{\lambda_1}(0) : \oplus \sum L_2(\mathbb{R}^1, d\lambda_1) = \int_{\mathbb{R}^1} H_{\lambda_1} d\lambda_1 \to \oplus \sum L_2(\mathbb{R}^1, d\lambda_1)$ gives a unitary equivalence between the considered representation and the direct sum of the representations (2). $\square$

Remark 1. For CCR of systems with finite degrees of freedom, the irreducibility of a representation is equivalent to the simplicity of the joint spectrum of the CSO $(Q_k)_{k=1}^n$, and is also equivalent to the simplicity of the joint spectrum of the CSO $(P_k)_{k=1}^n$.

4.2. Representations of CCR of systems with countable degrees of freedom. Measures and cocycles.

Consider now the group

$$G_0^\infty = j - \lim_{\to} G^n, \quad j = (j_{n+1}^n),$$

$$j_{n+1}^n \; G^n \hookrightarrow G^{n+1}, \; j_{n+1}^n((t_1, s_1, \alpha_1), \ldots, (t_n, s_n, \alpha_n)) = ((t_1, s_1, \alpha_1), \ldots, (t_n, s_n, \alpha_n), (0,0,1))$$

DEFINITION 3. The unitary representation of the group G_0^∞

$$((t_1, s_1, \alpha_1), \ldots, (t_n, s_n, \alpha_n), \ldots) \mapsto U_{((t_1,s_1,\alpha_1), \ldots, (t_n,s_n,\alpha_n), \ldots)}$$

on H such that

$$U_{((0,0,\alpha_1), \ldots, (0,0,\alpha_n), \ldots)} = W_{(\alpha_1, \ldots, \alpha_n, \ldots)} = \alpha_1 \cdots \alpha_n \cdots I$$

is a representation of the CCR of a system with countable degrees of freedom in Weyl form.

The generators of the representations of the commutative subgroups

$$(\mathbb{R}_0^\infty, (0, \ldots, 0, \ldots), (1, \ldots, 1, \ldots)) \ni$$

$$((t_1, 0, 1), \ldots, (t_n, 0, 1), \ldots) \mapsto U_{((t_1,0,1), \ldots, (t_n,0,1), \ldots)} = U_{(t_1, \ldots, t_n, \ldots)} = U_t$$

and

$$((0, \ldots, 0, \ldots), \mathbb{R}_0^\infty, (1, \ldots, 1, \ldots)) \ni$$

$$((0, s_1, 1), \ldots, (0, s_n, 1), \ldots) \mapsto U_{((0,s_1,1), \ldots, (0,s_n,1), \ldots)} = V_{(s_1, \ldots, s_n, \ldots)} = V_s \quad (t,s \in \mathbb{R}_0^\infty)$$

is a countable collection of self-adjoint operators $(Q_k, P_k)_{k=1}^\infty$.

The CCR (1) are realized on the dense set, $H^\infty(P_1^2+Q_1^2, \ldots, P_n^2+Q_n^2, \ldots)$, which is invariant with respect to the operators of the collection.

THEOREM 2. The representation of the CCR in Weyl form of a system with countable degrees is unitarily equivalent to the representation on

$$H = \oplus \int_{\mathbb{R}^\infty} H_\lambda \, d\rho(\lambda) \ni \vec{f}(\cdot),$$

where $\rho(\cdot)$ is a $\mathbb{R}_0^\infty$ quasi-invariant measure on $\mathbb{R}^\infty$, established by the operators U_t $(t \in \mathbb{R}_0^\infty)$ and V_s $(s \in \mathbb{R}_0^\infty)$ which are defined by

$$(U_t \vec{f})(\lambda) = e^{i(t,\lambda)} \vec{f}(\lambda) \tag{3}$$

$$(V_s \vec{f})(\lambda) = C_s(\lambda) \sqrt{\frac{d\rho_s(\lambda)}{d\rho(\lambda)}} \vec{f}(\lambda+s). \tag{4}$$

The measurable unitary operator functions, $C_s(\lambda) : H_{\lambda+s} \to H_\lambda$ satisfy the equations

$$C_{s+s'}(\lambda) = C_s(\lambda) \, C_{s'}(\lambda+s). \tag{5}$$

Proof. Suppose first that the joint spectrum of the CSO $(Q_k)_{k=1}^\infty$ is simple, i.e. that the representation $\mathbb{R}_0^\infty \ni t \mapsto U_t$ is cyclic. Then the representation space of the CCR for countable degrees of freedom is unitarily equivalent to the space $H = L_2(\mathbb{R}^\infty, d\rho(\lambda))$. Under this equivalence, the operators U_t are realized by the operators of multiplication by $e^{i(t,\lambda)}$ in $L_2(\mathbb{R}^\infty, d\rho(\lambda))$.

On the dense set of trigonometric polynomials

$$T(\lambda) = \sum_{k=1}^{n} c_k e^{i(\lambda, s^{(k)})} = \sum_{k=1}^{n} c_k U_{s^{(k)}} 1 ,$$

the operators V_s ($s \in \mathbb{R}_0^\infty$) act as:

$$(V_s T)(\lambda) = \sum_{k=1}^{n} c_k V_s U_{s^{(k)}} = \sum_{k=1}^{n} c_k e^{i(s, s^{(k)})} U_{s^{(k)}} V_s =$$

$$= \sum_{k=1}^{n} c_k e^{i(\lambda+s, s^{(k)})} a_s(\lambda) ,$$

where $a_s(\lambda) = (V_s 1)(\lambda)$.

Because the operators V_s are unitary, for any function $f(\cdot) \in L_2(\mathbb{R}^\infty, d\rho(\lambda))$ we have

$$\int_{\mathbb{R}^\infty} |f(\lambda)|^2 d\rho(\lambda) = \int_{\mathbb{R}^\infty} |f(\lambda+s)|^2 |a_s(\lambda)|^2 d\rho(\lambda).$$

Consequently the measure $\rho(\cdot)$ is $\mathbb{R}_0^\infty$-quasi-invariant and

$$\frac{d\rho_s(\lambda)}{d\rho(\lambda)} = |a_s(\lambda)|^2 \; (|a_s(\lambda)| > 0 \; (\mathrm{mod}\, \rho(\cdot))).$$

Because $V_{s+s'} = V_s V_{s'}$, we have $a_{s+s'}(\lambda) = a_s(\lambda)\, a_{s'}(\lambda+s)$ (mod $\rho(\cdot)$).

Introduce $\mathbf{B}(\mathbb{R}^\infty)$-measurable functions $\alpha_s(\lambda) = \dfrac{a_s(\lambda)}{|a_s(\lambda)|}$ such that

a) $|\alpha_s(\lambda)| = 1$

b) $\alpha_{s+s'}(\lambda) = \alpha_s(\lambda)\, \alpha_{s'}(\lambda+s)$.

Then the operators V_s on $L_2(\mathbb{R}^\infty, d\rho(\lambda))$ are given by the formulas (4):

$$(V_s f)(\lambda) = \alpha_s(\lambda) \sqrt{\frac{d\rho_s(\lambda)}{d\rho(\lambda)}}\, f(\lambda+s)$$

and for the case of a simple spectrum of the CSO $(Q_k)_{k=1}^\infty$ Theorem 2 is proved.

If the joint spectrum of the CSO $(Q_k)_{k=1}^\infty$ is not simple, the space of the representation is unitarily equivalent to the direct sum $H = \oplus \sum L_2(\mathbb{R}^\infty, d\rho_n(\lambda)) = \int_{\mathbb{R}^\infty} H_\lambda \, d\rho(\lambda)$ where $\rho(\cdot) = \rho_1(\cdot) \gg \rho_2(\cdot) \gg \cdots$.

Let Ω_n be a vector in H with component in $L_2(\mathbb{R}^\infty, d\rho_n(\lambda))$ equal to one, and other components equal to zero. Denote the component of $V_s \Omega_n$ in $L_2(\mathbb{R}^\infty, d\rho_m(\lambda))$ by $a_{s,n}^m(\lambda)$. As before, for $\vec{f}(\cdot) = (f_1(\cdot), f_2(\cdot), \ldots) \in \oplus \sum L_2(\mathbb{R}^\infty, d\rho_n(\lambda))$, we get

$$(V_s \vec{f})(\lambda) = (\sum_k f_k(\lambda+s)\, a_{s,k}^1(\lambda), \sum_k f_k(\lambda+s)\, a_{s,k}^2(\lambda), \ldots).$$

Here we use induction to prove that the measures $\rho_k(\cdot)$ ($k = 1,2,\ldots$) are $\mathbb{R}_0^\infty$ quasi-invariant and so the dimension function, $\dim H_\lambda$ is $\mathbb{R}_0^\infty$-invariant.

The matrix function $C_s(\lambda) = \sqrt{\dfrac{d\rho(\lambda)}{d\rho_s(\lambda)}}\,(a_{s,n}^m(\lambda))_{m,n}$ defines a unitary operator for all $s \in \mathbb{R}_0^\infty$ and $\rho(\cdot)$-almost all $\lambda \in \mathbb{R}^\infty$. It satisfies

$$C_{s+s'}(\lambda) = C_s(\lambda)\, C_{s'}(\lambda + s)$$

for $\rho(\cdot)$-almost all $\lambda \in \mathbb{R}^\infty$. With this, the operators V_s act according to (4)

$$(V_s \vec{f})\,(\lambda) = C_s(\lambda) \sqrt{\frac{d\rho_s(\lambda)}{d\rho(\lambda)}}\, \vec{f}(\lambda + s).$$

□

So, the study of CCR of systems with countable degrees of freedom can be reduced to the study of:

a) $\mathbb{R}_0^\infty$-quasi-invariant probability measures $\rho(\cdot)$ on $(\mathbb{R}^\infty, \mathbf{B}(\mathbb{R}^\infty))$;

b) measurable vector functions $(C_s(\lambda))_{s\in \mathbb{R}_0^\infty}$ with values in the unitary operators, defined on $H_{\lambda+s} = H_\lambda$ with values in H_λ satisfying (5) modulo $\rho(\cdot)$.

4.3. Representations with one-dimensional cocycle.

Under the assumption that the CSO $(Q_k)_{k=1}^\infty$ have a simple joint spectrum, the representation space of the CCR can be realized as the space $L_2(\mathbb{R}^\infty, d\rho(\lambda)) \ni f(\cdot)$, where $\rho(\cdot)$ is a $\mathbb{R}_0^\infty$-quasi-invariant measure on $(\mathbb{R}^\infty, \mathbf{B}(\mathbb{R}^\infty))$.
The operators $U_t = \exp(i \sum_{k=1}^\infty t_k Q_k)$ in this representation are the multiplication operators:

$$(U_t f)\,(\lambda) = \exp(i \sum_{k=1}^\infty t_k \lambda_k)\, f(\lambda) = \exp(i(t, \lambda))\, f(\lambda)$$

$$(t \in \mathbb{R}_0^\infty, \lambda \in \mathbb{R}^\infty). \qquad (6)$$

The operators $V_s = \exp(i \sum_{k=1}^\infty s_k P_k)$ can be realized as

$$(V_s f)\,(\lambda) = \alpha_s(\lambda) \sqrt{\frac{d\rho(\lambda + s)}{d\rho(\lambda)}}\, f(\lambda + s)$$

$$(s \in \mathbb{R}_0^\infty, \lambda \in \mathbb{R}^\infty). \qquad (7)$$

Here the function $\alpha_s(\lambda)$ is $\mathbf{B}(\mathbb{R}^\infty)$-measurable, with absolute value equal to one $\rho(\cdot)$-almost everywhere, depends on the parameter $s \in \mathbb{R}_0^\infty$, and satisfies the functional equation

$$\alpha_{s^{(1)}+s^{(2)}}(\lambda) = \alpha_{s^{(1)}}(\lambda)\, \alpha_{s^{(2)}}(\lambda + s^{(1)}) \pmod{\rho(\cdot)}. \tag{8}$$

In what follows we will call such functions cocycles (1-cocycles).

The simplest cocycle is of the following form: $\alpha_s(\lambda) = \dfrac{\phi(\lambda+s)}{\phi(\lambda)}$ where $\phi(\cdot)$ is $B(\mathbb{R}^\infty)$-measurable function, $|\phi(\lambda)| = 1 \pmod{\rho(\cdot)}$. A cocycle of this kind is called trivial. It should be noted that in the finite-dimensional space $\mathbb{R}^n$, every cocycle $\alpha_{s_1,\dots,s_n}(\lambda_1,\dots,\lambda_n)$ is trivial. In the case of infinitely many variables, there are cocycles which are not trivial. We consider some examples.

Example 1. A product-cocycle is called a cocycle $\alpha_s(\lambda) = \prod_{n=1}^{\infty} \phi_n(\lambda_n + s_n)/\phi_n(\lambda_n)$ where $\phi_n(\lambda_n)$ is a Borel function and $|\phi_n(\lambda_n)| = 1 \pmod{d\lambda_n}$ $(n \in \mathbb{N})$. It is evident that the definition of $\alpha_s(\lambda)$ is correct because for every $s \in \mathbb{R}_0^\infty$ only a finite number of factors is different from the identity.

Particular examples of nontrivial product-cocycles can easily be constructed if $\rho(\cdot)$ is a product-measure. For example, let $\rho(\cdot)$ be the standard Gaussian measure $g(\cdot)$. Because $\sum_{n=1}^{\infty} \lambda_n^2 = \infty \pmod{g(\cdot)}$, the corresponding product-cocycle,

$$\alpha_s(\lambda) = \exp\{i \sum_{n=1}^{\infty} (2\lambda_n s_n + s_n^2)$$

is nontrivial. □

Example 2. A cocycle of the form

$$\alpha_s(\lambda) = \prod_{n=1}^{\infty} \frac{\phi_n(\lambda_{m_n} + s_{m_n}, \dots, \lambda_{m_n+p_n} + s_{m_n+p_n})}{\phi_n(\lambda_{m_n}, \dots, \lambda_{m_n+p_n})}$$

where $\phi_n(\cdot)$ is a Borel function with the absolute value equal to one, $m_n \to \infty$ for $n \to \infty$, is called a locally dependent cocycle. A locally dependent cocycle satisfies the condition: for every $p \in \mathbb{N}$ there exist $n_p \geq p$ and a Borel function $F_{n_p}(\lambda_1, \dots, \lambda_{n_p})$ with $|F_{n_p}(\cdot)| = 1 \pmod{d\lambda_1 \cdots d\lambda_{n_p}}$ such that for every $s \in \mathbb{R}^p \times \{0,0,\dots\}$, $\alpha_s(\lambda) = F_{n_p}(\lambda_1 + s_1, \dots, \lambda_{n_p} + s_{n_p}) F_{n_p}^{-1}(\lambda_1, \dots, \lambda_{n_p})$. By defining $F_{n_p}(\lambda) = \prod_{k=1}^{n_p} \phi_k(\lambda)$, we can take n_p to be the first number such that $m_{n_p+k} > p$ $(k = 1,2,\dots)$.

In what follows we will consider the cocycles which satisfy the following continuity condition with respect to s:

$$\int_{\mathbb{R}^\infty} |\alpha_s(\lambda) - 1|\, d\rho(\lambda) \to 0 \ \text{ for } s \to 0 \text{ in } \mathbb{R}_0^\infty. \tag{9}$$

It follows from the functional equation (8) for $\alpha_s(\lambda)$ that (9) holds if 0 is replaced by any point $s_0 \in \mathbb{R}_0^\infty$. The meaning of (9) is clarified by the following lemma.

LEMMA 1. Let $\alpha_s(\lambda)$ satisfy the condition (9). Then the mapping $\mathbb{R}_0^\infty \ni s \mapsto \alpha_s(\lambda) \in L_1(\mathbb{R}^\infty, d\rho(\lambda))$ is continuous.

Proof. It follows from (8) that $\forall s, s_0 \in \mathbb{R}_0^\infty$

$$\int_{\mathbb{R}^\infty} |\alpha_s(\lambda) - \alpha_{s_0}(\lambda)|\, d\rho(\lambda) = \int_{\mathbb{R}^\infty} |\alpha_{s_0}(\lambda)\, \alpha_{s-s_0}(\lambda+s_0) -$$

$$- \alpha_{s_0}(\lambda)|\, d\rho(\lambda) = \int_{\mathbb{R}^\infty} |\alpha_{s-s_0}(\lambda+s_0) - 1|\, d\rho(\lambda) =$$

$$= \int_{\mathbb{R}^\infty} |\alpha_{s-s_0}(\lambda) - 1| \frac{d\rho_{-s_0}(\lambda)}{d\rho(\lambda)}\, d\rho(\lambda)$$

tends to zero if $s \to s_0$ by (9) and Lebesgue's theorem. □

We have the following lemma.

LEMMA 2. Each locally dependent cocycle satisfies the condition (9).

Proof. Let $s^{(k)} \to 0$ in $\mathbb{R}_0^\infty$. Because $\mathbb{R}_0^\infty$ has the inductive limit topology, there is N such that

$$s^{(k)} = (s_1^{(k)}, \ldots, s_N^{(k)}, 0, 0, \ldots) \quad (k = 1,2,\ldots)$$

$$\text{and } s_j^{(k)} \to 0,\ k \to \infty \quad (j = 1, \ldots, N).$$

According to the definition of a locally dependent cocycle,

$$\alpha_{s^{(k)}}(\lambda) = \frac{F_{m_N}(\lambda+s^{(k)})}{F_{m_N}(\lambda)}$$

and, consequently,

$$\int_{\mathbb{R}^\infty} |\alpha_{s^{(k)}}(\lambda) - 1|\, d\rho(\lambda) = \int_{\mathbb{R}^{m_N}} \left| \frac{F_{m_N}(\lambda+s^{(k)})}{F_{m_N}(\lambda)} - 1 \right| d\rho_{m_N}(\lambda) =$$

$$= \int_{\mathbb{R}^{m_N}} |F_{m_N}(\lambda+s^{(k)}) - F_{m_N}(\lambda)|\, d\rho_{m_N}(\lambda).$$

The last integral tends to zero as $s^{(k)} \to 0$ because of the continuity in the mean of the function $F_{m_N}(\cdot) \in L_1(\mathbb{R}^{m_N}, d\rho_{m_N}(\lambda))$. □

We call the cocycles $\alpha_s(\cdot)$ and $\beta_s(\cdot)$ equivalent if their ratio (evidently, being a cocycle) is trivial.

PROPOSITION 2. Let $\alpha_s(\cdot)$ satisfy the condition (9) and let $\beta_s(\cdot)$ be equivalent to $\alpha_s(\cdot)$. Then $\beta_s(\cdot)$ also satisfies (9).

Proof. We have $\forall s \in \mathbb{R}_0^\infty$

$$\int_{\mathbb{R}^\infty} |\beta_s(\lambda)-1|\,d\rho(\lambda)=\int_{\mathbb{R}^\infty} |\alpha_s(\lambda)\,\frac{\beta_s(\lambda)}{\alpha_s(\lambda)}-1|\,d\rho(\lambda)\leq$$

$$\leq \int_{\mathbb{R}^\infty} |\phi(\lambda+s)-\phi(\lambda)|\,d\rho(\lambda)+\int_{\mathbb{R}^\infty} |\alpha_s(\lambda)-1|\,d\rho(\lambda).$$

For $s\to 0$, the first summand tends to zero because of the continuity in the mean of the function $\phi(\cdot)\in L_2(\mathbb{R}^\infty, d\rho(\lambda))$ and the second summand because of the imposed conditions. ▯

THEOREM 3. Suppose that the cocycle $\alpha_s(\lambda)$ in (7) satisfies the condition (9). Then the unitary group $(V_s)_{s\in\mathbb{R}_0^\infty}$ is strongly continuous.

Proof. We give a direct proof of this fact. It is based on the following property of a $\mathbb{R}_0^\infty$-quasi-invariant measure $\rho(\cdot)$: $\forall n\in\mathbb{N}$ $\rho(\cdot)\sim\rho_n(\cdot)\otimes\rho^n(\cdot)$, where $\rho_n(\cdot)$ is the projection of $\rho(\cdot)$ on $\mathbf{B}(\mathbb{R}^n)\otimes\mathbb{R}^\infty$ and $\rho^n(\cdot)$ is the projection on $\mathbb{R}^n\otimes\mathbf{B}(\mathbb{R}^\infty)$ (see Chapter 3, §3.1). Denote $p_n(\lambda_1,\ldots,\lambda_n)=\dfrac{d\rho_n(\lambda_1,\ldots,\lambda_n)}{d\lambda_1\cdots d\lambda_n}$, $q_n(\lambda)=\dfrac{d\rho}{d(\rho_n\otimes\rho^n)}$. For $s\in\mathbb{R}^n\times\{0,0,\ldots\}$ we have

$$\frac{d\rho_s}{d\rho}(\lambda)=\frac{q_n(\lambda+s)}{q_n(\lambda)}\,\frac{p_n(\lambda_1+s_1,\ldots,\lambda_n+s_n)}{p_n(\lambda_1,\ldots,\lambda_n)}. \tag{10}$$

To prove that V_s is strongly continuous it is sufficient to show that $\forall g(\cdot)\in L_2(\mathbb{R}^\infty, d\rho(\lambda))$ $((V_s-I)\,g,\,g)\to 0$ for $s\to 0$ in $\mathbb{R}_0^\infty$. We have

$$|((V_s-I)\,g,\,g)|\leq \int_{\mathbb{R}^\infty} |\alpha_s(\lambda)\sqrt{\frac{d\rho_s}{d\rho}}(\lambda)\,g(\lambda+s)-g(\lambda)|\times$$

$$\times|g(\lambda)|\,d\rho(\lambda)\leq \int_{\mathbb{R}^\infty} |\sqrt{\frac{d\rho_s}{d\rho}}(\lambda)-1|\;|g(\lambda+s)|\;|g(\lambda)|\,d\rho(\lambda)+$$

$$+\int_{\mathbb{R}^\infty} |\alpha_s(\lambda)-1|\;|g(\lambda+s)|\;|g(\lambda)|\,d\rho(\lambda)+$$

$$+\int_{\mathbb{R}^\infty} |g(\lambda+s)-g(\lambda)|\;|g(\lambda)|\,d\rho(\lambda). \tag{11}$$

For $s\to 0$, the second term in (11) tends to zero by the condition (9) and by Lebesgue's theorem, and the third term by the continuity in the mean of the function $g(\cdot)$. Consider the first term in (11). It is sufficient to show that

$$\int_{\mathbb{R}^\infty} |\sqrt{\frac{d\rho_s}{d\rho}}(\lambda)-1|\,d\rho(\lambda)\to 0 \quad\text{for}\quad s\to 0.$$

Let $s\in\mathbb{R}^n\times\{0,0,\ldots\}$. Using (10), we get

$$\int_{\mathbb{R}^\infty} \left| \frac{d\rho_s}{d\rho}(\lambda) - 1 \right| d\rho(\lambda) \le \int_{\mathbb{R}^\infty} \left| \frac{q_n(\lambda+s)}{q_n(\lambda)} \frac{p_n(\lambda_1+s_1, \ldots, \lambda_n+s_n)}{p_n(\lambda_1, \ldots, \lambda_n)} - \right.$$

$$\left. - 1 \right| d\rho(\lambda) \le \int_{\mathbb{R}^\infty} \left| \frac{q_n(\lambda+s)}{q_n(\lambda)} - 1 \right| \frac{p_n(\lambda_1+s_1, \ldots, \lambda_n+s_n)}{p_n(\lambda_1, \ldots, \lambda_n)} d\rho(\lambda) +$$

$$+ \int_{\mathbb{R}^\infty} \left| \frac{p_n(\lambda_1+s_1, \ldots, \lambda_n+s_n)}{p_n(\lambda_1, \ldots, \lambda_n)} - 1 \right| d\rho(\lambda) =$$

$$= \int_{\mathbb{R}^\infty} |q_n(\lambda+s) - q_n(\lambda)| \frac{p_n(\lambda_1+s_1, \ldots, \lambda_n+s_n)}{p_n(\lambda_1, \ldots, \lambda_n)} d(\rho_n \otimes \rho^n)(\lambda) +$$

$$+ \int_{\mathbb{R}^n} |p_n(\lambda_1+s_1, \ldots, \lambda_n+s_n) - p_n(\lambda_1, \ldots, \lambda_n)| \, d\rho_n(\lambda_1, \ldots, \lambda_n).$$

Because $q_n(\lambda)$ and $p_n(\lambda)^{-1} p_n(\lambda+s) \in L_1(\mathbb{R}^\infty, d(p_n \otimes p^n)(\lambda))$ and $p_n(\lambda_1, \ldots, \lambda_n) \in L_1(\mathbb{R}^n, d\lambda_1, \ldots, d\lambda_n)$,

$$\int_{\mathbb{R}^\infty} \left| \sqrt{\frac{d\rho_s}{d\rho}}(\lambda) - 1 \right| d\rho(\lambda) \to 0 \quad \text{as} \quad s \to 0$$

in $\mathbb{R}^\infty$ by the continuity in the mean of the functions $q_n(\lambda)$ and $p_n(\lambda)$ and by Lebesgue's theorem. □

Note that the equivalence of the cocycles $\alpha_s(\cdot)$ and $\beta_s(\cdot)$, and the equivalence of the measures $\rho(\cdot)$ and $\mu(\cdot)$ on $(\mathbb{R}^\infty, \mathrm{B}(\mathbb{R}^\infty))$ is necessary and sufficient for the representations of the CCR constructed using (6) and (7) with the aid of the cocycle $\alpha_s(\cdot)$ and the measure $\rho(\cdot)$ and the cocycle $\beta_s(\cdot)$ and the measure $\mu(\cdot)$, to be unitarily equivalent.

We give another condition for the irreducibility of a representation of the CCR of the systems with countable degrees of freedom under the assumption that $(Q_k)_{k=1}^\infty$ have a simple joint spectrum.

THEOREM 4. If the CSO $(Q_k)_{k=1}^\infty$ have a simple joint spectrum, a representation of the CCR is irreducible if and only if the measure $\rho(\cdot)$ is ergodic with respect to the shifts by vectors in $\mathbb{R}_0^\infty$.

Proof. Let $\rho(\cdot)$ be ergodic with respect to shifts from $\mathbb{R}_0^\infty$. We show that a bounded operator A which commutes with the operators of the representation is a multiple of the identity. Indeed, by Theorem 1 and the commutativity of A with the operators U_t it follows that A is an operator of multiplication by an essentially bounded function $A(\lambda)$. The commutativity of A with V_s implies that the function satisfies the equality $A(\lambda+s) = A(\lambda)$ for $\rho(\cdot)$-almost all $\lambda \in \mathbb{R}^\infty$ and for all $s \in \mathbb{R}_0^\infty$. Because the measure $\rho(\cdot)$ is ergodic, we conclude that $A(\lambda) \equiv$ const $\rho(\cdot)$-almost everywhere.

Conversely, let the representation be irreducible. Then we show that the measure $\rho(\cdot)$ is ergodic relatively to the shifts from $\mathbb{R}_0^\infty$. For this, it is sufficient to show that an arbitrary measurable bounded function $\psi(\lambda)$ satisfying $\psi(\lambda+s) = \psi(\lambda)$ is constant for all $s \in \mathbb{R}_0^\infty$ and $\rho(\cdot)$-almost all $\lambda \in \mathbb{R}^\infty$. Consider the operator L_ψ defined by the following

$$(L_\psi f)(\lambda) = \psi(\lambda) f(\lambda) \quad (f(\cdot) \in L_2(\mathbb{R}^\infty, d\rho(\lambda))).$$

It is clear that the operator L_ψ commutes with the operators U_t and because $\psi(\lambda+s) = \psi(\lambda)$, it also commutes with the operators V_s. So, because the representation is irreducible, it is a multiple of the identity, i.e. $\psi(\lambda) \equiv$ const. ▯

The class of measures $\rho(\cdot)$ on $\mathbb{R}^\infty$ ergodic with respect to shifts from $\mathbb{R}_0^\infty$ is described by A.V. Skorokhod (see Chapter 3, §3.1).

Remark 2. For the representation (3), (4) of CCR to be a factor-representation, it is necessary that the measure be $\mathbb{R}_0^\infty$-ergodic. ▯

Remark 3. Unlike the CCR of systems with finite degrees of freedom, for which the irreducibility of a representation and the simplicity of the spectrum of the CSO $(Q_k)_{k=1}^\infty$ are equivalent, for the CCR with countable degrees of freedom there exist irreducible representations with any finite or infinite multiplicity of the spectrum of the CSO $(Q_k)_{k=1}^\infty$ (see V.Ya. Golodets [1,3]). ▯

In the simplest irreducible representations of the CCR with countable degrees of freedom constructed as tensor products of irreducible representations (2) of the group ψ, the spectrum of the CSO $(Q_k)_{k=1}^\infty$ is simple. For such representations constructed using (6) and (7), the measure $\rho(\cdot)$ can be chosen to be a product-measure and the cocycle to be a product-cocycle.

4.4. Gaussian representations of CCR.

On the space $L_2(\mathbb{R}^\infty, dg_B(\lambda))$, where $g_B(\cdot)$ is a $\mathbb{R}_0^\infty$-quasi-invariant Gaussian measure on $(\mathbb{R}^\infty, \mathbf{B}(\mathbb{R}^\infty))$ (see Chapter 3, §3.1), consider the following representation of the CCR with infinite degrees of freedom:

$$(U_t f)(\lambda) = e^{i(t,\lambda)} f(\lambda),$$

$$(V_s f)(\lambda) = \alpha_s^{(C)}(\lambda) \sqrt{\frac{dg_{B,s}(\lambda)}{dg(\lambda)}}\, f(\lambda+s),$$

$$f \in L_2(\mathbb{R}^\infty, dg_B(\lambda)).$$

Here the cocycle

$$\alpha_s^{(C)}(\lambda) = \exp(-i\,[2(Cs,\lambda) + (Cs,s)]), \tag{12}$$

is given by an infinite real symmetric matrix C such that

a) $\alpha_s^{(C)}$ exists for $g_B(\cdot)$-almost all λ;

b) the mapping $\mathbb{R}_0^\infty \ni s \mapsto \alpha_s^{(C)}(\cdot) \in L_2(\mathbb{R}^\infty, dg_B(\lambda))$ is continuous in zero.

Such cocycles will be called Gaussian.

As a Gaussian cocycle satisfies (9), the unitary groups U_t and V_s are weakly continuous, and generate a Weil representation of the CCR of a system with countable degrees of freedom.

First of all, we will study the existence conditions for a Gaussian cocycle. The existence of a function $\alpha_s^{(C)}(\cdot)$ is equivalent with the existence of a $g_B(\cdot)$-measurable linear functional (Cs, λ).

The set of $g_B(\cdot)$-measurable linear functionals can be described as the completion of the space $\mathbb{R}_0^\infty$ with respect to the norm $\|s\|_{H_B} = \|As\|_{l_2}$ (here $A : \mathbb{R}_0^\infty \to \mathbb{R}_0^\infty$ is the operator defined in Chapter 3, §3.1, $B = A^* A$). Indeed, every $g_B(\cdot)$-measurable linear functional is represented by a sequence of continuous linear functionals convergent with respect to the quadratic mean. It follows from the formula

$$\int_{\mathbb{R}^\infty} |(s^{(1)}, \lambda) - (s^{(2)}, \lambda)|^2 \, dg_B(\lambda) = \|A s^{(1)} - A s^{(2)}\|_{l_2}^2, \quad s^{(1)}, s^{(2)} \in \mathbb{R}_0^\infty ,$$

that there is a one-to-one correspondence between the set of $g_B(\cdot)$-measurable linear functionals and the elements of the space H_B obtained from $\mathbb{R}_0^\infty$ by taking the completion with respect to the inner product

$$(s^{(1)}, s^{(2)})_{H_B} = (As^{(1)}, As^{(2)})_{l_2} = (Bs^{(1)}, s^{(2)})_{l_2}.$$

Now we formulate a necessary and sufficient condition on the matrix C which allows to define a Gaussian cocycle $\alpha_s^{(C)}(\cdot)$.

LEMMA 3. A function $\alpha_s^{(C)}(\cdot)$ will be a Gaussian cocycle if and only if $C(\mathbb{R}_0^\infty) \subset H_B$ and the operator $C : \mathbb{R}_0^\infty \to H_B$ is continuous.

Proof. The functional (Cs, λ) is $g_B(\cdot)$-measurable if and only if $Cs \in H_B$, i.e. the condition $C(\mathbb{R}_0^\infty) \subset H_B$ is necessary to define $\alpha_s^{(C)}(\cdot)$.

Suppose this condition holds. Then

$$\int_{\mathbb{R}^\infty} |\alpha_s^{(C)}(\lambda) - 1|^2 \, dg_B(\lambda) = \int_{\mathbb{R}^\infty} |\exp(-i\,[2(Cs, \lambda) + (Cs,s)]) - 1|^2 \, dg_B(\lambda)$$

$$= 2 \int_{\mathbb{R}^\infty} (1 - \cos[2\,(Cs, \lambda) + (Cs, s)]) \, dg_B(\lambda) =$$

$$= \int_{\mathbb{R}^\infty} \sin^2 \tfrac{1}{2} \,[2(Cs, \lambda) + (Cs, s)] \, dg_B(\lambda) \le$$

$$\le 4 \int_{\mathbb{R}^\infty} (Cs, \lambda)^2 \, dg_B(\lambda) + 4(Cs, s) \int_{\mathbb{R}^\infty} (Cs, \lambda) \, dg_B(\lambda) + (Cs, s) \le$$

$$\le 4 \, \|Cs\|_{H_B}^2 + 2 \int_{\mathbb{R}^\infty} ((Cs, s)^2 + (Cs, \lambda)^2) \, dg_B(\lambda) + (Cs, s)^2 =$$

$$= 6 \, \|Cs\|_{H_B}^2 + 3(Cs, s)_{H_B}^2.$$

Thus, $\int_{\mathbb{R}^\infty} |\alpha_s^{(C)}(\lambda) - 1|^2 \, dg_B(\lambda) \to 0$ for $s \to 0$ if and only if $\|Cs\|_{H_B} \to 0$ for $s \to 0$.

Consequently, for condition b) to hold it is necessary and sufficient that $C : \mathbb{R}_0^\infty \to H_B$ be continuous. □

We give an example of a Gaussian cocycle.

Example 3. For a $\mathbb{R}_0^\infty$-quasi-invariant measure $g_B(\cdot)$, the function $\alpha_s^{(B^{-1})}(\cdot)$ is a Gaussian cocycle. Indeed, because $B^{-1} = A^{-1}(A^*)^{-1}$, and because from the $\mathbb{R}_0^\infty$-quasi-invariance of $g_B(\cdot)$ it follows that $(A^*)^{-1}(\mathbb{R}_0^\infty) \subset l_2$ and $A^{-1}(l_2) = H_B$ (see Chapter 3, Proposition 9) we have that $B^{-1}(\mathbb{R}_0^\infty) = A^{-1}((A^*)^{-1}(\mathbb{R}_0^\infty)) \subset H_B$. □

For the standard Gaussian measure we have $H_I = l_2$, and it follows from Lemma 3 that the Gaussian cocycles $\alpha_s^{(C)}(\cdot) \in L_2(\mathbb{R}^\infty, dg(\lambda))$ are defined by the continuous operators $C : \mathbb{R}_0^\infty \to l_2$. For this case, we give a criteria for a Gaussian cocycle to be trivial.

LEMMA 4. A Gaussian cocycle $\alpha_s^{(C)}(\cdot) \in L_2(\mathbb{R}^\infty, dg(\lambda))$ is trivial if and only if the closure of C in l_2 is a Hilbert-Schmidt operator.

Proof. Let C be a Hilbert-Schmidt operator on l_2 and

$$P_n : \mathbb{R}^\infty \ni (\lambda_1, \ldots, \lambda_n, \lambda_{n+1}, \ldots) \to (\lambda_1, \ldots, \lambda_n, 0, \ldots) \in \mathbb{R}_0^\infty$$

be the projection on the subspace $\mathbb{R}^n \times (0,0,\ldots)$. Define a sequence of functions:

$$f_n(\lambda) = (P_n C P_n \lambda, \lambda) - sp\, P_n C P_n.$$

Because C is a Hilbert-Schmidt operator on l_2 the sequence $(f_n(\cdot))_{n=1}^\infty$ converges with respect to the measure $g(\cdot)$ to a measurable function $f(\cdot)$ (see, for example, the book I.I. Gikhman and A.V. Skorokhod [2]). But

$$\alpha_s^{(C)}(\lambda) = \exp(i\,[f(\lambda+s) - f(\lambda)])$$

and so the cocycle $\alpha_s^{(C)}(\cdot)$ is trivial.

Conversely, consider a trivial cocycle $\alpha_s^{(C)}(\cdot)$. Its restrictions:

$$\alpha_s^{(C)}(P_n \lambda) = \frac{\exp(-i\,[(P_n C P_n(\lambda+s), \lambda+s) + a_n])}{\exp(-i\,[(P_n C P_n \lambda, \lambda) + a_n])}$$

are functions defined everywhere on $\mathbb{R}^\infty$ and the sequence $\exp(-i\,[(P_n C P_n \lambda, \lambda) + a_n])$ converges with respect to the measure to a measurable function. From this convergent sequence of functions choose a subsequence $\exp(-i\,[(P_{n_k} C P_{n_k} \lambda, \lambda) + a_{n_k}])$ which is pointwise convergent almost everywhere. But then the subsequence $[(P_{n_k} C P_{n_k} \lambda, \lambda) + a_{n_k}]$, where a_{n_k} is a number sequence, is also pointwise convergent almost everywhere. As known (again see the book I.I. Gikhman and A.V. Skorokhod [2]), this is possible if and only if C is a Hilbert-Schmidt operator. □

4.5. The Gårding domain for representations of CCR.

In this section, for every representation of the CCR with countable degrees of freedom we will construct a Gårding domain, i.e. a dense domain D in the representation space H which is invariant relatively to all the operators $(Q_k, P_k)_{k=1}^{\infty}$ and which is even a core for all these operators (see Chapter 2, §2.5). Using Gårding-Wightman's representation of the CCR, M. Reed [2] and G. Hegerfeldt [2] constructed a Gårding domain consisting correspondingly of analytic and entire vectors for the collection $(Q_k, P_k)_{k=1}^{\infty}$. We will construct a Gårding type domain directly following Gårding's construction (see Chapter 2, §2.5). However, we will be using a quasi-invariant measure on the group G_0^{∞} instead of an invariant measure, and a space of entire minimal type functions of order 2 on the group G_0^{∞} instead of $C_0^{\infty}(G_0^{\infty})$.

First we give a construction of a representation of the CCR for one degree of freedom, so we can generalize later on for a representation of the CCR with countable degrees of freedom.

The space $A(\mathbb{R}^2) = A(\mathbb{R}^1) \otimes A(\mathbb{R}^1)$ is a nuclear linear topological space of functions on $\mathbb{R}^2$. These functions can be extended to entire minimal type functions of order 2 defined on $\mathbb{C}^2$.

Now, on H let there be defined the cyclic Weyl representation of the CCR with one degree of freedom, let there be two strongly continuous one-parameter groups, U_{t_1} and V_{s_1}, such that

$$V_{s_1} U_{t_1} = e^{it_1 s_1} U_{t_1} V_{s_1} \tag{13}$$

and a vector $\Omega \in H$ such that $\mathrm{CLS}\{U_{t_1} V_{s_1} \Omega \mid t_1, s_1 \in \mathbb{R}^1\} = H$.

Define the set

$$D = \{ \int_{\mathbb{R}^2} a(t_1, s_1)\, U_{t_1} V_{s_1} \Omega \frac{1}{\pi} \exp(-(t_1^2 + s_1^2))\, dt_1\, ds_1 \mid$$

$$a(t_1, s_1) \in A(\mathbb{R}^2)\}.$$

The set D is linear. Consider the mapping F_0

$$A(\mathbb{R}^2) \ni a(t_1, s_1) \mapsto (F_0 a) = \int_{\mathbb{R}^2} a(t_1, s_1)\, U_{t_1} V_{s_1} \Omega\, dg(t_1, s_1) \in H.$$

From the inequality

$$\|F_0 a\|_H = \| \int_{\mathbb{R}^2} a(t_1, s_1)\, U_{t_1} V_{s_1} \Omega\, dg(t_1, s_1) \|_H \leq$$

$$\leq \int_{\mathbb{R}^2} | a(t_1, s_1) |\, dg(t_1, s_1) \leq \|a\|_{L_2(\mathbb{R}^2, dg(t_1, s_1))} =$$

$$= \|a\|_{A_1(\mathbb{R}^1)\otimes A_1(\mathbb{R}^1)} \tag{14}$$

it follows that $D \subset H$ and that F_0 is continuous. The mapping F_0 induces the topology of the space $\mathbf{A}(\mathbb{R}^2)$ on D. If F_0 does not map $\mathbf{A}(\mathbb{R}^2)$ on D bijectively then the topology on D coincides with the factor topology of the space $\mathbf{A}(\mathbb{R}^2)/\operatorname{Ker} F_0$.

THEOREM 5. The following statements hold:

1) D is a nuclear linear topological space;

2) D is topologically densely imbedded in H;

3) $U_{t_1} V_{s_1}, V_{s_1} U_{t_1} \in L(D \to D)$;

4) D is contained in the domains of generators of representations of one-parameter subgroups of the group G and these generators map D into itself;

5) D consists of entire vectors for generators of representations of one-parameter subgroups of the group G;

6) the function $G \ni (t_1, s_1, \alpha_1) \to U_{t_1} V_{s_1} W_{\alpha_1} f \in D$ is continuous for all $f \in D$.

Proof. 1. D is nuclear being a factor space $\mathbf{A}(\mathbb{R}^2)/\operatorname{Ker} F_0$ of the nuclear space $\mathbf{A}(\mathbb{R}^2)$.

2. We show that D is topologically densely imbedded in H. Because F_0 is continuous, the imbedding $D \hookrightarrow H$ is topological. To prove that D is dense in H we need to prove several lemmas.

Let U_{t_1} be a strongly continuous unitary representation of the additive group $\mathbb{R}^1$ on a separable Hilbert space H:

$$\mathbb{R}^1 \ni t_1 \mapsto U_{t_1} \in \mathbf{U}(H),$$

let A_1 be the self-adjoint operator on H such that $U_{t_1} = e^{it_1A_1}$, let $E_1(\Delta)$ be the resolution of the identity of the operator A_1 where $\Delta \in \mathbf{B}(\mathbb{R}^1)$ are Borel subsets of $\mathbb{R}^1$. Then we have the following lemma:

LEMMA 5. Let $f \in H$. Let there exists $\varepsilon > 0$ such that $f \in H_\varepsilon^\omega(A_1)$. Then $\operatorname{CLS}\{A_1^n f \mid n = 0, 1, \ldots\} = \operatorname{CLS}\{E_1(\Delta) f \mid \Delta \in \mathbf{B}(\mathbb{R}^1)\} = \operatorname{CLS}\{U_{t_1} f \mid t_1 \in \mathbb{R}^1\}$.

Proof. The proof of this fact is based on the uniqueness theorem for the Fourier transform of complex measures on $(\mathbb{R}^1, \mathbf{B}(\mathbb{R}^1))$ and the properties of this transform.

Recall that a finite complex measure $\omega(\cdot)$ on $(\mathbb{R}^1, \mathbf{B}(\mathbb{R}^1))$ is a σ-additive function of the sets $\mathbf{B}(\mathbb{R}^1) \ni \Delta \mapsto \omega(\Delta) \in \mathbb{C}^1$. It is clear that $\omega_1(\Delta) = \operatorname{Re}\omega(\Delta)$ and $\omega_2(\Delta) = \operatorname{Im}\omega(\Delta)$ are sign changing measures on $(\mathbb{R}^1, \mathbf{B}(\mathbb{R}^1))$. For sign changing measures $\mu(\cdot)$ on $(\mathbb{R}^1, \mathbf{B}(\mathbb{R}^1))$ the Jordan decomposition, $\mu(\cdot) = \mu_+(\cdot) - \mu_-(\cdot)$ holds. Here $\mu_+(\cdot)$, $\mu_-(\cdot)$ are already measures, and $|\mu(\cdot)| = \mu_+(\cdot) + \mu_-(\cdot)$ is also a measure. Define the Fourier transform of a complex measure $\omega(\cdot)$ as follows:

$$g(t_1) = \int_{\mathbb{R}^1} e^{it_1\lambda_1} d\omega(\lambda_1) = \int_{\mathbb{R}^1} e^{it_1\lambda_1} d\omega_1(\lambda_1) + i \int_{\mathbb{R}^1} e^{it_1\lambda_1} d\omega_2(\lambda_1).$$

For the Fourier transform of a complex measure $\omega(\cdot)$, the uniqueness theorem holds: if $g(t_1) \equiv 0 \, (t_1 \in \mathbb{R}^1)$, then $\omega(\Delta) = 0 \, (\Delta \in \mathbf{B}(\mathbb{R}^1))$. Indeed, for a measure ω on $(\mathbb{R}^1, \mathbf{B}(\mathbb{R}^1))$ this is equivalent to the fact that the characteristic function of a random variable uniquely defines its distribution function. If $\omega(\cdot) = \omega_+(\cdot) - \omega_-(\cdot)$ is a sign changing measure on $(\mathbb{R}^1, \mathbf{B}(\mathbb{R}^1))$ and $g(t_1) = g_+(t_1) - g_-(t_1) = \int_{\mathbb{R}^1} e^{it_1\lambda_1} d\omega_+(\lambda_1) - \int_{\mathbb{R}^1} e^{it_1\lambda_1} d\omega_-(\lambda_1) = 0$ then $g_+(t_1) = g_-(t_1)$ and so $\omega_+(\Delta) = \omega_-(\Delta)$ and hence $\omega(\Delta) = 0$.

For an arbitrary complex measure $\omega(\cdot) = \omega_1(\cdot) + i\,\omega_2(\cdot)$ it follows from the equality $g(t_1) \equiv 0$ that $\int_{\mathbb{R}^1} e^{it_1\lambda_1} d\omega_1(\lambda_1) = \dfrac{g(t_1) + \overline{g(-t_1)}}{2} = 0$, $\int_{\mathbb{R}^1} e^{it_1\lambda_1} d\omega_2(\lambda_1) = \dfrac{g(t_1) - \overline{g(-t_1)}}{2i} = 0$ and so $\omega_1(\Delta) = \omega_2(\Delta) = 0 \;\; (\Delta \in \mathbf{B}(\mathbb{R}^1))$.

As known, the full variation of a complex measure $\omega(\cdot)$ on Δ denoted by $v(\omega(\cdot), \Delta)$, equals, by definition,

$$v(\omega(\cdot), \Delta) = \sup \sum_{i=1}^{n} |\omega(\Delta_i)|$$

where the supremum is taken with respect to all finite systems (Δ_i) of mutually disjoint sets of $\mathbf{B}(\mathbb{R}^1)$ with $\Delta_i \subset \Delta$. For a finite complex measure $\omega(\cdot)$ on $(\mathbb{R}^1, \mathbf{B}(\mathbb{R}^1))$ its full variation is countably additive and bounded.

Suppose that for a finite complex measure $\omega(\cdot)$ there exists $\delta > 0$ such that

$$\int_{\mathbb{R}^1} e^{\delta|\lambda_1|} dv(\omega(\cdot), \lambda_1) < \infty.$$

Then the function $g(t_1) = \int_{\mathbb{R}^1} e^{it_1\lambda_1} d\omega(\lambda_1) \;\; (t_1 \in \mathbb{R}^1)$ can be analytically extended to the strip $\{z \in \mathbb{R}^1 \mid |\mathrm{Im}\, z| < \delta\}$.

Now consider $h \in H$. The following three requirements are equivalent

(α) $(A_1^n f, h) = 0, \quad n = 0, 1, \ldots,$

(β) $(E_1(\Delta) f, h) = 0, \quad \Delta \in \mathbf{B}(\mathbb{R}^1),$

(γ) $(U_{t_1} f, h) = 0, \quad t_1 \in \mathbb{R}^1.$

To prove this consider a finite complex measure $\omega(\Delta) = (E_1(\Delta) f, h) \;\; (\Delta \in \mathbf{B}(\mathbb{R}^1))$. Because the requirements $f \in H_\varepsilon^\omega(A_1)$, $f \in D(e^{\varepsilon|A_1|})$ and $\int_{\mathbb{R}^1} e^{2t_1|\lambda_1|} d(E_1(\lambda_1) f, f) < \infty \; (0 \le t_1 < \varepsilon)$ are equivalent, the Fourier transform of the complex measure ω can be analytically extended to the strip $\{z \in \mathbb{C}^1 \mid |\mathrm{Im}\, z| < 2\varepsilon\}$:

$$g(z) = \int_{\mathbb{R}^1} e^{iz\lambda_1} \, d\omega(\lambda_1).$$

Suppose (α) holds. Then

$$\left. \frac{d^n g(z)}{dz^n} \right|_{z=0} = \int_{\mathbb{R}^1} (i\lambda_1)^n \, d\omega(\lambda_1) = ((iA_1)^n f, h) = 0, \quad n = 0, 1, \ldots . \tag{15}$$

So $g(t_1) \equiv 0$ and from the uniqueness theorem it follows that $\omega(\Delta) = (E_1(\Delta) f, h) = 0$ $(\Delta \in \mathbf{B}(\mathbb{R}^1))$. Thus, ($\alpha$) implies ($\beta$). The fact that ($\beta$) implies ($\gamma$) follows from the relation

$$(U_{t_1} f, h) = \int_{\mathbb{R}^1} e^{it_1\lambda_1} \, d(E_1(\lambda_1) f, h). \tag{16}$$

It follows from (15), (16) and the uniqueness theorem that (γ) implies (α). We have shown the equivalence of the requirement (α), (β) and (γ) and so the lemma is proved. ▯

COROLLARY 1. Let A_1, B_1 be self-adjoint operators, possibly unbounded and noncommuting, which act in a Hilbert space H. Let $E_1(\cdot), F_1(\cdot)$ be their resolutions of the identity, $U_{t_1} = e^{it_1A_1}$, $V_{s_1} = e^{is_1B_1}$.

For $f \in H$ let there exist $\varepsilon > 0$ such that $f \in H_\varepsilon^\omega(A_1, B_1)$ i.e.

$$\sum_{n=0}^{\infty} \frac{1}{n!} \sum_{\substack{k_1, \ldots, k_n, \\ j_1, \ldots, j_n = 0 \\ k_1 + j_1 + \cdots + k_n + j_n = n}}^{n} \|A_1^{k_1} B_1^{j_1} \cdots A_1^{k_n} B_1^{j_n} f\| s^n < \infty \quad (0 \leq s < \varepsilon).$$

Then

$$\mathrm{CLS}\, \{A_1^k B_1^j f \mid k, j = 0, 1, \cdots\} = \mathrm{CLS}\, \{E_1(\Delta) F_1(\Delta') f \mid$$

$$\Delta, \Delta' \in \mathbf{B}(\mathbb{R}^1)\} = \mathrm{CLS}\, \{U_{t_1} V_{s_1} \mid t_1, s_1 \in \mathbb{R}^1\}.$$

Proof. Let $h \in H$. It is sufficient to show that the following three requirements are equivalent

(α') $(A_1^k B_1^j f, h) = 0, \quad k, j = 0, 1, \ldots;$

(β') $(E_1(\Delta) F_1(\Delta') f, h) = 0, \quad \Delta, \Delta' \in \mathbf{B}(\mathbb{R}^1);$

(γ') $(U_{t_1} V_{s_1} f, h) = 0, \quad t_1, s_1 \in \mathbb{R}^1.$

From $f \in H_\varepsilon^\omega(A_1, B_1)$ it follows that $f \in H_\varepsilon^\omega(A_1) \cap H_\varepsilon^\omega(B_1)$ and $B_1^j f \in H_\varepsilon^\omega(A_1)$. Thus from ($\alpha'$) it follows from Lemma 5 that

$$(E_1(\Delta)B_1^j f, h) = (B_1^j f, E_1(\Delta)h) = 0 \ \ \forall j = 0,1,\dots, \text{ and } \forall \Delta \in \mathrm{B}(\mathbb{R}^1).$$

Again from Lemma 5 we get that

$$(F_1(\Delta')f, E_1(\Delta)h) = (E_1(\Delta)F_1(\Delta')f, h) = 0 \ \ \forall \Delta, \Delta' \in \mathrm{B}(\mathbb{R}^1).$$

So, (β′) follows from (α′).

Now we show that (γ′) follows from (β′). Let

$$(E_1(\Delta)F_1(\Delta')f, h) = (F_1(\Delta')f, E_1(\Delta)h) = 0, \quad \forall \Delta, \Delta' \in \mathrm{B}(\mathbb{R}^1).$$

Then from Lemma 5, $(V_{s_1} f, E_1(\Delta)h) = (E_1(\Delta)V_{s_1} f, h) = 0$. Again using Lemma 5, we get $(U_{t_1} V_{s_1} f, h) = 0$ $(t_1, s_1 \in \mathbb{R}^1)$.

Similarly one can show that (α′) follows from (γ′). □

Remark 4. The corollary holds for any finite collection of operators. □

COROLLARY 2. Let A_1 be a self-adjoint operator acting in a Hilbert space H, let $U_{t_1} = e^{it_1 A_1}$ $(t_1 \in \mathbb{R}^1)$ and let $E_1(\cdot)$ be a resolution of the identity of the operator A_1. By $\mathrm{A}(\mathbb{R}^1)$ we denote the space of functions introduced above, we take $dg(t_1) = \pi^{-1/2} \exp(-t_1^2)\, dt_1$ a Gaussian measure on $(\mathbb{R}^1, \mathrm{B}(\mathbb{R}^1))$, and $f \in H$ an arbitrary vector. Then

$$\mathrm{CLS}\ \{\int_{\mathbb{R}^1} a(t_1)\, U_{t_1}\, dg(t_1) f \mid a(\cdot) \in \mathrm{A}(\mathbb{R}^1)\} =$$

$$= \mathrm{CLS}\ \{E_1(\Delta) f \mid \Delta \in \mathrm{B}(\mathbb{R}^1)\} = \mathrm{CLS}\ \{U_{t_1} f \mid t_1 \in \mathbb{R}^1\}.$$

Proof. Let $h \in H$. It is sufficient to show that the requirements

$$(\alpha'') \quad (\int_{\mathbb{R}^1} a(t_1)\, U_{t_1}\, dg(t_1) f, h) = 0, \quad a(\cdot) \in \mathrm{A}(\mathbb{R}^1);$$

$$(\beta'') \quad (E_1(\Delta) f, h) = 0, \quad \Delta \in \mathrm{B}(\mathbb{R}^1)$$

are equivalent. Denote $U_{a(\cdot)} = \int_{\mathbb{R}^1} a(t_1)\, U_{t_1}\, dg(t_1)$. Because

$$U_{t_1} = \int_{\mathbb{R}^1} e^{it_1\lambda_1}\, dE_1(\lambda_1), \quad U_{a(\cdot)} = \int_{\mathbb{R}^1} a(t_1) \int_{\mathbb{R}^1} e^{it_1\lambda_1}\, dE_1(\lambda_1)\, dg(t_1) =$$

$$= \int_{\mathbb{R}^1} (\int_{\mathbb{R}^1} a(t_1) e^{it_1\lambda_1}\, dg(t_1))\, dE_1(\lambda_1) = \int_{\mathbb{R}^1} b(\lambda_1) e^{-\frac{\lambda_1^2}{4}}\, dE_1(\lambda_1).$$

(We have used that

$$\{\int_{\mathbb{R}^1} a(t_1) e^{it_1\lambda_1}\, dg(t_1) \mid a(\cdot) \in \mathrm{A}(\mathbb{R}^1)\} = \{b(\lambda_1) e^{-\lambda_1^2/4} \mid b(\cdot) \in \mathrm{A}(\mathbb{R}^1)\}$$

see Chapter 2, §2.5). So (α″) can be rewritten as

$$(U_{a(\cdot)}f,\, h) = \int_{\mathbb{R}^1} b(\lambda_1)\, e^{-\lambda_1^2/4}\, d(E_1(\lambda) f,\, h) = 0.$$

Because $A(\mathbb{R}^1)$ contains all polynomials, it follows from (α'') that

$$\int_{\mathbb{R}^1} \lambda_1^n\, e^{-\lambda_1^2/4}\, d(E_1(\lambda_1) f,\, h) = 0.$$

Consider the complex measure, $\omega(\Delta) = \int_{\Delta} e^{-\lambda_1^2/4}\, d(E_1(\lambda_1) f,\, h)$ and its Fourier transform $g(t_1) = \int_{\mathbb{R}^1} e^{it_1\lambda_1}\, e^{-\lambda_1^2/4}\, d(E_1(\lambda_1) f,\, h)$. The function $g(t_1)$ can be analytically extended to the whole complex plane by

$$g(z) = \int_{\mathbb{R}^1} \exp(i z \lambda_1 - \tfrac{1}{4}\lambda_1^2)\, d(E_1(\lambda_1) f,\, h).$$

Because

$$\left.\frac{d^n g}{dz^n}\right|_{z=0} = \int_{\mathbb{R}^1} (i\lambda_1)^n \exp(-\lambda_1^2/4)\, d(E_1(\lambda_1) f,\, h) = 0\,,$$

we see that $g(z) \equiv 0$, which means that $\omega(\Delta) = \int_{\Delta} \exp(-\lambda_1^2/4)\, d(E_1(\lambda_1) f,\, h) = 0$ and, consequently, that $(E_1(\Delta) f,\, h) = 0$.

The fact that (β'') implies (α'') follows from the equality

$$\left(\int_{\mathbb{R}^1} a(t_1)\, U_{t_1}\, dg(t_1) f,\, h\right) = \int_{\mathbb{R}^1} b(\lambda_1)\, e^{-\lambda_1^2/4}\, d(E_1(\lambda_1) f,\, h).$$

□

The following corollary finishes the proof that $D \subset H$ is dense.

COROLLARY 3. Let A_1, B_1 be self-adjoint operators acting in a Hilbert space H, let $U_{t_1} = e^{it_1 A_1}$, $V_{s_1} = e^{is_1 B_1}$, let $E_1(\cdot)$, $F_1(\cdot)$ be resolutions of the identity of the operators A_1 and B_1, let $A(\mathbb{R}^2)$ be the introduced above space of functions, let $dg(t_1, s_1) = dg(t_1) \otimes dg(s_1)$ be a Gaussian measure on $(\mathbb{R}^2, B(\mathbb{R}^2))$, and let $f \in H$ be an arbitrary vector.

Then

$$\text{CLS}\,\{\int_{\mathbb{R}^2} a(t_1, s_1)\, U_{t_1} V_{s_1}\, dg(t_1, s_1) f \mid a(t_1, s_1) \in A(\mathbb{R}^2)\} =$$

$$= \text{CLS}\,\{E_1(\Delta_1) F_1(\Delta_2) f \mid \Delta_1, \Delta_2 \in B(\mathbb{R}^1)\} =$$

$$= \text{CLS}\,\{U_{t_1} V_{s_1} f \mid t_1, s_1 \in \mathbb{R}^1\}.$$

Proof. Let $h \in H$. We are ready if we can show the equivalence of the requirements

$$(\alpha''')\quad (\int_{\mathbb{R}^2} a(t_1,s_1)\,U_{t_1}\,V_{s_1}\,dg(t_1,s_1)f,\,h)=0,\quad a(\cdot)\in \mathrm{A}(\mathbb{R}^2),$$

$$(\beta''')\quad (E_1(\Delta_1)\,F_1(\Delta_2)f,\,h)=0,\quad \Delta_1,\Delta_2\in \mathrm{B}(\mathbb{R}^1),$$

where we use Corollary 1. Take the functions $a(t_1,s_1)\in \mathrm{A}(\mathbb{R}^2)$ of the form $a(t_1,s_1)=a_1(t_1)\,a_2(s_1)$, $a_1(\cdot),a_2(\cdot)\in \mathrm{A}(\mathbb{R}^1)$.
Denote $U_{a_1(\cdot)}=\int_{\mathbb{R}^1} a_1(t_1)\,U_{t_1}\,dg(t_1)$, $V_{a_2(\cdot)}=\int_{\mathbb{R}^1} a_2(s_1)\,V_{s_1}\,dg(s_1)$.

Then (α''') takes the form

$$(U_{a_1(\cdot)}\,V_{a_2(\cdot)}f,\,h)=0,\quad a_1(\cdot),a_2(\cdot)\in \mathrm{A}(\mathbb{R}^1).$$

By Corollary 2, this is equivalent to

$$(E_1(\Delta_1)\,V_{a_2(\cdot)}f,\,h)=(V_{a_2(\cdot)}f,\,E_1(\Delta_1)h)=0,\quad \Delta_1\in \mathrm{B}(\mathbb{R}^1).$$

Using Corollary 2 once more, we get

$$(F_1(\Delta_2)f,\,E_1(\Delta_1)h)=(E_1(\Delta_1)\,F_1(\Delta_2)f,\,h)=0,\quad \Delta_1,\Delta_2\in \mathrm{B}(\mathbb{R}^1).$$

$\square$

3. We show that $U_{t_1}\,V_{s_1}\in L(D\to D)$. Taking into consideration the way the operators $U_{\tau_1}\,V_{\sigma_1}$ of the representation act on the vectors from D

$$U_{\tau_1}\,V_{\sigma_1}\int_{\mathbb{R}^2} a(t_1,s_1)\,U_{t_1}\,V_{s_1}\,dg(t_1,s_1)\Omega=$$

$$=\int_{\mathbb{R}^2} a(t_1,s_1)\,U_{\tau_1}\,V_{\sigma_1}\,U_{t_1}\,V_{s_1}\,dg(t_1,s_1)\Omega=$$

$$=\int_{\mathbb{R}^2} a(t_1,s_1)\,e^{it_1\sigma_1}\,U_{t_1+\tau_1}\,V_{s_1+\sigma_1}\,dg(t_1,s_1)\Omega=$$

$$=\int_{\mathbb{R}^2} e^{i(t_1-\tau_1)\sigma_1}\,a(t_1-\tau_1,\,s_1-\sigma_1)\,U_{t_1}\,V_{s_1}\times$$

$$\times\frac{dg(t_1-\tau_1,\,s_1-\sigma_1)}{dg(t_1,s_1)}\,dg(t_1,s_1)\Omega$$

and recalling the topology of the space D we find that the property 3) follows from the continuity of the operator $T_{(\tau_1,\sigma_1)}:\mathrm{A}(\mathbb{R}^2)\to \mathrm{A}(\mathbb{R}^2)$ operating as

$$(T_{(\tau_1,\sigma_1)}\,a)\,(t_1,s_1)=e^{i(t_1-\tau_1)\sigma_1}\,\frac{dg(t_1-\tau_1,\,s_1-\sigma_1)}{dg(t_1,s_1)}\times$$

$$\times\, a(t_1-\tau_1,\,s_1-\sigma_1).$$

LEMMA 6. $T_{(\tau_1,\sigma_1)}\in L(\mathrm{A}(\mathbb{R}^2)\to \mathrm{A}(\mathbb{R}^2))$.

Proof. Denote by T_{τ_1} the operator, acting on $A(\mathbb{R}^1)$:

$$(T_{\tau_1} a)(t_1) = \frac{dg(t_1 - \tau_1)}{dg(t_1)} a(t_1 - \tau_1)).$$

It is continuous: $T_{\tau_1} \in L(A(\mathbb{R}^1) \to A(\mathbb{R}^1))$ (see Chapter 2, §2.5, Lemma 2). Introduce similar operators in $A(\mathbb{R}^2)$

$$(T_{(\tau_1,0)} a)(t_1, s_1) = \frac{dg(t_1 - \tau_1, s_1)}{dg(t_1, s_1)} a(t_1 - \tau_1, s_1)$$

$$(T_{(0,\sigma_1)} a)(t_1, s_1) = e^{it_1\sigma_1} \frac{dg(t_1, s_1 - \sigma_1)}{dg(t_1, s_1)} a(t_1, s_1 - \sigma_1)$$

which are also continuous. The relation

$$T_{(\tau_1, \sigma_1)} = T_{(\tau_1,0)} T_{(0,\sigma_1)} \tag{17}$$

finishes the proof.

4,5. We show that D consists of entire vectors for generators of one-parameter subgroups of the group $U_{(t_1,s_1,\alpha_1)}$. From this, in particular, the statement 4 will follow.

The generators of the one-parameter subgroups U_{t_1} and V_{s_1} are Q_1 and P_1 i.e. $U_{t_1} = e^{it_1 Q_1}$ and $V_{s_1} = e^{is_1 P_1}$. Denote the resolutions of the identity of the operators Q_1 and P_1 by $E_1(\cdot)$ and $F_1(\cdot)$ respectively.

We need to show the inclusions $D \subset H^c(Q_1)$ and $D \subset H^c(P_1)$. This is equivalent to $D \subset \bigcap_{\varepsilon>0} D(e^{\varepsilon|Q_1|})$ and $D \subset \bigcap_{\varepsilon>0} D(e^{\varepsilon|P_1|})$ (see Chapter 1, §1.6). And these inclusions hold because the operators

$$Q_{1,\varepsilon} = e^{\varepsilon|Q_1|} \int_{\mathbb{R}^2} a(t_1, s_1) U_{t_1} V_{s_1} dg(t_1, s_1),$$

$$P_{1,\varepsilon} = e^{\varepsilon|P_1|} \int_{\mathbb{R}^2} a(t_1, s_1) U_{t_1} V_{s_1} dg(t_1, s_1)$$

are bounded on H.

Indeed, transform the expression for $Q_{1,\varepsilon}$ as follows

$$Q_{1,\varepsilon} = e^{\varepsilon|Q_1|} \int_{\mathbb{R}^2} a(t_1, s_1) e^{it_1 Q_1} V_{s_1} dg(t_1) dg(s_1) =$$

$$= e^{\varepsilon|Q_1|} \int_{\mathbb{R}^2} a(t_1, s_1) \int_{\mathbb{R}^1} e^{it_1\lambda_1} dE_1(\lambda_1) V_{s_1} dg(t_1) dg(s_1) =$$

$$= e^{\varepsilon|Q_1|} \int_{\mathbb{R}^2} (\int_{\mathbb{R}^1} a(t_1, s_1) e^{it_1\lambda_1} dg(t_1)) dE_1(\lambda_1) V_{s_1} dg(s_1) =$$

$$= e^{\varepsilon|Q_1|} \int_{\mathbb{R}^1} \int_{\mathbb{R}^1} b(\lambda_1, s_1)\, e^{-\lambda_1^2/4}\, dE_1(\lambda_1)\, V_{s_1}\, dg(s_1) =$$

$$= \int_{\mathbb{R}^1} \int_{\mathbb{R}^1} e^{\varepsilon|\lambda_1|}\, b(\lambda_1, s_1)\, e^{-\lambda_1^2/4}\, dE_1(\lambda_1)\, V_{s_1}\, dg(s_1).$$

We have used the fact that $a(t_1, s_1)$ is a function of $A(\mathbb{R}^1)$ for a fixed s_1. So $b(\lambda_1, s_1)$ is again a function of $A(\mathbb{R}^1)$ such that

$$|b(\lambda_1, s_1)| \le C_\delta \exp(-\delta(\lambda_1^2 + s_1^2)) \text{ for any } \delta > 0.$$

From the estimate
$|e^{\varepsilon|\lambda_1|}\, b(\lambda_1, s_1) \exp(-\lambda_1^2/4)| \le e^{\varepsilon|\lambda_1|}\, C_\delta \exp(-\delta(\lambda_1^2 + s_1^2)) \exp(-\lambda_1^2/4) \le K_{\varepsilon,\delta} \exp(-\delta s_1^2)$ $(\varepsilon > 0)$,
it follows that

$$\|Q_{1,\varepsilon}\| \le K_{\varepsilon,\delta} \int_{\mathbb{R}^1} \exp(-\delta s_1^2)\, dg(s_1) < \infty.$$

Since $U_{t_1} V_{s_1} = e^{-t_1 s_1} V_{s_1} U_{t_1}$ we can similarly prove that the operator $P_{1,\varepsilon}$ is bounded. ▯

6. We prove that not only the vector-function $\mathbb{R}^2 \ni (t_1, s_1) \mapsto U_{t_1} V_{s_1} f \in D$ $(f \in D)$ is continuous, but that, in addition, the vector function $\mathbb{R}^2 \ni (t_1, s_1) \mapsto T_{(t_1, s_1)}\, a(\cdot)$ $(a(\cdot) \in A(\mathbb{R}^2))$ is also continuous for all $a \in A(\mathbb{R}^2)$. Using (17), we see that this is a consequence of Lemma 3 in Chapter 2. ▯

If the representation $G \ni (t_1, s_1, \alpha_1) \mapsto U_{(t_1, s_1, \alpha_1)}$ is not cyclic, then there exists a sequence of cyclic vectors Ω_n in H_n, such that $H = \oplus \sum H_n$. The set $D = \oplus \sum D_n$ will be the set in question. ▯

Let now U_t, V_s be the representation of $\mathbb{R}_0^\infty \ni t,s$ on H such that

$$V_s U_t = \exp(i \sum_{k=1}^{\infty} t_k s_k) U_t V_s = e^{i(t,s)} U_t V_s.$$

Suppose that a given representation of the CCR of a system with countable degrees of freedom is cyclic, and that Ω is a cyclic vector, i.e.

$$\text{CLS}\, \{U_t V_s \Omega \mid t,s \in \mathbb{R}_0^\infty\} = H.$$

Let $E(\cdot)$, $F(\cdot)$ be resolutions of the identity defined on $(\mathbb{R}^\infty, \mathrm{B}(\mathbb{R}^\infty))$ such that

$$U_t = \int_{\mathbb{R}^\infty} e^{i(t,\lambda)}\, dE(\lambda), \quad V_s = \int_{\mathbb{R}^\infty} e^{i(s,\lambda)}\, dF(\lambda).$$

Among the measures $\mu_f(\Delta) = (E(\Delta) f, f)$ choose a measure with respect to which all others are absolutely continuous. Denote it by $\mu(\Delta)$. Similarly, define $\nu(\Delta)$. Then $\mu_f(\cdot) \ll \mu(\cdot)$ and $\nu_f(\cdot) \ll \nu(\cdot)$.

As known (see Chapter 1, §1.3), for any probability measure on $(\mathbb{R}^\infty, \mathrm{B}(\mathbb{R}^\infty))$, there exists a Hilbert space $l_2(a_n)$ of full measure. For the measures $\mu(\cdot)$ and $\nu(\cdot)$ choose a Hilbert space $l_2(a_n) \subset \mathbb{R}^\infty$ $(a_n > 0)$ such that $\mu(l_2(a_n)) = \nu(l_2(a_n)) = 1$. On the space $l_2\left(\frac{1}{a_n}\right)$

construct a Gaussian measure $dg_\alpha(\lambda) = \bigotimes_{k=1}^{\infty} \sqrt{\frac{\alpha_k}{\pi}} \exp(-\alpha_k \lambda_k^2) d\lambda_k$ such that $g_\alpha(l_2\left[\frac{1}{a_n}\right]) = 1$. To do this, it is necessary to find a sequence $(\alpha_k)_{k=1}^{\infty}$ such that $\sum_{k=1}^{\infty} a_k / \alpha_k < \infty$.

Define a Gaussian measure $dg_\alpha(t) \otimes dg_\alpha(s)$ on $(\mathbb{R}^\infty \times \mathbb{R}^\infty, \mathrm{B}(\mathbb{R}^\infty \times \mathbb{R}^\infty))$, and use it to construct the function space $\mathrm{A}^\alpha(\mathbb{R}^\infty \times \mathbb{R}^\infty)$ (see Chapter 2, §2.5).

Define the set D to be:

$$D = \{ \int_{\mathbb{R}^\infty \times \mathbb{R}^\infty} a(t,s)\, U_t V_s\, dg_\alpha(t,s)\, \Omega \mid a(t,s) \in \mathrm{A}^\alpha(\mathbb{R}^\infty \times \mathbb{R}^\infty)\}.$$

Its topology is the topology induced by the space $\mathrm{A}^\alpha(\mathbb{R}^\infty \times \mathbb{R}^\infty)$.

THEOREM 6. The following statements hold:

1) D is a nuclear topological linear space;

2) D is topologically densely imbedded in H;

3) $U_t V_s \in L(D \to D)$;

4) D is contained in the domains of generators of one-parameter subgroups of the group G_0^∞ and these generators map D into itself;

5) D consists of entire vectors for the generators of representations of one-parameter subgroups of the group G_0^∞;

6) the function $G_0^\infty \ni (t,s,\alpha) \to U_{(t,s,\alpha)} f \in D$ is continuous for all $f \in D$.

The proof is similar to the proof of Theorem II in Chapter 2. If the representation is not cyclic, then $D = \bigoplus_{n=1}^{\infty} \sum D_n$ satisfies the conditions of the theorem.

Comments to Chapter 4.

1. The classic Stone-von Neumann uniqueness theorem for representations of the CCR of systems with finite degrees of freedom in Weyl's form is given. The proof follows the A.A. Kirillov [1].

2. Gårding-Wightman's form for representations of systems with countable degrees of freedom is given (see L. Gårding and A. Wightman [2], F.A. Berezin [3], G. Emch [1]). The decomposition method with respect to a countable family of the CSO $(Q_k)_{k=1}^{\infty}$ using at later stages the commutation relations, follows I.M. Gel'fand and N.Ya. Vilenkin [1], G. Hegerfeldt, O. Melsheimer [1] and others.

3. The construction of cocycles in this section follows the article V.I. Kolomytsev and Yu.S. Samoĭlenko [2].

4. For Gaussian representations of the CCR see, for example, A.S. Holevo [2,5,6]. Conditions for quasi-equivalence of Gaussian representations of the CCR are obtained in H. Araki [4], A.S. Holevo [2,3], A. van Daele [1]. Gaussian representations of the CCR with the simple joint spectrum of the CSO $(Q_k)_{k=1}^{\infty}$ can be obtained by using a Gaussian measure and a Gaussian cocycle. In the section, the study of such CCR follows A.Yu. Daletskiĭ.

5. The given construction of a Gårding domain for representations of CCR with countable degrees of freedom is due to A.V. Kosyak [2].

Chapter 5.
UNITARY REPRESENTATIONS OF THE GROUP OF FINITE SU(2)-CURRENTS ON A COUNTABLE SET

For the "prelimiting" objects considered in Chapter 4 there existed an irreducible representation unique up to unitary equivalence. In this chapter, we consider skew-adjoint representation of the real Lie algebra $SU(2)_0^\infty$ and the corresponding unitary representations of the group $SU(2)_0^\infty$. The "prelimiting" groups $SU(2)^n$ have countably many nonequivalent unitary representations. But being limited to the irreducible (factor) representations of the group $SU(2)_0^\infty$ we will construct their commutative models.

At the beginning we will give preliminaries of the theory of unitary representations of the groups $SU(2)$ and $SU(2)^n$. Then we find the Gårding-Wightman form for the irreducible (factor) representations of the group $SU(2)_0^\infty$ and use it to study particular classes of unitary representations of the group $SU(2)_0^\infty$ and the Lie algebra $su(2)_0^\infty$.

5.1. Unitary representations of the groups $SU(2)$ and $SU(2)^n$

One of the simplest compact non-commutative Lie groups is the group $SU(2)$ of unitary complex 2×2 matrices. By $su(2)$ we denote its real Lie algebra. A basis, a_1, a_2, a_3 in $su(2)$ can be chosen such that

$$[a_1, a_2]=a_3\,,\ [a_2,a_3]=a_1\,,\ [a_3,a_1]=a_2. \tag{1}$$

For unitary representations of the group, there correspond skew-adjoint representations of the relation (1). Consider the complex extension $su(2)_C$ of the real Lie algebra $su(2)$ with a basis e_-, e_0, e_+:

$$e_+=i(a_1+i\,a_2),\ \ e_0=i\,a_3\,,\ \ e_-=i(a_1-i\,a_2)$$

satisfying the commutation relations

$$[e_0,e_+]=e_+\,,\ \ [e_0,e_-]=-e_-\,,\ \ [e_+,e_-]=2\,e_0.$$

As all irreducible representations of the connected, simply connected compact group $SU(2)$ are finite-dimensional, there is a one-to-one correspondence between the representation and an irreducible triple of operators, E_-, E_0, E_+ on a finite dimensional Hilbert space H satisfying the commutation relations

$$[E_0,E_+]=E_+\,,\ \ [E_0,E_-]=-E_-\,,\ \ [E_+,E_-]=2E_0 \tag{2}$$

and the adjoint relations

$$E_0^* = E_0\,,\; E_+^* = E_-\,,\; E_-^* = E_+. \tag{3}$$

The irreducible unitary representations of the group $SU(2)$ are determined uniquely up to unitary equivalence by an integer or semi-integer l and can be realized on a space H_l of dimension $2l+1$. In an orthonormal basis of H_l, e_ν ($\nu=-l, -l+1, \ldots, l$) consisting of the eigenvectors of the self-adjoint operator E_0, the operators E_-, E_0, E_+ act according to the formulas

$$\begin{aligned} E_- e_\nu &= \sqrt{(l+\nu)\,(l-\nu+1)}\; e_{\nu-1}\,, \\ E_0\, e_\nu &= \nu\, e_\nu\,, \\ E_- e_\nu &= \sqrt{(l-\nu)\,(l+\nu-1)}\; e_{\nu+1}. \end{aligned} \tag{4}$$

Denote by $U^{(l)}$ the irreducible representation of the group $SU(2) \ni g_1 \mapsto U_{g_1}^{(l)}$ on H_l corresponding to the representation of the algebra $su(2)$ on H_l.

Together with the irreducible representations also consider factor representations of $SU(2)$, i.e. representations $SU(2) \ni g_1 \mapsto U_{g_1}$ on H such that the $*$-algebra (W^*-algebra, von Neumann algebra) which is closed in the weak operator topology and generated by the representation operators, is a factor, i.e. a W^*-algebra with center consisting only of scalar multiples of the identity.

All the factor representations of the group $SU(2)$ are direct sums of a finite or infinite number of the same irreducible representation associated to a given factor, i.e. they are realized on $H_l \oplus H_l \oplus \cdots$ by the formulas $SU(2) \ni g_1 \mapsto U_{g_1}^{(l)} \oplus U_{g_1}^{(l)} \oplus \cdots$.

Irreducible unitary representations of the group $SU(2)^n \ni (g_1, \ldots, g_n)$ ($g_k \in SU(2)$; $k = 1, \ldots, n$) are determined uniquely up to unitary equivalence by the collections of integers or semi-integers $(l_1, \ldots, l_n)$ and they are realized on the space $H_{l_1} \otimes \ldots \otimes H_{l_n}$ according to the formulas $SU(2)^n \ni (g_1, \ldots, g_n) \mapsto U_{g_1}^{(l_1)} \otimes \cdots \otimes U_{g_n}^{(l_n)}$.

As it was for the group $SU(2)$, all the factor representations of the group $SU(2)^n$ are direct sums of the same fixed irreducible representation of the group $SU(2)^n$.

5.2. Unitary representations of the group $SU(2)_0^\infty$. Measures and cocycles.

Now consider the group $SU(2)_0^\infty$ of $SU(2)$-valued sequences $(g_1, \ldots, g_n, \ldots)$ ($g_n \in SU(2)$) such that only a finite number of g_n's is different from the identity e of the group $SU(2)$.

First of all we construct a series of irreducible unitary representations of the group $SU(2)_0^\infty$. For a given sequence of numbers $(l_k)_{k=1}^\infty$ (non-negative integers or semi-integers) define the representation

$$SU(2)_0^\infty \ni (g_1, \ldots, g_n, \ldots) \mapsto U_{g_1}^{(l_1)} \otimes \cdots \otimes U_{g_n}^{(l_n)} \otimes \cdots \tag{5}$$

on the space $H = \bigotimes_{k=1;(f_k)}^{\infty} H_{l_k}$. In the sequel such representations will be called product representations.

PROPOSITION 1. Product representations are irreducible.

Proof. On H consider the projections P_n into the subspaces $K_n = H_{l_1} \otimes \cdots \otimes H_{l_n} \otimes f_{n+1} \otimes \cdots \subset H$. For any bounded operator A on H commuting with the representation (5), the operators $P_n A P_n$ on K_n commute with the irreducible unitary representation $SU(2)^n \ni (g_1, \ldots, g_n) \mapsto U_{g_1}^{(l_1)} \otimes \cdots \otimes U_{g_n}^{(l_n)} \otimes I \otimes \cdots$. So, consequently, for any $n = 1,2,\ldots$, $P_n A P_n = \lambda_n I$. But because $s-\lim P_n = I$, we have that $A = \lambda I$ and so the representation (5) on H is irreducible. □

PROPOSITION 2. Two product representations of $SU(2)_0^\infty$

$$U^{(l_1, \ldots, l_n, \ldots)} = U^{(l_1)} \otimes \cdots \otimes U^{(l_n)} \otimes \cdots \quad \text{on } H = \bigotimes_{k=1;(f_k)}^{\infty} H_{l_k}$$

and

$$\tilde{U}^{(m_1, \ldots, m_n, \ldots)} = U^{(m_1)} \otimes \cdots \otimes U^{(m_n)} \otimes \cdots \quad \text{on } \tilde{H} = \bigotimes_{k=1;(g_k)}^{\infty} H_{m_k}$$

are unitarily equivalent if and only if both $l_k = m_k$ $(k = 1,2,\ldots)$ and the stabilizations $(f_k)_{k=1}^\infty$ and $(g_k)_{k=1}^\infty$ $(f_k, g_k \in H_{l_k} = H_{m_k},\ \|f_k\|_{H_{l_k}} = \|g_k\|_{H_{m_k}} = 1)$ are weakly equivalent, i.e.

$$\sum_{k=1}^{\infty} (1 - |(f_k, g_k)|) < \infty.$$

Proof. The condition $l_k = m_k$ $(k = 1,2,\ldots)$ is necessary for the representations $U^{(l_1, \ldots, l_n, \ldots)}$ and $\tilde{U}^{(m_1, \ldots, m_n, \ldots)}$ of $SU(2)_0^\infty$ to be unitarily equivalent because it is necessary for the unitary equivalence of the representations

$$SU(2)^n \ni (g_1, \ldots, g_n) \mapsto U_{g_1}^{(l_1)} \otimes \cdots \otimes U_{g_n}^{(l_n)} \otimes U_e^{(l_{n+1})} \otimes \cdots \quad \text{on } H$$

and

$$SU(2)^n \ni (g_1, \ldots, g_n) \mapsto U_{g_1}^{(m_1)} \otimes \cdots \otimes U_{g_n}^{(m_n)} \otimes U_e^{(m_{n+1})} \otimes \cdots \quad \text{on } \tilde{H}.$$

Now, because the operator $U : H \to \tilde{H}$ that realizes the equivalence of the representations $U^{(l_1, \ldots, l_n, \ldots)}$ and $U^{(m_1, \ldots, m_n, \ldots)}$ is a product operator, i.e. it has the form $U = U_1 \otimes \cdots \otimes U_n \otimes \cdots$, where $U_n : H_{l_n} \to H_{m_n}$ commutes with the irreducible representation $SU(2) \ni g_n \mapsto U_{g_n}^{(l_n)}$ we see that

$$U = \lambda_1 I_{H_{l_1}} \otimes \cdots \otimes \lambda_n I_{H_{l_n}} \otimes \cdots \quad (|\lambda_n| = 1, n \in \mathbb{N}).$$

But then the stabilizations $(\lambda_k f_k)_{k=1}^\infty$ and $(g_k)_{k=1}^\infty$ belong to $\tilde{H}$ and are strongly equivalent, and so $\sum_{k=1}^{\infty}(1-\lambda_k(f_k, g_k))<\infty$. The latter is equivalent to weak equivalence of $(f_k)_{k=1}^\infty$ and $(g_k)_{k=1}^\infty$.

Thus, the conditions of Proposition 2 are necessary. To prove that the conditions are sufficient, one constructs an operator U which realizes the unitary equivalence between the representations $U^{(l_1,\ldots,l_n\cdots)}$ and $\tilde{U}^{(m_1,\ldots,m_n,\ldots)}$ in the form

$$U=\lambda_1 I_{H_{l_1}}\otimes\cdots\otimes\lambda_n I_{H_{l_n}}\otimes\cdots$$

where $\lambda_k(|\lambda_k|=1;\ k=1,2,\ldots)$ are chosen so that the stabilizations $(\lambda_k f_k)_{k=1}^\infty$ and $(g_k)_{k=1}^\infty$ are strongly equivalent. □

Although the representations $U^{(l_1)}\otimes\cdots\otimes U^{(l_n)}$ on $H_{l_1}\otimes\cdots\otimes H_{l_n}$ for $l_1,\ldots,l_n=0,\frac{1}{2},1,\ldots$ form a complete collection of irreducible representations of the group $SU(2)^n$, the collection of product representations is only a small part of all irreducible representations of $SU(2)_0^\infty$.

Consider the real Lie algebra $su(2)_0^\infty$. The vectors $(a_1^{(k)}, a_2^{(k)}, a_3^{(k)})_{k=1}^\infty$ satisfying the commutation relations

$$[a_1^{(k)}, a_2^{(j)}]=\delta_{kj}a_3^{(k)}\ ,\ [a_3^{(k)}, a_3^{(j)}]=\delta_{kj}a_1^{(k)}\ ,\ [a_3^{(k)}, a_1^{(j)}]=\delta_{kj}a_2^{(k)}$$

form a basis in it. A basis in $(su(2)_0^\infty)_C$ can be chosen accordingly. The vectors $(e_-^{(k)}, e_0^{(k)}, e_+^{(k)})_{k=1}^\infty$ must satisfy the commutation relations:

$$[e_0^{(k)}, e_+^{(j)}]=\delta_{kj}e_+^{(k)}\ ,\ [e_0^{(k)}, e_-^{(j)}]=-\delta_{kj}e_-^{(k)}\ ,\ [e_+^{(k)}, e_-^{(j)}]=2\delta_{kj}e_0^{(k)}.$$

For every unitary representation of the group $SU(2)$ there is a corresponding triple of, generally speaking, unbounded operators (E_-, E_0, E_+) which satisfy the commutation relations (2), and the adjointness relations (3), on the common invariant set of their joint analytic vectors. The closure of the operator $\frac{1}{2}(E_-E_++E_+E_-)+E_0^2$ defined on this set of vectors, is a self-adjoint operator Δ. Using the CSO $(\Delta^{(k)})_{k=1}^\infty$ we determine the domain, $\Psi=H^\omega(\Delta^{(1)},\ldots,\Delta^{(n)},\ldots)$ dense in H, invariant with respect to the operators $(E_-^{(k)}, E_0^{(k)}, E_+^{(k)})_{k=1}^\infty$ where these operators satisfy the commutation relations

$$[E_+^{(k)}, E_-^{(j)}]=2\delta_{kj}E_0^{(k)}\ ,\ [E_0^{(k)}, E_+^{(j)}]=\delta_{kj}E_+^{(k)}\ ,$$
$$[E_0^{(k)}, E_-^{(j)}]=-\delta_{kj}E_-^{(k)} \tag{6}$$

and the adjointness relation

$$E_0^{(k)*}=E_0^{(k)}\ ,\ E_+^{(k)*}=E_-^{(k)}\ ,\ E_-^{(k)*}=E_+^{(k)}\quad (k=1,2,\ldots). \tag{7}$$

Now suppose that the unitary representation of $su(2)_0^\infty$ on H is irreducible. Then $\Delta^{(k)}=l_k(l_k+1)I$ with l_k an integer or semi-integer, $k=1,2,\ldots$, and the $E_0^{(k)}$ are CSO with joint spectrum $M(l_1,\ldots,l_n,\ldots)=\prod_{k=1}^{\infty}M_{l_k}$ where $M_{l_k}=\{-l_k, -l_k+1,\ldots, l_k-1, l_k\}$. Using the spectral theorem for the CSO $(E_0^{(k)})_{k=1}^{\infty}$, we decompose the representation space H into the direct integral

$$H = \oplus \int_{M(l_1,\ldots,l_n,\ldots)} H_\lambda \, d\rho(\lambda) \quad (\lambda \in M(l_1,\ldots,l_n,\ldots) \subset \mathbb{R}^\infty).$$

Here the operators $(E_0^{(k)})_{k=1}^{\infty}$ act on vectors $\vec{f}(\cdot) \in \oplus \int_{M(l_1,\ldots,l_n,\ldots)} H_\lambda \, d\rho(\lambda)$ by the rule: $(E_0^{(k)} f)(\lambda) = \lambda_k f(\lambda)$ $(k = 1,2,\ldots)$. From the relation (6) we have:

$$(E_+^{(k)} \vec{f})(\lambda) = C_+^{(k)}(\lambda) \, \vec{f}(\lambda - 1_k),$$

$$(E_-^{(k)} \vec{f})(\lambda) = C_-^{(k)}(\lambda) \, f(\lambda + 1_k),$$

where $C_+^{(k)}(\lambda) : H_{\lambda - 1_k} \to H_\lambda$, $C_-^{(k)}(\lambda) : H_{\lambda + 1_k} \to H_\lambda$, $1_k = (\underbrace{0, \ldots, 0, 1}_{k\text{–th place}},0,\ldots).$,

Let us find out what properties of the operator functions $(C_+^{(k)}(\lambda), C_-^{(k)}(\lambda))_{k=1}^{\infty}$ follow from the relations (6) and (7). From the commutation relation (6) we have

$$C_+^{(k)}(\lambda) \, C_-^{(k)}(\lambda - 1_k) - C_-^{(k)}(\lambda) \, C_+^{(k)}(\lambda + 1_k) = 2\lambda_k. \tag{8}$$

Because the operator $\Delta^{(k)} = l_k(l_k + 1) I$ is scalar, we get

$$C_+^{(k)}(\lambda) \, C_-^{(k)}(\lambda - 1_k) + C_-^{(k)}(\lambda) \, C_+^{(k)}(\lambda + 1_k) = 2\lambda_k(l_k + 1) - 2\lambda_k^2. \tag{9}$$

So the equalities (8) and (9) yield:

$$\begin{aligned} C_+^{(k)}(\lambda) \, C_-^{(k)}(\lambda - 1_k) &= l_k(l_k + 1) - \lambda_k(\lambda_k - 1) \\ C_-^{(k)}(\lambda) \, C_+^{(k)}(\lambda + 1_k) &= l_k(l_k + 1) - \lambda_k(\lambda_k + 1). \end{aligned} \tag{10}$$

Now, by the adjointness relation (7)

$$C_+^{(k)}(\lambda) = C_-^{(k)}(\lambda) \, \frac{d\rho(\lambda - 1_k)}{d\rho(\lambda)} \tag{11}$$

and so from (10) and (11), we get

$$C_-^{(k)}(\lambda) \, C_-^{(k)*}(\lambda) = C_-^{(k)*}(\lambda) \, C_-^{(k)}(\lambda) =$$

$$= [l_k(l_k + 1) - \lambda_k(\lambda_k + 1)] \, \frac{d\rho(\lambda + 1_k)}{d\rho(\lambda)}$$

and

$$C_+^{(k)}(\lambda) \, C_+^{(k)*}(\lambda) = C_+^{(k)*}(\lambda) \, C_+^{(k)}(\lambda) =$$

$$= [l_k(l_k+1) - \lambda_k(\lambda_k-1)] \frac{d\rho(\lambda-1_k)}{d\rho(\lambda)}.$$

Setting

$$C_+^{(k)}(\lambda) = \sqrt{[l_k(l_k+1)-\lambda_k(\lambda_k-1)] \frac{d\rho(\lambda-1_k)}{d\rho(\lambda)}}\; U_+^{(k)}(\lambda)$$

$$C_-^{(k)}(\lambda) = \sqrt{[l_k(l_k+1)-\lambda_k(\lambda_k+1)] \frac{d\rho(\lambda+1_k)}{d\rho(\lambda)}}\; U_-^{(k)}(\lambda)$$

we get a collection of unitary operator functions

$$(U_+^{(k)}(\lambda),\, U_-^{(k)}(\lambda))_{k=1}^\infty : U_+^{(k)} : H_{\lambda-1_k} \to H_\lambda,\; U_-^{(k)} : H_{\lambda+1_k} \to H_\lambda$$

Denote $U_+^{(k)}(\lambda) = U(\lambda)$. The commutativity of $E_+^{(j)}$ and $E_-^{(j)}$ with $E_+^{(k)}$ and $E_-^{(k)}$ $(j \neq k)$ leads to the equality

$$U_k(\lambda)\, U_j(\lambda-1_k) = U_j(\lambda)\, U_k(\lambda-1_j). \tag{12}$$

So we have proved the following theorem.

THEOREM 1. Any irreducible representation of the group $SU(2)_0^\infty$ acts on $H = \oplus \int_{M(l_1,\ldots,l_n,\ldots)} H_\lambda\, d\rho(\lambda)$ where $\rho(\cdot)$ is a probability measure on $M(l_1,\ldots,l_n,\ldots)$ which is quasi-invariant with respect to the group of finite shifts (i.e. the group of shifts generated by the transformations

$$\lambda = (\lambda_1,\ldots,\lambda_{k-1},\lambda_k,\lambda_{k+1},\ldots) \mapsto (\lambda_1,\ldots,\lambda_{k-1},\lambda_k \tilde{+} 1,\lambda_{k+1},\ldots),$$

$$\lambda_k \tilde{+} 1 = \begin{cases} \lambda_k+1, & \lambda_k \neq l_k \\ -l_k, & \lambda_k = l_k \end{cases} \quad (k = 1,2,\ldots)).$$

Here the operators $(E_0^{(k)}, E_+^{(k)}, E_-^{(k)})_{k=1}^\infty$ are defined by the formulas

$$(E_0^{(k)} \vec{f})(\lambda) = \lambda_k(\lambda),$$

$$(E_+^{(k)} \vec{f})(\lambda) = \sqrt{[l_k(l_k+1)-\lambda_k(\lambda_k-1)] \frac{d\rho(\lambda-1_k)}{d\rho(\lambda)}} \times U_k(\lambda)\, \vec{f}(\lambda-1_k), \tag{13}$$

and

$$(E_-^{(k)} \vec{f})(\lambda) = \sqrt{[l_k(l_k+1)-\lambda_k(\lambda_k+1)] \frac{d\rho(\lambda+1_k)}{d\rho(\lambda)}} \times U_k^*(\lambda+1_k)\, f(\lambda+1_k) \quad (k = 1,2,\ldots).$$

The operators $U_k(\lambda)$ are unitary operators from $H_{\lambda-1_k}$ into H_λ $(k = 1,2,\ldots)$ satisfying the relations

$$U_k(\lambda)\, U_j(\lambda - 1_k) = U_j(\lambda)\, U_k(\lambda - 1_j). \tag{14}$$

Since the Laplace-Casimir operators $(\Delta^{(k)})_{k=1}^{\infty}$ in the given representation (13) of the algebra $(su(2)_0^{\infty})_C$ are bounded, the representation constructed in the theorem can be integrated to get a unitary representations of the group.

Remark 1. The irreducible representations of $SU(2)_0^{\infty}$ are always realized according to the formulas (13). However, not all the representations constructed by (13) are irreducible. The irreducibility of such a representation depends on the measure $\rho(\cdot)$ on $M(l_1, \ldots, l_n, \ldots) = \prod_{k=1}^{\infty} M_{l_k} \ni \lambda$ and the unitary operator functions $(U_k(\lambda))_{k=1}^{\infty}$ which satisfy the relations (14). A necessary condition for a representation to be irreducible is ergodicity of the measure on $M(l_1, \ldots, l_n, \ldots)$ with respect to the finite shifts.

The operators $U_k(\lambda)$ establish an isomorphism between H_λ and $H_{\lambda - 1_k}$. So the dimension of H_λ is constant on the orbits of the group of finite shifts. However, the spectrum of the operators $(E_0^{(k)})_{k=1}^{n}$ is simple in any irreducible representation of $SU(2)^n$ but for the group $SU(2)_0^{\infty}$ there exist irreducible representations with any dimension (even infinite) of H_λ. □

Remark 2. Any factor representation of $SU(2)_0^{\infty}$ can also be given using quasi-invariant measures on $M(l_1, \ldots, l_n, \ldots)$ and unitary operator functions $(U_k(\lambda))_{k=1}^{\infty}$ satisfying the relations (14). For a representation to be a factor it is necessary that the measure is ergodic with respect to the finite shifts. □

Also note that the constructed representations of the Lie algebra $su(2)_0^{\infty}$ for integer $l_k (k = 1,2,\ldots)$ can be integrated to get a representation of the group $SO(3)_0^{\infty}$.

5.3. Representations with one-dimensional cocycle.

In this section we suppose that the CSO $(E_0^{(k)})_{k=1}^{\infty}$ has a simple joint spectrum, i.e. $\dim H_\lambda = 1$ for all $\lambda (\mathrm{mod}\, \rho(\cdot))$.

THEOREM 2. If $\dim H_\lambda = 1 (\mathrm{mod}\, \rho(\cdot))$, a necessary and sufficient condition for the representation (13) to be irreducible is the ergodicity of the measure $\rho(\cdot)$ on $M_{(l_1, \ldots, l_n, \ldots)}$.

The proof follows the proof of Theorem 4 in Chapter 4. □

If the CSO $(E_0^{(k)})_{k=1}^{\infty}$ have a simple joint spectrum, then $U_k(\lambda) = \alpha_k(\lambda)$ $(k = 1,2,\ldots)$ are measurable functions with absolute values equal to one $\rho(\cdot)$-almost everywhere and with

$$\alpha_k(\lambda)\, \alpha_l(\lambda - 1_k) = \alpha_l(\lambda)\, \alpha_k(\lambda - 1_l).$$

Using the functions $\alpha_k(\lambda)$ $(k = 1,2,\ldots)$ we can uniquely reconstruct a 1-cocycle, viz. a function $\alpha_t(\lambda)$ $(t \in \bigcup_{n=1}^{\infty} \mathbb{Z}_{l_1} \times \cdots \times \mathbb{Z}_{l_n} \times 0 \times 0 \cdots)$ for which

a) its absolute value is equal to 1 $\rho(\cdot)$-almost everywhere;

b) $\alpha_{t_1+t_2}(\lambda) = \alpha_{t_1}(\lambda)\,\alpha_{t_2}(\lambda + t_1)$;

c) $\alpha_{1_k}(\lambda) = \alpha_k(\lambda)$.

Following Chapter 4, Section 4.3, we introduce the notion of trivial and equivalent cocycles. Similarly we can prove the proposition.

PROPOSITION 3. For $\dim H_\lambda = 1$ the two representations given by a measure $\rho(\cdot)$ on $M(l_1, \ldots, l_n, \ldots)$ and a cocycle $\alpha_t(\cdot)$ and a measure $\mu(\cdot)$ on $M(s_1, \ldots, s_n, \ldots)$ and a cocycle $\beta_t(\cdot)$ are unitarily equivalent if and only if:

1) $l_k = s_k$ $(k = 1, 2, \ldots)$;

2) the measures $\rho(\cdot)$ and $\mu(\cdot)$ are equivalent;

3) the cocycles $\alpha_t(\cdot)$ and $\beta_t(\cdot)$ are equivalent.

For product representations of $SU(2)_0^\infty$, the joint spectrum of the CSO $(E_0^{(k)})_{k=1}^\infty$ is simple. The measures corresponding to product representations are product measures and the associated cocycles are product cocycles which can be calculated using the stabilization $(f_k)_{k=1}^\infty$.

As in Chapter 4, Section 4.3, one can construct Markov and locally dependent cocycles.

5.4. $SU(2)_{\pi(2)}^\infty$ and its representations.

We will outline the construction of only a few irreducible representations of the group $SU(2)_{\pi(2)}^\infty$ of step $SU(2)$-currents $g(\cdot)$, on the interval $[0,1]$ with discontinuities in binary-rational points. We consider representations ("integrals"), with the property that for all binary-rational shifts r of the interval $[0,1]$ (mod 1), the representation $T_{g(\cdot)}^r = T_{g(\cdot + r \bmod 1)}$ is equivalent to the representation $T_{g(\cdot)}$. Here, we will consider the representations of $SU(2)_{\pi(2)}^\infty$ that can be extended to a representation of $SO(3)_{\pi(2)}^\infty$.

As before, denote the $(2l+1)$-dimensional irreducible representation of the group $SO(3)$ on the space H_l by $U^{(l)}$. It is known that $U^{(l_1)} \otimes U^{(l_2)} = \sum_{k=|l_1-l_2|}^{l_1+l_2} U^{(k)}$. To a sequence of indices $l_0, l_1, \ldots, l_k, \ldots$, such that for all $k = 0, 1, \ldots, l_k \leq 2l_{k+1}$ there corresponds the Hilbert space $H = i - \lim\limits_{\rightarrow} (\otimes H_{l_k}^{2^k})$, where the inclusion

$$i_{n+1}^n : K_n = \otimes H_{l_n}^{2^n} \hookrightarrow K_{n+1} = \otimes H_{l_{n+1}}^{2^{n+1}} =$$

$$= \otimes (H_{l_{n+1}} \otimes H_{l_{n+1}})^{2^n} = \otimes \Big(\sum_{k=0}^{2l_{n+1}} H_k\Big)^{2^n} =$$

$$= (\otimes H_{l_n})^{2^n} \oplus (H_0 \oplus \cdots \oplus H_{l_n - 1} \oplus H_{l_n + 1} \oplus \cdots \oplus H_{2l_{n+1}})^{2^n}$$

is determined by the mapping K_n into the first isomorphic summand.

Since the diagram

$$\begin{array}{ccc} K_n = \otimes H_{l_n}^{2^n} & \xrightarrow{(\bigotimes_{k=1}^{2^n} U^{(l_n)})g(\cdot)} & K_n = \otimes H_{l_n}^{2^n} \\ i_{n+1}^n \downarrow & & \downarrow i_{n+1}^n \\ K_{n+1} = \otimes H_{l_{n+1}}^{2^{n+1}} & \xrightarrow{(\bigotimes_{k=1}^{2^{n+1}} U^{(l_{n+1})})g(\cdot)} & K_{n+1} = \otimes H_{l_{n+1}}^{2^{n+1}} \end{array}$$

is commutative for all $g(\cdot) \in SO(3)^{2^n}$, and since the representation of this group realized on K_n is irreducible, the representation

$$SO(3)^{\infty}_{\pi(2)} \ni g(\cdot) \mapsto T_{g(\cdot)} = \lim_{\rightarrow} (U^{(l_k)} \otimes \cdots \otimes U^{(l_k)})\, g(\cdot)$$

realized on $H = i - \lim\limits_{\rightarrow} K_n$ is also irreducible. For the binary-rational shift $r = \dfrac{k_0}{2^n}$, the unitary operator which establishes the equivalence of the representations $T_{g(\cdot)}$ and $T^r_{g(\cdot)}$ operates as follows: for $m > n$ and $f_k \in H_{l_m}$

$$U f_1 \otimes f_2 \otimes \cdots \otimes f_{2^m} =$$

$$= U(f_1 \otimes \cdots \otimes f_{m/n}) \otimes \cdots \otimes (f_{n-1/n \cdot m+1} \otimes \cdots \otimes f_{2^m}) =$$

$$= U\, \phi_1 \otimes \cdots \otimes \phi_n = \phi_{1+k_0} \otimes \cdots \otimes \phi_{k_0}$$

where $\phi_k \in \otimes H_{l_m}^{m/n}$.

Comments to Chapter 5.

1. Structure of irreducible unitary representations of the groups $SU(2)$ and $SU(2)^n$ is described in different book and textbooks on representation theory (see, for example, D.P. Zhelobenko [1]).

2. The exposition scheme is parallel to Chapter 4, Section 4.2. The exposition follows Yu.S. Samiĭlenko and B.L. Tsigan.

3. The exposition is parallel to Chapter 4, Section 4.3.

4. The construction of irreducible representations of $SO(3)^{\infty}_{\pi(2)}$ given in this section follows Yu.S. Samoĭlenko [6].

Chapter 6.
REPRESENTATIONS OF THE GROUP OF UPPER TRIANGULAR MATRICES

6.1. The group $B_0(\mathbb{N}, \mathbb{R})$ and its completion. $B_0(\mathbb{N}, \mathbb{R})$-quasi-invariant measures.

Let $B(n, \mathbb{R})$ be the group of real upper triangular matrices of order n with unity on the main diagonal. The subgroup of matrices which have zeros in the last column above the main diagonal is isomorphic to the group $B(n-1, \mathbb{R})$. This allows to define the imbedding of the groups:

$$B(n-1, \mathbb{R}) \ni g \;\mapsto\; \begin{bmatrix} g & 0 \\ 0 & 1 \end{bmatrix} \in B(n, \mathbb{R}).$$

To the increasing sequence of imbedded groups $B(n, \mathbb{R})$, $n \geq 2$, there naturally corresponds the group $B_0(\mathbb{N}, \mathbb{R}) = \lim\limits_{\rightarrow} B(n, \mathbb{R})$ of infinite upper triangular matrices of the form $I + A$, where A is a finite strictly upper triangular matrix. The convergence of the sequence $g_n = I + A_n \rightarrow g = I + A$ in the inductive limit topology supposes that the matrices A_n are equi-finite and that the matrix elements converge.

If we denote the Lie algebra of the group $B(n, \mathbb{R})$ by $\mathbf{b}(n, \mathbb{R})$, then the Lie algebra $\mathbf{b}_0(\mathbb{N}, \mathbb{R})$ of the group $B_0(\mathbb{N}, \mathbb{R})$ is the inductive limit, $\mathbf{b}_0(\mathbb{N}, \mathbb{R}) = \lim\limits_{\rightarrow} \mathbf{b}(n, \mathbb{R})$. It is the Lie algebra of the infinite strictly upper triangular matrices having only a finite number of nonzero elements. The matrix units $(E_{jk})_{1 \leq j < k}$ for which the commutation relation

$$[E_{jk}, E_{lm}] = \delta_{kl}\, E_{jm}\,, \quad j < m$$

holds form a basis for the Lie algebra $\mathbf{b}_0(\mathbb{N}, \mathbb{R})$.

The exponential mapping $\mathbf{b}_0(\mathbb{N}, \mathbb{R}) \ni u \mapsto \exp u \in B_0(\mathbb{N}, \mathbb{R})$ is a homeomorphism between the group $B_0(\mathbb{N}, \mathbb{R})$ and its Lie algebra, and for the elements $t E_{jk} \in \mathbf{b}_0(\mathbb{N}, \mathbb{R})$ $(t \in \mathbb{R}^1, j < k)$, we have $\exp t E_{jk} = I + t E_{jk}$.

Introducing the product topology in the space $B_0(\mathbb{N}, \mathbb{R})$ and completing $B_0(\mathbb{N}, \mathbb{R})$ we get the set $B(\mathbb{N}, \mathbb{R})$ of all (without the finiteness condition) upper triangular matrices with unity on the main diagonal. $B(\mathbb{N}, \mathbb{R})$ is a topological group and matrix multiplication and taking inverses make sense because the matrices of $B(\mathbb{N}, \mathbb{R})$ are upper triangular. Its Lie algebra $\mathbf{b}(\mathbb{N}, \mathbb{R})$ is the Lie algebra of all real strictly upper triangular matrices, $B_0(\mathbb{N}, \mathbb{R})$ is a dense subgroup of $B(\mathbb{N}, \mathbb{R})$. Next we describe the class of "intermediate" Hilbert-Lie groups, $B_a(\mathbb{N}, \mathbb{R})$ $(B_0(\mathbb{N}, \mathbb{R}) \subset B_a(\mathbb{N}, \mathbb{R}) \subset B(\mathbb{N}, \mathbb{R}))$, which are separated by growth conditions on the matrix elements. For an arbitrary strictly upper triangular matrix $a = (\alpha_{jk})$ $(\alpha_{jk} > 0, j < k)$ (the weight matrix), on the Lie algebra $\mathbf{b}_0(\mathbb{N}, \mathbb{R})$ impose the Hilbert norm

$$u \mapsto \|u\|_a^2 = \sum_{j<k} u_{jk}^2 \alpha_{jk},$$

and consider the Hilbert space $\mathbf{b}_a(\mathbb{N}, \mathbb{R})$ which is a completion of $\mathbf{b}_0(\mathbb{N}, \mathbb{R})$ with respect to the norm $\|\cdot\|_a$. In this case the operation of matrix commutation cannot be extended in general to get a continuous operation on $\mathbf{b}_a(\mathbb{N}, \mathbb{R})$, so that $\mathbf{b}_a(\mathbb{N}, \mathbb{R})$ is not, in general, a Lie algebra.

LEMMA 1. (A.V. Kosyak [3]). The set $\mathbf{b}_a(\mathbb{N}, \mathbb{R})$ is an associative Hilbert algebra with respect to matrix multiplication if and only if

$$\alpha_{jk} \le c^2 \alpha_{jm} \alpha_{mk}, \quad j < m < k. \tag{1}$$

In this case $\mathbf{b}_a(\mathbb{N}, \mathbb{R})$ is a Lie algebra corresponding to the Hilbert-Lie group of matrices of the form $I + A$, $A \in \mathbf{b}_a(\mathbb{N}, \mathbb{R})$.

Let $\mathbf{U}$ denote the set of arbitrary weights, $a = (\alpha_{jk})$, and $\mathbf{U}_0$ the set of the weights which satisfy (1). Partially order $\mathbf{U}$ by setting for $a = (\alpha_{jk})$, $a' = (\alpha_{jk}')$, $a \le a'$ if for some $c > 0$, $\alpha_{jk} \le c\, \alpha_{jk}'$, $\forall j < k$. If $a < a'$, then there is a continuous imbedding $\mathbf{b}_{a'}(\mathbb{N}, \mathbb{R}) \hookrightarrow \mathbf{b}_a(\mathbb{N}, \mathbb{R})$. Indeed, in this case we have the inequality $\|\cdot\|_a \le \sqrt{c}\, \|\cdot\|_{a'}$.

PROPOSITION 1. The following relations hold

$$\mathbf{b}_0(\mathbb{N}, \mathbb{R}) = \bigcap_{a \in \mathbf{U}_0} \mathbf{b}_a(\mathbb{N}, \mathbb{R}) \subset \bigcup_{a \in \mathbf{U}_0} \mathbf{b}_a(\mathbb{N}, \mathbb{R}) = \mathbf{b}(\mathbb{N}, \mathbb{R}).$$

Proof. First of all note that if we replace $\mathbf{U}_0$ by $\mathbf{U}$, the given relations become obvious. So, it is sufficient to prove that for any weight $a \in \mathbf{U}$ there exist weights $a_0, a_1 \in \mathbf{U}_0$ such that $a_0 \le a \le a_1$. Let $a \in \mathbf{U}$. Construct the weight a_0 recursively by setting

$$\alpha_{jj+1}^0 = \alpha_{jj+1}$$

$$\alpha_{jj+k}^0 = \min(\alpha_{jj+k}, \min_{j<l<j+k} (\alpha_{jl}^0 \alpha_{lj+k}^0)) \quad j,k = 1,2,\ldots.$$

To construct the weight a_1 we set

$$\alpha_{1k}^1 = \alpha_{1k}$$

$$\alpha_{1k}^1 = \max(\alpha_{jk}, \max_{l<j} (\alpha_{lk}^1 / \alpha_{lj}^1)), \quad 2 \le j < k.$$

It is obvious from the construction that the weights a_0 and a_1 satisfy all the necessary requirements. □

We present one more property of the group $B_0(\mathbb{N}, \mathbb{R})$ that will be used later on. For an arbitrary n consider the subgroups of $B_0(\mathbb{N}, \mathbb{R})$, $B(n, \mathbb{R})$, H_n, $B_{0,n}(\mathbb{N}, \mathbb{R})$ formed by the corresponding matrices:

$$g_n = \begin{pmatrix} * & 0 \\ 0 & I \end{pmatrix}, \quad h_n = \begin{pmatrix} I & * \\ 0 & I \end{pmatrix}, \quad g^{(n)} = \begin{pmatrix} I & 0 \\ 0 & * \end{pmatrix}.$$

The matrices are devided into blocks such that the upper left block has the dimension

$n \times n$.

PROPOSITION 2. For any $n \in \mathbb{N}$, the group $B_0(\mathbb{N}, \mathbb{R})$ can be decomposed into the semi-direct product

$$B_0(\mathbb{N}, \mathbb{R}) = H_n \wedge (B(n, \mathbb{R}) \times B_{0,n}(\mathbb{N}, \mathbb{R})).$$

Proof. From the directly verified equality

$$\begin{bmatrix} A_n & H_n \\ 0 & A^{(n)} \end{bmatrix} = \begin{bmatrix} I & 0 \\ 0 & A^{(n)} \end{bmatrix} \begin{bmatrix} I & H_n \\ 0 & I \end{bmatrix} \begin{bmatrix} A_n & 0 \\ 0 & I \end{bmatrix}$$

it follows that any element of the group $B_0(\mathbb{N}, \mathbb{R})$ can be represented as the product $g = g^{(n)} h_n g_n$, $g^{(n)} \in B_{0,n}(\mathbb{N}, \mathbb{R})$, $h_n \in H_n$, $g_n \in B(n, \mathbb{R})$. To complete the proof it remains to show that H_n is a normal subgroup of $B_0(\mathbb{N}, \mathbb{R})$. Indeed, decompose g into the product $g = g^{(n)} h_n g_n$, then $g^{-1} = g_n^{-1} h_n^{-1} (g^{(n)})^{-1}$ and

$$\begin{bmatrix} A_n & H_n \\ 0 & A^{(n)} \end{bmatrix} \begin{bmatrix} I & H_n \\ 0 & I \end{bmatrix} \begin{bmatrix} A_n & H_n \\ 0 & A^{(n)} \end{bmatrix}^{-1} = \begin{bmatrix} I & A_n H_n' A^{(n)^{-1}} \\ 0 & I \end{bmatrix} \in H_n.$$

Remark 1. Because for $g_n \in B(n, \mathbb{R})$ and $g^{(n)} \in B_{0,n}(\mathbb{N}, \mathbb{R})$, $g_n g^{(n)} = g^{(n)} g_n$, we also have the decompositions

$$B_0(\mathbb{N}, \mathbb{R}) = (H_n \otimes B(n, \mathbb{R})) \otimes B_{0,n}(\mathbb{N}, \mathbb{R}) =$$

$$= (H_n \otimes B_{0,n}(\mathbb{N}, \mathbb{R})) \otimes B(n, \mathbb{R}).$$

Remark 2. The obvious analog of Proposition 2 holds for the groups $B_a(\mathbb{N}, \mathbb{R})$ and $B(\mathbb{N}, \mathbb{R})$ as well.

As the group $B_0(\mathbb{N}, \mathbb{R})$ is not locally compact, it does have neither a $B_0(\mathbb{N}, \mathbb{R})$-invariant measure (A. Weil [1]), nor a $B_0(\mathbb{N}, \mathbb{R})$-quasi-invariant measure (Xia Dao Xing [1]). This complicates the use of the standard constructions of representation theory in the study of the group $B_0(\mathbb{N}, \mathbb{R})$. However, it turns out that there exist Borel probability measures quasi-invariant relatively to the right action of $B_0(\mathbb{N}, \mathbb{R})$, on suitable completions of the group $B_0(\mathbb{N}, \mathbb{R})$, in particular, on the groups $B(\mathbb{N}, \mathbb{R})$ and $B_a(\mathbb{N}, \mathbb{R})$.

Define the right and left actions of the group $B_0(\mathbb{N}, \mathbb{R})$ on the Hilbert-Lie algebra $\mathbf{b}_a(\mathbb{N}, \mathbb{R})$ by setting for $g = I + u_0 \in B_0(\mathbb{N}, \mathbb{R})$, $u \in \mathbf{b}_a(\mathbb{N}, \mathbb{R})$

$$R_g(u) = ug = u(I + u_0) \in \mathbf{b}_a(\mathbb{N}, \mathbb{R})$$

$$L_g(u) = gu = (I + u_0)u \in \mathbf{b}_a(\mathbb{N}, \mathbb{R}).$$

On the Hilbert space $X = \mathbf{b}_a(\mathbb{N}, \mathbb{R})$, consider a Gaussian measure $\mu_c(\cdot)$ with a zero mean and a positive nuclear correlation operator C for which the matrix units are eigen-vectors:

$$C E_{jk} = \lambda_{jk} E_{jk}, \quad \lambda_{jk} > 0, \quad 1 \le j < k.$$

THEOREM 1. The measure $\mu_c(\cdot)$ is quasi-invariant with respect to the right action of $B_0(\mathbb{N}, \mathbb{R})$. The condition

$$\sum_{m=n+1}^{\infty} \frac{a_{km}}{\lambda_{km}} \frac{\lambda_{nm}}{a_{nm}} < \infty$$

for all k,n $(k<n)$ is necessary and sufficient for the measure $\mu_c(\cdot)$ to be quasi-invariant with respect to the left action of the group $B_0(\mathbb{N}, \mathbb{R})$.

Proof. Let us find the correlation operator of the measure $\mu_c^R(\cdot) = \mu_c(R_g^{-1}(\cdot))$

$$\int_X \exp(i(x,u))\, d\mu_c(R_{g^{-1}}(u)) = \int_X \exp((x, R_g(u)))\, d\mu_c(u) =$$

$$= \int_X \exp(i(R_g^* x, u))\, d\mu_c(u) = \exp(-\tfrac{1}{2}(R_g C R_g^* x, x).$$

So, under the transformation $R_{g^{-1}} : X \to X$ the measure $\mu_c(\cdot)$ goes into the measure $\mu_{R_g C R_g^*}(\cdot)$. Similarly, under the transformation $L_{g^{-1}} : X \to X$ the measure $\mu_c(\cdot)$ transforms into the measure $\mu_{L_g C L_g^*}(\cdot)$. According to the Gajek-Feldman's criterion (see Chapter 3), Gaussian measures with correlation operators C and C_1 are equivalent if and only if $C^{-1/2} C_1 C^{-1/2} - I$ is a Hilbert-Schmidt operator. There, it is assumed that $C^{-1/2} C_1 C^{-1/2}$ is invertible. Introduce the operators $r_{u_0} u = u_0 u$, $l_{u_0} u = u u_0$, then $R_g = I + r_{u_0}$, $L_g = I + l_{u_0}$. We have that

$$C^{-1/2} R_g C R_g^* C^{-1/2} = C^{-1/2}(I + r_{u_0}) C (I + r_{u_0}^*) C^{-1/2} =$$

$$= I + C^{-1/2} r_{u_0} C^{1/2} + (C^{-1/2} r_{u_0} C^{1/2})^* + (C^{-1/2} r_{u_0} C^{1/2})(C^{-1/2} r_{u_0} C^{1/2})^*,$$

$$C^{-1/2} L_g C L_g C^{-1/2} =$$

$$= I + C^{-1/2} l_{u_0} C^{1/2} + (C^{-1/2} l_{u_0} C^{1/2})^* + (C^{-1/2} l_{u_0} C^{1/2})(C^{-1/2} l_{u_0} C^{1/2})^*$$

are invertible.

Let $u_0 = t E_{jk}$. The Hilbert-Schmidt norm of the operator $C^{-1/2} r_{u_0} C^{1/2}$,

$$| C^{-1/2} r_{u_0} C^{1/2} |^2 = \sum_{\substack{l<m \\ p<q}} |(C^{-1/2} r_{u_0} C^{1/2} E_{lm}, E_{pq})|^2 =$$

$$= \sum_{\substack{l<m \\ p<q}} \left| \left(\frac{\lambda_{lm}}{\lambda_{pq}} \right)^{1/2} (r_{u_0} E_{lm}, E_{pq}) \right|^2 =$$

$$= \sum_{\substack{l<m \\ p<q}} \left| \left(\frac{\lambda_{lm}}{\lambda_{pq}} \right)^{1/2} + \delta_{jm} \left(\frac{a_{lk}}{a_{lm}} \right)^{1/2} (E_{lk}, E_{pq}) \right|^2 =$$

$$= \sum_{l=1}^{m-1} t^2 \frac{\lambda_{lm}\, a_{lk}}{\lambda_{lk}\, a_{lm}} < \infty.$$

We also have that $\| C^{-1/2} R_g C R_g^* C^{-1/2} - I \| < \infty$ which means that the measure is quasi-invariant with respect to the right action.

Similarly, we have for the left action:

$$\| C^{-1/2} l_{u_0} C^{1/2} \|^2 = t^2 \sum_{m=k+1}^{\infty} \frac{\lambda_{km}}{\lambda_{jm}} \frac{a_{jm}}{a_{km}},$$

and thus the condition of the theorem is sufficient for the measure $\mu_c(\cdot)$ to be quasi-invariant. For $U_0 = E_{jk}$ we find after some calculations that

$$\| C^{-1/2} l_{u_0} C^{1/2} + (C^{-1/2} l_{u_0} C^{1/2})^* + (C^{-1/2} l_{u_0} C^{1/2})(C^{-1/2} l_{u_0} C^{1/2})^* \|^2 =$$

$$= 2 \sum_{m=k+1}^{\infty} \frac{\lambda_{km}}{\lambda_{jm}} \frac{a_{jm}}{a_{km}} + \sum_{m=k+1}^{\infty} \left[\frac{\lambda_{km}}{\lambda_{jm}} \frac{a_{jm}}{a_{km}} \right]^2 > 2 \sum_{m=k+1}^{\infty} \frac{\lambda_{km}}{\lambda_{jm}} \frac{a_{jm}}{a_{km}}.$$

From this it follows that the conditions are necessary. □

COROLLARY 1. There exists a $B_0(\mathbb{N}, \mathbb{R})$-right quasi-invariant measure on any Hilbert Lie group $B_a(\mathbb{N}, \mathbb{R})$.

Indeed, one can take the image of the measure on $\mathrm{b}_a(\mathbb{N}, \mathbb{R})$ under the mapping $\mathrm{b}_a(\mathbb{N}, \mathbb{R}) \ni u \mapsto I + u \in B_a(\mathbb{N}, \mathbb{R})$.

Remark 3. Each right quasi-invariant measure on a finite-dimensional Lie group is equivalent to a left quasi-invariant measure. In the case of the group $B_0(\mathbb{N}, \mathbb{R})$ this is not so: there exist measures on $B_a(\mathbb{N}, \mathbb{R})$ which behave well under the right action but not under the left action.

6.2. Commutative models for representations of the group $B_0(\mathbb{N}, \mathbb{R})$.

In what follows we construct commutative models of unitary representations of the group $B_0(\mathbb{N}, \mathbb{R})$, i.e. we give the form of the operators of the representations of the group in the space of Fourier-images, which are constructed on the basis of a commutative family of operators of the representation of a given commutative subgroup. The term "commutative models" was first introduced in A.M. Vershik, I.M. Gel'fand and M.I. Graev [6]. Here as the commutative subgroup we choose $\mathbf{H}_n$.

THEOREM 2. Let $B_0(\mathbb{N}, \mathbb{R}) \ni g \mapsto T_g$ be a strongly continuous unitary representation of the group $B_0(\mathbb{N}, \mathbb{R})$ on a Hilbert space H. The space H can be decomposed into the direct integral

$$H = \oplus \int_{\mathfrak{h}_n'} H_x \, d\mu(x)$$

so that the representation T has the form

$$(T_{(\exp h)} f)\,(x) = \exp(i(h,x))\,f(x)\,, \quad \exp h \in H_n\,;$$

$$(T_g f)\,(x) = \psi(g,x)\sqrt{\frac{d\mu(Ad_g^*(x))}{d\mu(x)}}\; f(Ad_g^*(x))\,,$$

$$g \in B(n,\ \mathbb{R}) \times B_{0,n}(\mathbb{N},\ \mathbb{R}).$$

Here $\mu(\cdot)$ is a $B_0(\mathbb{N},\ \mathbb{R})$-quasi-invariant probability measure on the space $\mathbf{h}_n'$ adjoint to the Lie algebra $\mathbf{h}_n$ of the group $H_n \subset B_0(\mathbb{N},\ \mathbb{R})$ and $\psi(\cdot,\cdot)$ is a weakly measurable unitary operator-valued function satisfying

$$\psi(e,x) = I$$

$$\psi(g_1 g_2, x) = \psi(g_1, x)\,\psi(g_2, Ad_{g_1}^*(x))$$

for all $g_1, g_2 \in B_0(\mathbb{N},\ \mathbb{R})$ and almost all $x \in \mathbf{h}_n'$ with respect to $\mu(\cdot)$. The dimension function $x \mapsto \dim H_x$ is constant on the orbits of the group $B_0(\mathbb{N},\ \mathbb{R})$ in $\mathbf{h}_n'$.

Proof. Let us fix an arbitrary n. The group H_n is isomorphic to the additive group of the nuclear space $\mathbb{R}_0^\infty$.

First, suppose that the family of operators $(T_{(\exp h)}\,,\ h \in \mathbf{h}_n)$ is cyclic, and $\Omega \in H\,, \|\Omega\| = 1$ is a corresponding cyclic vector. In this case the family of operators $(T_{(\exp h)}\,,\ h \in \mathbf{h}_n)$ is unitarily equivalent to the family of operators of multiplication by $\exp(i(h,x))\,,\ h \in \mathbf{h}_n$ in the space $L_2(\mathbf{h}_n', d\mu(x))$ ($\mu(\cdot)$ is a spectral measure of the family $(T_{(\exp h)}\,,\ h \in \mathbf{h}_n)$. For the total set of vectors $f_h \in H$ of the form

$$f_h = T_{(\exp h)}\,\Omega = \exp(i(h,x))\,,$$

it follows from the group law that

$$T_g f_h = T_g\,T_{(\exp h)}\,\Omega = T_g\,T_{(\exp h)}\,T_g^*\,T_g\,\Omega =$$

$$= T_{(g \exp h g^{-1})}\,T_g\,\Omega.$$

Because H_n is a normal subgroup, $g \exp h\, g^{-1} = \exp Ad_g(h) \in H_n$ and

$$(T_g f_h)\,(x) = \exp(i(Ad_g(h), x))\,\alpha(g,x) =$$

$$= \exp(i(h,\ Ad_g^*(x)))\,\alpha(g,x) = \alpha(g,x)\,f_h(Ad_g^*(x))$$

where $\alpha(g,x) = T_g\,\Omega$. Using linearity and continuity, we can extend the last equality to the whole space $L_2(\mathbf{h}_n', d\mu(\cdot))$.

We show that the spectral measure $\mu(\cdot)$ is quasi-invariant with respect to the group action of $B_0(\mathbb{N},\ \mathbb{R})$. Indeed,

$$k(h) = (T_{(\exp h)}\,\Omega,\ \Omega)\,, \quad h \in \mathbf{h}_n$$

is the Fourier transform of the measure $\mu(\cdot)$. Because the operators T_g are unitary

$$k(h) = (T_{(\exp h)}\,\Omega, \Omega) = (T_g\, T_{(\exp h)}\,\Omega, T_g\,\Omega) =$$

$$= (T_{(\exp Ad_g(h))}\, T_g\,\Omega, T_g\,\Omega) = \int_{\mathrm{h}_n'} \exp(i(h, Ad_g^*(x)))\ |\alpha(g,x)|^2\, d\mu(x) =$$

$$= \int_{\mathrm{h}_n'} \exp(i(h,x))\ |\alpha(g, Ad_{g^{-1}}^*(x))|^2\, d\mu(Ad_{g^{-1}}^*(x))$$

and so from the uniqueness of the Fourier transform we find that the measure $\mu(\cdot)$ is $B_0(\mathbb{N}, \mathbb{R})$-quasi-invariant, and $\dfrac{d\mu(Ad_g^*(x))}{d\mu(x)} = |\alpha(g,x)|^2$. Setting $\psi(g,x) = \dfrac{\alpha(g,x)}{|\alpha(g,x)|}$ we derive

$$(T_g f)(x) = \psi(g,x)\ |\alpha(g,x)|\ f(Ad_g^*(x)) =$$

$$= \psi(g,x) \sqrt{\frac{d\mu(Ad_g^*(x))}{d\mu(x)}}\ f(Ad_g^*(x)).$$

From $T_{g_1 g_2} = T_{g_1} T_{g_2}$ we find the relations for $\psi(\cdot,\cdot)$

$$\psi(g_1 g_2, x) = \psi(g_1, x)\, \psi(g_2, Ad_{g_1}^*(x))$$

which hold for all $g_1, g_2 \in B_0(\mathbb{N}, \mathbb{R})$ and almost all $x \in \mathrm{h}_n'$ with respect to $\mu(\cdot)$.

In the case of a multiple spectrum of the family $(T_{(\exp h)}, h \in f_n)$ we follow the proof of the theorem on the commutative models for representation of the CCR (see Chapter 4).

6.3. Continuous extension of representations of $B_0(\mathbb{N}, \mathbb{R})$.

In this section we show that any strongly continuous unitary representation of the group $B_0(\mathbb{N}, \mathbb{R})$ can be continuously extended to a representation of a Hilbert-Lie group $B_a(\mathbb{N}, \mathbb{R})$ which depends on the representation. The extensions of representations will be used to construct the Gårding domain for representations of the group $B_0(\mathbb{N}, \mathbb{R})$.

THEOREM 3. Any strongly continuous unitary representation of the group $B_0(\mathbb{N}, \mathbb{R})$ can be continuously extended to a strongly continuous unitary representation of a Hilbert-Lie group $B_a(\mathbb{N}, \mathbb{R})$ where a depends on the representation.

Proof. Let us look at the convergence of infinite products of unitary operators.

Let $(U_n(t))_{n\in \mathbb{N}}$ be an arbitrary family of strongly continuous unitary one-parameter groups in H. On $\mathbb{R}_0^\infty$ we define an operator-valued function

$$\mathbb{R}_0^\infty \ni t \;\mapsto\; U_1(t_1)\, U_2(t_2) \cdots U_n(t_n) = U(t).$$

Here $t = (t_1, \ldots, t_n, 0, \ldots) \in \mathbb{R}_0^\infty$. The function $U(t)$ is continuous on $\mathbb{R}_0^\infty$.

LEMMA 2. The function $\mathbb{R}_0^\infty \ni t \mapsto U(t)$ can be continuously extended to a function that is continuous at zero and defined on a topological subspace $R(U)$ which properly contains $\mathbb{R}_0^\infty$.

Proof. Consider an arbitrary sequence of unitary operators $(U_n)_{n\in \mathbb{N}}$. Suppose that for $\forall f \in H$, the series

$$\sum_{n=1}^{\infty} \|(U_{n+1}-U_n)f\| = \sum_{n=1}^{\infty} \|(U_n^{-1}U_{n+1}-I)f\|$$

and

$$\sum_{n=1}^{\infty} \|(U_{n+1}^{-1}-U_n^{-1})f\| = \sum_{n=1}^{\infty} \|(U_n U_{n+1}^{-1}-I)f\|$$

converge.
Then there exists the limiting unitary operator $U = s-\lim U_n$. Indeed, for all $f \in H$

$$\|(U_{m+p}-U_m)f\| = \|\sum_{n=m}^{m+p-1}(U_{n+1}-U_n)f\| \le \sum_{n=m}^{m+p-1}\|(U_{n+1}-U_n)f\|$$

and so the sequence $(U_n f)_{n\in \mathbb{N}}$ is fundamental, and we can define $Uf = s-\lim U_n f$. The map $f \mapsto Uf$ is a linear unitary operator. The linearity is clear. Also U is isometric, $\|Uf\| = \lim_{n\to\infty} \|U_n f\| = \|f\|$. Hence, it is sufficient to show that $\mathbf{R}(U) = H$. But the second formula in (2) implies that the operator $U^{-1} = s-\lim U_n^{-1}$ exists.

It should be noted that it is sufficient to require that the conditions (2) only hold on a dense subset T in H but not on all of H. Indeed, in this case the isometric operator $Uf = \lim U_n f$ $(f\in T)$ is defined on T. So U can be continuously extended to the whole space H. Doing the same for the operator U^{-1}, we find that the operator U is unitary.

Fix a countable dense set $T = (f_m)_{m=1}^{\infty}$ in H and $t^{(n)} = (t_1, \ldots, t_n, 0, \ldots) \in \mathbb{R}_0^\infty$ for $t = (t_1, \ldots, t_n, t_{n+1}, \ldots) \in \mathbb{R}^\infty$. According to what has been said before, for the existence of the unitary operator $U(t) = s-\lim U(t^{(n)})$ it is sufficient that the conditions

$$\gamma_{1,m}^{(t)} = \sum_{n=0}^{\infty} \|(U(t^{(n+1)}) - U(t^{(n)}))f_m\| < \infty$$

$$\gamma_{2,m}^{(t)} = \sum_{n=0}^{\infty} \|(U^{-1}(t^{(n+1)}) - U^{-1}(t^{(n)}))f_m\| < \infty$$

hold for all $m \in \mathbb{N}$.

Define a topological space

$$R(U) = \{t \in \mathbb{R}^\infty \mid \gamma_{k,m}^{(t)} < \infty,\ k=1,2;\ m \in \mathbb{N}\}.$$

Its topology is generated by the family of sets

$$W(t,k,m,\varepsilon) = \{s \in R(U) \mid \gamma_{k,m}(t-s) < \varepsilon\ (t\in R(U);\ k=1,2;\ m\in \mathbb{N};\ \varepsilon<0)\}.$$

It is clear that $R(U)$ contains $\mathbb{R}_0^\infty$, and that the mapping $t \mapsto U(t)$ is continuous in the topology defined on $R(U)$. □

Now we go back to the proof of the theorem. For $t = I + \sum_{k<n} t_{kn} E_{kn}$, we denote

$$t^{(n)} = I + \sum_{r<p\leq n} t_{rp} E_{rp}\,,\ t_n = I + \sum_{r=1}^{n-1} t_{rn} E_{rn}\,,\ t_{kn} = I + t_{kn} E_{kn}\,,\ t_1 = I.$$

Direct calculation shows that

$$t^{(n)} = t_n t_{n-1} \cdots t_2 t_1\,,\quad n = 1,2,\ldots.$$

Note that for a fixed n the set of the matrices t_n is isomorphic to the additive group $\mathbb{R}^{n-1}$. Denote the generators of the one-parameter groups $U(t_{kn})$ by A_{kn}. Because $B_0(\mathbb{N}, \mathbb{R})$ is the inductive limit of finite-dimensional Lie groups, there exists a dense set Φ in H which consists of infinitely differentiable vectors for the family of operators $A_{kn} (k<n)$ (see P. Richter [1]). Choose a countable dense set $T = (f_m)_{m=1}^{\infty} \subset \Phi$ in H. To prove the theorem it remains to show that in the set $R(U)$ constructed for T there exists a Hilbert-Lie algebra $B_a(\mathbb{N}, \mathbb{R})$. As in the proof of Proposition 1, it is easy to construct such a weight $a = (a_{jk}) \in \mathrm{U}_0$ so that for all m

$$B_m^2 = \sum_{k<n} \|A_{k,n} f_m\|^2 a_{kn}^{-1} < \infty.$$

Then we have

$$\gamma_{1,m}(t) = \sum_{n=1}^{\infty} \|(U((t_{n+1})^{-1}) - I) f_m\| \leq \sum_{n=2}^{\infty} \|(\sum_{k=1}^{n-1} t_{kn} A_{kn}) f_m\| \leq$$

$$\leq \sum_{n=2}^{\infty} \sum_{k=1}^{n-1} |t_{kn}|\, \|A_{kn} f_m\| \leq (\sum_{k<n} |t_{kn}|^2 a_{kn})^{1/2} (\sum_{k<n} \|A_{kn} f_m\|^2 a_{kn}^{-1})^{1/2} \leq$$

$$\leq \|t - I\|_a \cdot B_m < \infty\,,$$

$$\gamma_{2,m}(t) = \sum_{n=1}^{\infty} \|(U((t^{(n)})^{-1} t^{(n+1)}) - I) f_m\| \leq$$

$$\leq \sum_{n=2}^{\infty} \|(\sum_{k=1}^{n-1} \omega_{kn} A_{kn}) f_m\| \leq \sum_{n=2}^{\infty} \sum_{k=1}^{n-1} |\omega_{kn}|\, \|A_{kn} f_m\| \leq$$

$$\leq \|t^{-1} - I\|_a B_m < \infty$$

(here $t^{-1} = I + \sum_{j<k} \omega_{jk} E_{jk}$). So $B_a(\mathbb{N}, \mathbb{R}) \subset R(U)$ and the representation U can be continuously extended to a representation of the group $B_a(\mathbb{N}, \mathbb{R})$. □

6.4. Gårding domains for representations of the group $B_0(\mathbb{N}, \mathbb{R})$

Using the scheme of Chapter 3 to construct a Gårding domain for representations of $B_0(\mathbb{N}, \mathbb{R})$, we need to replace the space $\mathrm{A}(\mathbb{R}^{\infty})$ by the space $S_\alpha^\beta(B_a(\mathbb{N}, \mathbb{R}))$ which is an analogue of the Gel'fand-Shilov space $S_\alpha^\beta(\mathbb{R}^n)$. (see I.M. Gel'fand, G.E. Shilov [1]). This necessity arises because the space

$$D = \{ \int_{B_a(\mathbb{N},\mathbb{R})} f(g)\, T_g\, \phi_0\, d\mu(g) \mid f \in \mathbf{A}(\mathbb{R}^\infty)\}$$

($\mu(\cdot)$ is a Gaussian measure on $B_a(\mathbb{N}, \mathbb{R})$ see Section 6.1) is not invariant with respect to the operators of the representation.

We introduce the space

$$S_\alpha^\beta(B_a(\mathbb{N}, \mathbb{R}))\ (\alpha = (\alpha_{km})_{k<m},\ \beta = (\beta_{km})_{k<m},\ \alpha_{km} = 1/4,\ \beta_{km} = 3/4)$$

as follows. Consider the real Banach space

$$\mathbf{b}_a^{(4)} = \{u = \sum_{j<k} u_{jk} E_{jk} \mid \|u\|^4_{\mathbf{b}_a^{(4)}} = \sum_{j<k} |u_{jk}|^4 a_{jk}^2 < \infty\}.$$

It is clear that $\mathbf{b}_a^{(4)} \supset \mathbf{b}_a(\mathbb{N}, \mathbb{R})$. Denote by $\mathbf{b}_{a,C}^{(4)}$ the complexification of $\mathbf{b}_a^{(4)}$. Let $H(\mathbf{b}_{a,C}^{(4)})$ be the space of entire functions on $\mathbf{b}_{a,C}^{(4)}$ (see L. Nachbin [1]). Define the set

$$S_\alpha^\beta(\mathbf{b}_{a,C}^{(4)}) = \{f \in H(\mathbf{b}_{a,C}^{(4)}) \mid |f(u+iv)| \le c \exp(-a\|u\|^4_{\mathbf{b}_a^{(4)}} + b\|v\|^4_{\mathbf{b}_a^{(4)}}\}.$$

Next we introduce the topology in the constructed space. For $A,B > 0$ we set

$$S_{\alpha,A}^{\beta,B}(\mathbf{b}_{a,C}^{(4)}) = \{f \in S_\alpha^\beta(\mathbf{b}_{a,C}^{(4)}) \mid |f(u+iv)| \le$$

$$\le C \exp(-(A-\varepsilon)\|u\|^4_{\mathbf{b}_a^{(4)}} + (B+\delta)\|v\|^4_{\mathbf{b}_a^{(4)}}),\ \varepsilon, \delta > 0\}.$$

The topology in $S_{\alpha,A}^{\beta,B}(\mathbf{b}_{a,C}^{(4)})$ is given by the countable family of the norms

$$\|f\|_{\alpha,A,p}^{\beta,B,q} = \sup_{u+iv \in \mathbf{b}_{a,C}^{(4)}} |f(u+iv)| \exp((A - \frac{1}{p})\|u\|^4_{\mathbf{b}_a^{(4)}} - (B + \frac{1}{q})\|v\|^4_{\mathbf{b}_a^{(4)}})$$

$$p,q = 1,2,\dots.$$

For $A > A'$, $B < B'$ $S_{\alpha,A}^{\beta,B}(\mathbf{b}_{a,C}^{(4)}) \subset S_{\alpha,A'}^{\beta,B'}(\mathbf{b}_{a,C}^{(4)})$ and the topology in $S_\alpha^\beta(\mathbf{b}_{a,C}^{(4)})$ is the inductive limit generated by the spaces $S_{\alpha,A}^{\beta,B}(\mathbf{b}_{a,C}^{(4)})$. Finally, set $S_\alpha^\beta(\mathbf{b}_a(\mathbb{N}, \mathbb{R})) = S_\alpha^\beta(\mathbf{b}_{a,C}^{(4)}) \restriction \mathbf{b}_a(\mathbb{N}, \mathbb{R})$. Introduce $S_\alpha^\beta(B_a(\mathbb{N}, \mathbb{R}))$ as the image of $S_\alpha^\beta(\mathbf{b}_a(\mathbb{N}, \mathbb{R}))$ under the mapping $\pi : \mathbf{b}_a(\mathbb{N}, \mathbb{R}) \ni u \mapsto I + u \in B_a(\mathbb{N}, \mathbb{R})$.

Let there be given a strongly continuous unitary cyclic representation T of the group $B_0(\mathbb{N}, \mathbb{R})$ with cyclic vector $\phi_0 \in H$, $\|\phi_0\| = 1$. According to the results of the preceding section, this representation can be extended to a representation of a Hilbert-Lie group, $B_a(\mathbb{N}, \mathbb{R})$, dependent on the representation. On the group $B_a(\mathbb{N}, \mathbb{R})$ let there be constructed a $B_0(\mathbb{N}, \mathbb{R})$-quasi-invariant Gaussian measure $\mu(\cdot)$.

THEOREM 4. The set

$$D = \{ \int_{B_a(\mathbb{N},\mathbb{R})} f(g)\, T_{g^{-1}}\, \phi_0\, d\mu(g) \mid f \in S_\alpha^\beta(B_a(\mathbb{N}, \mathbb{R}))\}$$

has the following properties:

a) D is a linear topological space;

b) D is densely topologically imbedded in H;

c) T_g acts continuously from D into D , $g \in B_0(\mathbb{N}, \mathbb{R})$;

d) D is contained in the domain of the operators of the infinitesimal representation dT and is invariant under the action of these operators;

e) D consists of the analytic vectors for the representation T.

Proof. The topology in D is introduced via the mapping

$$S_\alpha^\beta(B_a(\mathbb{N}, \mathbb{R})) \ni f \mapsto \int_{B_a(\mathbb{N}, \mathbb{R})} f(g) T_{g^{-1}} \phi_0 \, d\mu(g) \in D.$$

Because $S_\alpha^\beta(B_a(\mathbb{N}, \mathbb{R}))$ is dense in $L_2(B_a(\mathbb{N}, \mathbb{R}), d\mu(\cdot))$ it follows that D is dense in H.

We show that $T_{g_0} : D \to D$ is continuous. Because the measure $\mu(\cdot)$ is $B_0(\mathbb{N}, \mathbb{R})$-quasi-invariant, we have

$$T_{g_0} \int_{B_a(\mathbb{N}, \mathbb{R})} f(g) T_{g^{-1}} \phi_0 \, d\mu(g) = \int_{B_a(\mathbb{N}, \mathbb{R})} f(g) T_{(gg_0^{-1})^{-1}} \phi_0 \, d\mu(g) =$$

$$= \int_{B_a(\mathbb{N}, \mathbb{R})} f(gg_0) \frac{d\mu(gg_0)}{d\mu(g)} T_{g^{-1}} \phi_0 \, d\mu(g).$$

It follows from the last formula that the invariance of D relatively to T_{g_0} is equivalent to the continuity in $S_\alpha^\beta(B_a(\mathbb{N}, \mathbb{R}))$ of the operator $U_{g_0} : f(g) \mapsto \frac{d\mu(gg_0)}{d\mu(g)} f(gg_0)$. This operator acts only on a finite number of variables, and so we can consider the same (corresponding) operator in $S_\alpha^\beta(B(n, \mathbb{R}))$.

Similarly, the problem of proving that D is invariant relatively to the operators of the representation dT can be reduced to showing that $S_\alpha^\beta(B_a(\mathbb{N}, \mathbb{R}))$ is invariant with respect to the operators

$$\frac{d}{dt} U(\exp t E_{jk}) \mid_{t=0} = \sum_{l=1}^{k-1} (u_{lk} \frac{\partial}{\partial u_{lk}} - 2 \frac{a_{lj}}{\lambda_{lj}} u_{lk} u_{lj}).$$

But because the last operator acts on a finite number of variables, it is equivalent to showing that the space $S_\alpha^\beta(B(n, \mathbb{R}))$ admits this invariance property. □

6.5. Irreducible representations of the group $B(\mathbb{N}, \mathbb{R})$.

To describe all irreducible representations of the group $B_0(\mathbb{N}, \mathbb{R})$ is a very difficult task. In particular, it contains the problem to describe all irreducible representations of the CCR of systems with infinite degree of freedom. One condition to get a class of irreducible representations that could be dealt with is the condition of continuity in the projective limit topology. Such representations can be continuously extended to representations of the group $B(\mathbb{N}, \mathbb{R})$.

We can get an irreducible representation of the group $B(\mathbb{N}, \mathbb{R})$ if we know an irreducible representation $B(n, \mathbb{R}) \ni g \mapsto U_g^{(n)}$ of the group $B(n, \mathbb{R})$. Indeed,

$$B(\mathbb{N}, \mathbb{R}) = (H_n \wedge B_n(\mathbb{N}, \mathbb{R})) \wedge B(n, \mathbb{R}),$$

$B(n, \mathbb{R})$ is a factor group, $B(n, \mathbb{R}) = B(\mathbb{N}, \mathbb{R}) / (H_n \wedge B_n(\mathbb{N}, \mathbb{R}))$. The formula

$$U_g = U^{(n)}_{p_n(g)} \qquad (g \in B(\mathbb{N}, \mathbb{R})) \tag{3}$$

($p_n : B(\mathbb{N}, \mathbb{R}) \to B(n, \mathbb{R})$ is the canonical projection from the group into the factor group) defines an irreducible representation of the group $B(\mathbb{N}, \mathbb{R})$. Note that the kernel of such a representation has finite codimension (it contains $H_n \wedge B_n(\mathbb{N}, \mathbb{R})$).

THEOREM 5. Every irreducible unitary representation of the group $B(\mathbb{N}, \mathbb{R})$ is unitarily equivalent to a representation of the form (3) for some n.

Proof. To prove the theorem we show that any irreducible representation of the group $B(\mathbb{N}, \mathbb{R})$ is trivial on a subgroup $H_n \otimes B_n(\mathbb{N}, \mathbb{R})$ for some $n \in \mathbb{N}$. Let $B(\mathbb{N}, \mathbb{R}) \ni g \mapsto U_g$ be an irreducible unitary representation of the group $B(\mathbb{N}, \mathbb{R})$ on H. According to Theorem 2, the spectral measure $\mu(\cdot)$ of the family of operators $\{U(\exp h) \mid h \in \mathbf{h}_1\}$ is quasi-invariant and ergodic with respect to the action of the group $B_0(\mathbb{N}, \mathbb{R})$.

LEMMA 3. The measure $\mu(\cdot)$ has its support in the measurable set $\mathbb{R}_0^\infty \subset \mathbb{R}^\infty$.

Proof. The Fourier transform of the measure $\mu(\cdot)$ is the function $k(h) = (U(\exp h)\Omega, \Omega)$, where Ω is a vector of maximal spectral type for the family of operators $\{U_{(\exp h)} \mid h \in \mathbf{h}_1\}$. Because U is a representation of the group $B(\mathbb{N}, \mathbb{R})$, $k(\cdot)$ is a continuous function on $\mathbf{h}_1 = \mathbb{R}^\infty$. But $\mathbb{R}^\infty$ with the projective limit topology induced by finite-dimensional subspaces is a nuclear space, so from Minlos theorem it follows that the function $k(\cdot)$ is the Fourier transform of some measure $\nu(\cdot)$ on $\mathbb{R}_0^\infty$ which is adjoint to $\mathbb{R}^\infty$. But the uniqueness of the Fourier transform implies that $\mu(\cdot) = \nu(\cdot)$ and $\mu(\mathbb{R}_0^\infty) = 1$. $\square$

Next we show that the measure $\mu(\cdot)$ has its support on a finite-dimensional subspace $\mathbb{R}^n \subset \mathbb{R}_0^\infty$, i.e.

$$\mu(\underbrace{\mathbb{R}^1 \times \cdots \times \mathbb{R}^1}_{n \text{ times}} \times \{0\} \times \cdots) = 1$$

for some n. Indeed, for an arbitrary n consider the subspace

$$\mathbb{R}^n \subset \mathbb{R}_0^\infty, \quad \mathbb{R}^n = \underbrace{\mathbb{R}^1 \times \cdots \times \mathbb{R}^1}_{n \text{ times}} \times \{0\} \times \ldots..$$

These subspaces are invariant with respect to the action of the group $B_0(\mathbb{N}, \mathbb{R})$. Because the measure $\mu(\cdot)$ is ergodic, the measure of such sets can be either zero or one. On the other hand $\mathbb{R}_0^\infty = \bigcup_{n=1}^{\infty} \mathbb{R}^n$ and $\mu(\mathbb{R}_0^\infty) = 1$, so $\mu(\mathbb{R}^n) = 1$ for some n.

Using the representation U of the group $B_0(\mathbb{N}, \mathbb{R})$, we construct the representation $u \mapsto A(u)$ of the Lie algebra $\mathbf{b}_0(\mathbb{N}, \mathbb{R})$ by unbounded self-adjoint operators on a dense Gårding domain D in H. Set $A_{jk} = A(E_{jk})$ $(j<k)$. Because the spectral measure of the family of operators $\{A_{1k} \mid k \geq 2\}$ has its support on some subspace $\mathbb{R}^n$, only a finite number of the operators A_{1k}, $k = 2,3,...$ is different from the zero operator. Similarly, using the commutative model of the representation with respect to the subgroup H_j we find that for any $j = 1,2,...$ only a finite number of the operators A_{jk}, $k > j$ is different from the zero operator. To prove the theorem it remains to show that for some n, $A_{jk} = 0$ for $j > n$. Let A_{k,l_k} be the last nonzero operator among the operators $(A_{kl})_{l=k+1}^{\infty}$ of the representation of the k-th row. If all $A_{kl} = 0$, we set $A_{kl_k} = A_{kk+1} = 0$. Consider the following sequence of the operators A_{jk}. Set $B_1 = A_{1l_1}$, $B_k = A_{kl_k}$ if $A_{k-1l_k} = 0$ and $B_k = A_{kl_k+1}$ if $A_{k-1l_k} \neq 0$. The family $(B_k)_{k=1}^{\infty}$ is a family of commuting skew-adjoint operators. As in Lemma 3, the spectral measure $\mu(\cdot)$ of this family has its support on $\mathbb{R}_0^\infty$. On the other hand, every operator from $(B_k)_{k=1}^{\infty}$ commutes with the irreducible representation U as it follows from the construction. So $B_k = \lambda_k I$, i.e. the spectral measure $\mu(\cdot)$ takes a nonzero value in a single point $\lambda \in \mathbb{R}_0^\infty$ and so starting with some n, $B_k = 0$, $k > n$. But then $A_{k,l_k} = 0$, $k > n+1$. □

It is known (see A.A. Kirillov [1]) that the irreducible unitary representations of the group $B(n, \mathbb{R})$, $n = 2,3,...$ can be described in terms of the orbits of the group $B(n, \mathbb{R})$ in the coadjoint representation. On the other hand every orbit in $\mathbf{b}'(\mathbb{N}, \mathbb{R})$ of the group $B(\mathbb{N}, \mathbb{R})$ coincides with an orbit of some group $B(n, \mathbb{R})$.

COROLLARY 2. There is a one-to-one correspondence between unitary representations of the group $B(\mathbb{N}, \mathbb{R})$ and its orbits in $\mathbf{b}'(\mathbb{N}, \mathbb{R})$. □

6.6. A class of exact irreducible representations of the group $B_0(\mathbb{N}, \mathbb{R})$.

In this chapter we construct a class of exact irreducible representations of the group $B_0(\mathbb{N}, \mathbb{R})$ and study the question of their extension to the Hilbert-Lie groups $B_a(\mathbb{N}, \mathbb{R})$.

Fix an arbitrary $\lambda \in \mathbb{R}^1$ $(\lambda \neq 0)$ and $n = 2,3 \cdots$ and construct a functional $\phi_\lambda \in \mathbf{b}'(n, \mathbb{R})$

$$\mathbf{b}(n, \mathbb{R}) \ni u \mapsto \lambda u_{1n}, \quad u = (u_{jk})_{j<k\leq n}.$$

Construct an irreducible unitary representation S_n^λ of the group $B(n, \mathbb{R})$ corresponding to the orbit which contains ϕ_λ. The representation S_n^λ can be realized in the space $L_2(\mathbb{R}^{n-2}, dx)$, $x = (x_2, \ldots, x_{n-1})$ using the formulas:

$$(S_n^\lambda(\exp t E_{1k})f)(x) = f(x_2, \ldots, x_k + t, \ldots, x_{n-1})$$

$$(S_n^\lambda(\exp t E_{1n})f)(x) = e^{it\lambda} f(x)$$

$$(S_n^\lambda(\exp t E_{jk})f)(x) = f(x_2, .., x_k + tx_j, \dots, x_{n-1})$$

$$(S_n^\lambda(\exp t E_{kn})f)(x) = e^{it\lambda x_k} f(x), \ 2 \le j < k \le n-1.$$

LEMMA 4. The restriction of the representation S_n^λ to the subgroup $B(n-1, \mathbb{R})$ does not depend on λ and can be decomposed into the direct integral of irreducible representations:

$$S_n^\lambda \restriction B(n-1, \mathbb{R}) = \oplus \int_{\mathbb{R}^1} S_{n-1}^\alpha \, d\alpha.$$

Proof. We get the necessary decomposition by using the Fourier transform. In the space of the Fourier images:

$$(S_n^\lambda(\exp t E_{1k})f)(x) = e^{itx_k} f(x)$$

$$(S_n^\lambda(\exp t E_{1n})f)(x) = e^{it\lambda} f(x)$$

$$(S_n^\lambda(\exp t E_{jk})f)(x) = f(x_2, \dots, x_j - tx_k, \dots, x_{n-1})$$

$$(S_n^\lambda(\exp t E_{kn})f)(x) = f(x_2, \dots, x_k - \lambda t, \dots, x_{n-1}), \ 2 \le j < k \le n-1.$$

□

In particular, it follows from Lemma 4 that none of the representations S_{n-1}^α is a subrepresentation of the representation $S_n^\lambda \restriction B(n-1, \mathbb{R})$.

For any group $B(n, \mathbb{R})$, $n \ge 2$, we define a reducible representation, $S_n = S_{n+1}^\lambda \restriction B(n, \mathbb{R})$ (by Lemma 4, it does not depend on $\lambda \in \mathbb{R}^1$). For the sequence of representations S_n, $n \ge 2$ we have:

LEMMA 5. The restriction of the representation S_n to the subgroup $B(n-1, \mathbb{R})$ is a multiple of S_{n-1}: $S_n \restriction B(n-1, \mathbb{R}) = S_{n-1} \otimes I$. Here $H_n = H_{n-1} \otimes L_2(\mathbb{R}^1, dx_n)$.

Proof. The proof follows directly from Lemma 4. □

Fix vectors $\phi_n (\|\phi_n\| = 1)$ in every space $V_n = L_2(\mathbb{R}^1, d\lambda_n)$ and consider the isometric imbedding of the spaces:

$$H_{n-1} \ni f \mapsto i_n^{n-1} f = f \otimes \phi_n \in H_n.$$

It follows from Lemma 5 that for all $n = 2, 3, \dots,\ g \in B(n, \mathbb{R})$

$$i_{n+1}^n \, S_n(g) = S_{n+1}(g) \, i_{n+1}^n.$$

This allows to define the representation $S^{(\phi)}$ of the group $B_0(\mathbb{N}, \mathbb{R})$ on the space $H = \overline{\bigcup_{n=2}^{\infty} H_n}$ to be the inductive limit of the representations S_n $(n = 2, 3, \dots)$ with respect to the inclusions $(i_{n+1}^n)_{n=2}^\infty$.

THEOREM 6. The representation $S^{(\phi)}$ is irreducible.

We prove this theorem after having found an explicit formula for $S^{(\phi)}$. Let the functions $\phi_n(\cdot) \in L_2(\mathbb{R}^1, dx_n)$ be nonzero almost everywhere for all $n = 2,3,\ldots$. In this case, the mapping $f(x_n) \mapsto \phi_n(x_n)^{-1} f(x_n)$ gives an isometric isomorphism between the spaces $L_2(\mathbb{R}^1, dx_n)$ and $L_2(\mathbb{R}^1, |\phi_n(x_n)|^2 dx_n)$. Here $|\phi_n(x_n)|^2 dx_n$ is a probability measure on $\mathbb{R}^1$. We let the space H_n of the representation S_n be $H_n = L_2(\mathbb{R}^{n-1}, d\mu_n(x))$ where $d\mu_n(x) = \bigotimes_{k=2}^{n} |\phi_k(\lambda_k)|^2 d\lambda_k$. The representation operators will be of the form:

$$(S_n(\exp t E_{1k})f)(x) = e^{itx_k} f(x), \quad 2 \le k \le n$$

$$(S_n(\exp t E_{jk})f)(x) = (\frac{\phi_j(x_j - tx_k)}{\phi_j(x)} f(x_2, \ldots, x_j - tx_k, \ldots, x_n) = \tag{4}$$

$$= \sqrt{\frac{d\mu_n(x_2, \ldots, x_j - tx_k, \ldots, x_n)}{d\mu_n(x)}} f(x_2, \ldots, x_j - x_j - tx_k, \ldots, x_n), \quad 2 \le j < k \le n.$$

The inclusions

$$L_2(\mathbb{R}^{n-1}, d\mu_n(\cdot)) \ni f(x_2, \ldots, x_n) \mapsto f(x_2, \ldots, x_n) \cdot 1 \in L_2(\mathbb{R}^n, d\mu_{n+1}(\cdot))$$

correspond to the inclusions $i^n_{n+1} : H_n \hookrightarrow H_{n+1}$. In this case the space $H = \overline{\bigcup_{n=2}^{\infty} H_n}$ is $L_2(\mathbb{R}^\infty, d\mu(x))$ where $d\mu(x) = \bigotimes_{k=2}^{\infty} |\phi_k(\lambda_k)|^2 d\lambda_k$ is the product measure on $\mathbb{R}^\infty$. So we have proved the following:

THEOREM 7. The representation $S^{(\phi)}$ can be realized in the space $L_2(\mathbb{R}^\infty, d\mu(x))$ ($\mathbb{R}^\infty = \mathbf{h}_c'$) as follows

$$(S^{(\phi)}(\exp u)f)(x) = e^{i(u,x)} f(x), \quad u \in \mathbf{h}_c$$

$$(S^{(\phi)}(g)f)(x) = \sqrt{\frac{d\mu(Ad_g^*(x))}{d\mu(x)}} f(Ad_g^*(x)), \quad g \in B_{0,1}(\mathbb{N}, \mathbb{R}).$$

$\square$

Irreducibility of the representation $S^{(\phi)}$ now is as a standard consequence of the fact that the product measure $d\mu(\cdot)$ is ergodic. This proves the theorem. $\square$

By virtue of Theorem 5 none of the representations $S^{(\phi)}$ can be continuously extended to a representation of the group $B(\mathbb{N}, \mathbb{R})$. On the other hand, $B(\mathbb{N}, \mathbb{R}) = \bigcup_{a \in \mathrm{U}_0} B_a(\mathbb{N}, \mathbb{R})$ and, as will be shown below, for every group $B_a(\mathbb{N}, \mathbb{R})$ there exists a representation $S^{(\phi)}$ of $B_0(\mathbb{N}, \mathbb{R})$ than can be extended to $B_a(\mathbb{N}, \mathbb{R})$.

THEOREM 8. For any weight $a \in \mathrm{U}_0$ there exists an irreducible representation $S^{(\phi)}$ which can be continuously extended to a representation of the Hilbert-Lie group $B_a(\mathbb{N}, \mathbb{R})$.

Proof. For a chosen weight $a \in U_0$ we can construct a Gaussian product measure $\gamma(\cdot)$ on $\mathbb{R}^\infty$ that has the following properties:

1) the action of the group $B_a(\mathbb{N}, \mathbb{R})$ is defined on a set of full-measure;

2) the constructed measure $\gamma(\cdot)$ is $B_a(\mathbb{N}, \mathbb{R})$-quasi-invariant;

3) the family of operators of multiplication by $e^{i(t,x)}$ in $L_2(\mathbb{R}^\infty, d\gamma(\cdot))$ is continuous with respect to $t \in l_2((a_{1k}))$. □

6.7. General position representations.

If U is a representation of the group $B_0(\mathbb{N}, \mathbb{R})$ then $U \upharpoonright B(n, \mathbb{R})$, $n = 2,3,...,$ is a representation of the group $B(n, \mathbb{R})$. Because for finite-dimensional nilpotent Lie groups, general position representations are the most interesting ones, it seems interesting to construct the irreducible representations of the group $B_0(\mathbb{N}, \mathbb{R})$ which can be decomposed into a direct integral of general position representations of the group $B(n, \mathbb{R})$. To construct such representations we will need some facts about general position representations of the groups $B(n, \mathbb{R})$.

The orbits in $\mathrm{b}'(n, \mathbb{R})$ of general position of the group $B(n, \mathbb{R})$ can be constructed as follows. The Lie algebra $\mathrm{b}(n, \mathbb{R})$ of the group $B(n, \mathbb{R})$ is the Lie algebra of strictly upper triangular matrices of dimension n. The adjoint space $\mathrm{b}'(n, \mathbb{R})$ is naturally identified with the space of strictly lower triangular matrices. Here the duality is of the form:

$$(u,x) = Tr\, ux, \quad u \in \mathrm{b}(n, \mathbb{R}), \quad x \in \mathrm{b}'(n, \mathbb{R}).$$

The coadjoint action of the group $B(n, \mathbb{R})$ in $\mathrm{b}'(n, \mathbb{R})$ is of the form

$$Ad_g^*(x) = (g^{-1} x g)_-$$

where $(g^{-1} x g)_-$ denotes the projection of a, generally speaking, not stricly lower triangular matrix $g^{-1} x g$ onto $\mathrm{b}'(n, \mathbb{R})$. Let $\Delta_{n,k}(x)$ denote the lower left corner minor of order k of the matrix x. The system of $B(n, \mathbb{R})$-invariant algebraic equations

$$\Delta_{n,k}(x) = \lambda_k, \; k = 1, \ldots, \left[\frac{n}{2}\right], \quad \lambda_1 \cdots \lambda_{[n/2]} \neq 0 \tag{5}$$

defines a unique orbit of the group $B(n, \mathbb{R})$ which is the general position orbit O_n^λ, $\lambda = (\lambda_1, \ldots, \lambda_m)$ $(m = [n/2])$ (see A.A. Kirillov [1]). Let Ω_n denote the set of general position orbits.

General position representations of the group $B(n, \mathbb{R})$ can be constructed as follows. Let O be an general position orbits of the group $B(n, \mathbb{R})$. The element

$$\lambda_\lambda^{(n)} = \begin{bmatrix} & & & 0 \\ & 0 & \mu_m & \\ 0 & & & \\ & \ddots & & 0 \\ \mu_1 & & 0 & \end{bmatrix}, \quad \lambda_l = (-1)^{l+1} \mu_1 \cdots \mu_l$$

can be chosen as the standard representative of a general position orbit.

In $\mathbf{b}(n, \mathbb{R})$ consider the commutative subalgebra $\mathbf{h}_n$ of matrices of the form

$$u = \begin{bmatrix} 0_{[n/2]} & * \\ 0 & 0_{[n+1/2]} \end{bmatrix}.$$

$\mathbf{h}_n$ is a subalgebra of maximal dimension submitted to the general position orbit, i.e. every functional $x \in O$ defines the one-dimensional representation $u \mapsto i(u,x)$ of the Lie algebra $\mathbf{h}_n$ and so we have the representation $U_x(\exp u) = \exp(i(u,x))$ of the group $H_n = \exp \mathbf{h}_n$. The irreducible general position representation T_n^x corresponding to the orbit O is a representation of the group $B(n, \mathbb{R})$ induced (in the sense of Mackey) by the representation U_x of the group H_n (see A.A. Kirillov [1]). $\square$

Let us study the relation between the general position representations of the group $B(n, \mathbb{R})$ and the representations of general position of the group $B(n-1, \mathbb{R})$.

LEMMA 6. The restriction of the representation T_n^x to the subgroup $B(n-1, \mathbb{R})$ does not depend on the choice of the orbit O and is equivalent to the direct integral of the representations of general position of the group $B(n-1, \mathbb{R})$:

$$T_n^x \restriction B(n-1, \mathbb{R}) = \oplus \int_{\Omega_{n-1}} T_{n-1}^y \, dy.$$

Proof. The restriction of the representation T_n^x to $B(n-1, \mathbb{R})$ can be decomposed into the direct integral of the representations that correspond to the orbits contained in the projection of $O_n^\lambda \in \Omega_n$ onto $\mathbf{b}'(n-1, \mathbb{R})$. A direct calculation using the formulas (6) shows that this projection is the union $\bigcup_{\Omega_{n-1}} O_{n-1}^\lambda$. This proves the lemma. $\square$

COROLLARY 3. Let T_n^x, T_{n+1}^y be general position representations of the groups $B(n, \mathbb{R})$ and $B(n+1, \mathbb{R})$ in the corresponding Hilbert spaces H_n and H_{n+1}. There are no isometric imbeddings $i_{n+1}^n : H_n \hookrightarrow H_{n+1}$ such that $T_{n+1}^y(g)\, i_{n+1}^n = i_{n+1}^n\, T_n^x(g)$ for all $g \in B(n, \mathbb{R})$.

For every group $B(n, \mathbb{R})$, define the reducible representation $U_n = T_{n+1}^y \restriction B(n, \mathbb{R})$ (as noted before, it does not depend on the choice of general position representation T_{n+1}^y). The following statement is a simple corollary of Lemma 6.

LEMMA 7. The restriction of the representation U_n to the subgroup $B(n-1, \mathbb{R})$ is a multiple of the representation U_{n-1}:

$$U_n \restriction B(n-1, \mathbb{R}) = U_{n-1} \otimes I.$$

Proof. The restriction of the representation U_n to the subgroup $B(n-1, \mathbb{R})$ has the form:

$$U_n \restriction B(n-1, \mathbb{R}) = \oplus \int_{\Omega_n} T_n^y \restriction B(n-1, \mathbb{R})\, dy = \oplus \int_{\Omega_n} U_{n-1}\, dy = U_{n-1} \otimes I,$$

where I is the identity operator on the space $V_n = L_2(\Omega_n, dy)$. □

Decompose the space H of the representation into the tensor product:

$$H_n = H_{n-1} \otimes V_n.$$

For every $n \geq 3$, choose $\phi_n \in V_n$, $\|\phi_n\| = 1$, and define isometric imbeddings of the spaces by setting $H_{n-1} \ni f \mapsto i_n^{n-1} f = f \otimes \phi_n \in H_n$. It easily follows from Lemma 7 that for any $n \geq 2$

$$i_{n+1}^n\, U_n(g) = U_{n+1}(g)\, i_{n+1}^n, \quad g \in B(n, \mathbb{R}). \tag{6}$$

This allows to construct a representation of the group $B_0(\mathbb{N}, \mathbb{R})$ that is the inductive limit of the sequence U_n, $n \geq 2$. The representation associated to the sequence $\phi = (\phi_n)_{n \geq 3}$ is denoted by $U^{(\phi)}$.

THEOREM 9. The representation $U^{(\phi)}$ is irreducible.

Proof. Let $H^{(\phi)} = \overline{\bigcup_{n=2}^{\infty} H_n}$ be the representation space for $U^{(\phi)}$ and let A be a bounded self-adjoint operator on $H^{(\phi)}$ commuting with the representation $U^{(\phi)}$. Denote by $P_n : H^{(\phi)} \to H_n$ the orthogonal projection onto the subspace H_n and set $A_n = P_n A P_n$. It follows from (6) that for any n the operator A_n commutes with the representation U_n of the group $B(n, \mathbb{R})$. According to the definition, the representation U_n can be decomposed into a direct integral $U_n = \oplus \int_{\Omega_n} T_n^y\, dy$. Because for different $y, y' \in \Omega_n$, the representations T_n^y and $T_n^{y'}$ are not equivalent, it follows that A_n is an operator of multiplication by a scalar essentially bounded measurable function $a_n(y)$. Decomposing the space H_n into the tensor product $H_n = H_{n-1} \otimes V_n$, we see that in this decomposition, $A_n = I \otimes a_n(\cdot)$. Taking into consideration the form of the imbeddings $i_n^{n-1} : H_{n-1} \hookrightarrow H_n$ we get that the operator A_{n-1} is a multiple of the identity. Indeed, for $f \in H_{n-1}$

$$A_{n-1} f = P_{n-1} A_n P_{n-1}(f \otimes \phi_n) = (a(\cdot)\, \phi_n(\cdot), \phi_n(\cdot))_{V_n} f.$$

As $P_n \to I$, A is a scalar operator. □

The following theorem presents conditions on representations $U^{(\phi)}$ and $U^{(\psi)}$ to be unitarily equivalent.

THEOREM 10. The representations $U^{(\phi)}$ and $U^{(\psi)}$ are unitarily equivalent if and only if

$$\sum_{n=3}^{\infty} (1 - |(\phi_n, \psi_n)|) < \infty. \tag{7}$$

Proof. Let the representations $U^{(\phi)}$ and $U^{(\psi)}$ be equivalent. There exists a unitary operator $S : H^{(\phi)} \to H^{(\psi)}$ such that $SU^{(\phi)} = U^{(\psi)} S$. Let $P_n : H^{(\phi)} \to H_n$ and $P_n' : H^{(\psi)} \to H_n$ be the orthogonal projections on H_n. Set $S_n = P_n' S P_n : H_n \to H_n$. The operator S_n commutes with the representation U_n. As shown in the proof of Theorem 9, it follows that S_n is a multiple of the identity, $S_n = \lambda I$, with $|\lambda| = 1$, because S is unitary. Represent the space $H^{(\phi)}$ as the infinite tensor product:

$$H^{(\phi)} = \bigotimes_{n=2;(\phi)}^{\infty} V_n \quad (V_2 = H_2).$$

Choose an arbitrary vector $\phi_2 \in V_2$ and consider the vector $f = \phi_2 \otimes \phi_3 \otimes \cdots \in H^{(\phi)}$. As $S_n \xrightarrow{s} S$, the sequence $S_n f$ converges in $H^{(\psi)}$ and $\lambda f = \lim S_n f \in H^{(\psi)}$. This is possible only if (7) holds.

Conversely, if (7) holds, then the spaces $H^{(\phi)}$ and $H^{(\psi)}$ coincide and the identity operator intertwines the representations $U^{(\phi)}$ and $U^{(\psi)}$. □

6.8. An analogue of the regular representation of the group $B_0(\mathbb{N}, \mathbb{R})$.

The regular representation of a finite-dimensional Lie group is a traditional tool to study representations. For infinite-dimensional Lie groups, it is difficult to construct regular representations because these groups do not have a Haar measure. In the case of the group $B_0(\mathbb{N}, \mathbb{R})$ it is possible to define an analogue of the regular representation if we choose a measure on the group $B(\mathbb{N}, \mathbb{R})$, quasi-invariant relatively to the right action of the group $B_0(\mathbb{N}, \mathbb{R})$. As shown in Section 6.1, such a measure can be taken to be the Gaussian measure on $B(\mathbb{N}, \mathbb{R})$ with diagonal correlation C. The operators of the representation act in the space $H = L_2(B(\mathbb{N}, \mathbb{R}), d\mu(g))$ according to

$$(U_{g_0} f)(g) = \sqrt{\frac{d\mu(gg_0)}{d\mu(g)}}\, f(gg_0).$$

Because there are many nonequivalent $B_0(\mathbb{N}, \mathbb{R})$-quasi-invariant measures on $B(\mathbb{N}, \mathbb{R})$, there exist many nonequivalent regular representations of the group $B_0(\mathbb{N}, \mathbb{R})$.

Let $\mu(\cdot)$ be a standard Gaussian measure on $B(\mathbb{N}, \mathbb{R})$ with correlation operator $C = I$ and let U be the corresponding regular representation of $B_0(\mathbb{N}, \mathbb{R})$.

THEOREM 11. The representation U is irreducible.

Proof. Let $H_n^0 \subset B_0(\mathbb{N}, \mathbb{R})$ be the subgroup of matrices of the form:

$$h = \begin{bmatrix} I_n & u \\ 0 & I_\infty \end{bmatrix}$$

where u is a finite matrix with n rows. Let $H_n \subset B(\mathbb{N}, \mathbb{R})$ be the closure of H_n^0 in $B(\mathbb{N}, \mathbb{R})$.

LEMMA 8.

$$\text{CLS}\,\{U_h 1 \mid h \in H_n^0\} = L_2(B(n, \mathbb{R}), d\mu_n(\cdot)) \otimes L_2(H_n, d\mu_n'(\cdot)),$$

where $\mu_n(\cdot)$, $\mu_n'(\cdot)$ are the projections of the measure on $B(n, \mathbb{R})$ and H_n.

Proof. H_n^0 is a commutative subgroup in $B_0(\mathbb{N}, \mathbb{R})$. Using the Fourier transform with respect to the variables from H_n, we find for $f(g,h) \in L_2(B(n, \mathbb{R}) \otimes H_n, d\mu_n(\cdot) \otimes d\mu_n'(\cdot))$

$$(U_{h_0} \hat{f})(g,h) = \exp(i\,Tr(g\,h_0\,h))\,\hat{f}(g,h), \quad h_0 \in H_n^0.$$

Using the law of large numbers, one can show that the family of functions $\{\exp(i\,Tr(g\,h_0\,h) \mid h \in H_n^0\}$ is total in $L_2(B(n, \mathbb{R}), d\mu_n(\cdot)) \otimes L_2(H_n, d\mu_n'(\cdot))^\infty$. Taking the inverse Fourier transform we get the statement. ▯

For $\phi(\cdot) \in L_\infty(B(n, \mathbb{R}), d\mu_n(\cdot))$ the operator of multiplication by $\phi(g)$ in $L_2(B(\mathbb{N}, \mathbb{R}), d\mu(\cdot))$ commutes with U_h, $h \in H_n^0$. Consequently, for a cylindrical function

$$f(g) \in L_2(B(n, \mathbb{R}), d\mu_n(\cdot) \subset L_2(B(\mathbb{N}, \mathbb{R}), d\mu(\cdot))$$

the function $(\phi f)(g)$ can be approximated by linear combinations of the functions $(U_h f)(g)$, $h \in H_n^0$. Because this is true for any $n = 1,2,...$ the operator of multiplication by $\phi(g)$ in $L_2(B(L\,\mathbb{N}, \mathbb{R}), d\mu(\cdot))$ can be weakly approximated by linear combinations of the operators U_g, $g \in B_0(\mathbb{N}, \mathbb{R})$ i.e. it is contained in the von Neumann algebra generated by the operators of the representation U. Thus, the bounded self-adjoint operator A which commutes with the representation U commutes with all the operators of multiplication by essentially bounded measurable cylindrical functions. So it is an operator of multiplication by a measurable essentially bounded function. Because the measure $\mu(\cdot)$ is ergodic relatively to the action of the group $B_0(\mathbb{N}, \mathbb{R})$, this function is a constant. ▯

Comments to Chapter 6.

1. The definition and the simplest properties of the group of infinite upper triangular matrices are given. The quasi-invariant Gaussian measures on $B(\mathbb{N}, \mathbb{R})$ have been studied in A.V. Kosyak and Yu.S. Samoĭlenko [4].

2. The theorem on commutative models is due to V.L. Ostrovskiĭ [2].

3,4. Results of A.V. Kosyak [4] are given.

5,6. A summary of the results of V.L. Ostrovskiĭ [2].

7. Irreducible representations of finite-dimensional Lie groups are studied in detail in A.A. Kirillov [1]. The results given in the section are contained in the thesis of V.L. Ostrovskiĭ.

8. The theorem on irreducibility of the regular representation is due to N.I. Nessonov [2].

Chapter 7.
A CLASS OF INDUCTIVE LIMITS OF GROUPS AND THEIR REPRESENTATIONS

7.1. The groups $G_0 = \lim\limits_{\rightarrow} G_n$, $G = \lim\limits_{\leftarrow} G_n$.

Let $G_1 \subset G_2 \subset \cdots \subset G_n \subset \cdots$ be an increasing sequence of imbedded locally compact separable groups. Suppose each of the groups G_n, $n \in \mathbb{N}$ can be decomposed into the semi-direct product $G_n = G^{(n)} \wedge G_{n-1}$. Here $G^{(n)} \subset G_n$ is a normal subgroup of G_n and

$$G_m = G^{(m,n)} \wedge G_n, \quad m > n, \tag{1}$$

where $G^{(m,n)} = G^{(m)} \wedge \cdots \wedge G^{(n+1)}$. In this chapter for the group $G_0 = \lim\limits_{\rightarrow} G_n$, we introduce a natural completion $G = \lim\limits_{\leftarrow} G_n$, we give a construction of G_0-quasi-invariant measures, and we describe irreducible unitary representations of G_0 which can be continuously extended to representations of G.

For arbitrary m,n $(m > n)$ define the projection $p_n^m : G_m \to G_n$ as canonical projection $G_m \to G_m / G^{(m,n)} = G_n$. It follows from (1) that the constructed projections form a projective system, i.e. for any $k < m < n$ $p_n^m p_k^n = p_k^m$. Consider the set G of all sequences $g = (g_1, g_2, \ldots)$, $g_n \in G_n$, $n \in \mathbb{N}$ with the property

$$p_n^m g_m = g_n, \quad m > n.$$

The topology of the projective limit in G is the weakest topology in which the projections $p_n : G \to G_n$, $n \in \mathbb{N}$ are continuous. The relations (1) allow to define the structure of a topological group on G by setting

$$g\, h = (g_n h_n)_{n \in \mathbb{N}}$$

for $g = (g_n)_{n \in \mathbb{N}}$, $h = (h_n)_{n \in \mathbb{N}}$. To show that the definition is correct it is necessary to show that

$$p_n^m(g_m h_m) = g_n h_n, \quad p_n^m(g_m^{-1}) = g_n^{-1}.$$

Indeed, because $G_m = G^{(m,n)} \wedge G_n$, any element $g_m \in G_m$ can be uniquely represented as $g_m = \tilde{g}\, g_n$, $\tilde{g} \in G^{(m,n)}$, $g_n \in G_n$. Then

$$p_n^m(g_m h_m) = p_n^m(\tilde{g}\, g_n \tilde{h}\, h_n) = p_n^m(\tilde{g}\, g_n \tilde{h}\, g_n^{-1} g_n h_n) = g_n h_n,$$

$$p_n^m(g_m^{-1}) = p_n^m(g_n^{-1} \tilde{g}^{-1}) = p_n^m(g_n^{-1} \tilde{g}^{-1} g_n g_n^{-1}) = g_n^{-1}$$

because $\tilde{g}\, g_n \tilde{h}\, g_n^{-1} \in G^{(m,n)}$, $g_n^{-1} \tilde{g}^{-1} g_n \in G^{(n,m)}$. The constructed group is the projective limit of the sequence of groups $(G_n)_{n \in \mathbb{N}}$.

The group $G_0 = \lim\limits_{\rightarrow} G_n$ contained in $G = \lim\limits_{\leftarrow} G_n$ is the dense subgroup of stabilizing sequences of the form $g = (g_1, \ldots, g_n, g_n, \ldots)$.

Example 1. Set $G_n = \underbrace{\phi \times \cdots \times \phi}_{n \text{ times}} = \phi_n$.

Define the inclusions $\phi^n \ni g = (g_1, \ldots, g_n) \mapsto (g_1, \ldots, g_n, e) \in \phi^{n+1}$. The projections $p_n^{n+1} : \phi^{n+1} \to \phi^n$ have the form $\phi^{n+1} \ni (g_1, \ldots, g_n, g_{n+1}) \mapsto (g_1, \ldots, g_n) \in \phi^n$. $G_0 = \phi_0^\infty$ is the group of finite sequences $(g_n)_{n \in \mathbb{N}}$ $(g_n \in \phi)$. The group G is isomorphic to the group ϕ^∞ of all sequences $(g_n)_{n \in \mathbb{N}}$ with product topology.

Example 2. Let $B(n, \mathbb{R})$ be the group of real upper triangular matrices of order n with the identity on the main diagonal; $B(n+1, \mathbb{R}) = \mathbb{R}^n \otimes B(n, \mathbb{R})$ where $B(n, \mathbb{R})$ is the subgroup of the matrices of $B(n+1, \mathbb{R})$ which have zeroes in the last column above the main diagonal, and $\mathbb{R}^n$ can be identified with the subgroup of the matrices in which the only non-zero elements are in the last column. The groups $B_0(\mathbb{N}, \mathbb{R}) = \lim\limits_{\rightarrow} B(n, \mathbb{R})$ and $B(\mathbb{N}, \mathbb{R}) = \lim\limits_{\leftarrow} B(\mathbb{N}, \mathbb{R})$ are the groups of infinite real matrices of the form $I + A_0$ and $I + A$ correspondingly, where A_0 is a finite strictly upper triangular matrix and A is any strictly upper triangular matrix. These groups have been considered in the previous chapter.

Example 3. Let ϕ be a locally compact Lie group. Set

$$G_1 = \phi, G_2 = \phi \times \phi = \phi^2, \ldots, G_n = G_{n-1}^2 = \phi^{2^{n-1}}$$

and define the inclusions by $G_n \ni g \mapsto (g, g) \in G_{n+1}$. The projections p_n^{n+1} are of the form $G_{n+1} \ni (g_1, g_2) \mapsto g_1 \in G_n$. $G_n = \lim\limits_{\rightarrow} G_n$ is the group $\phi_{\pi(2)}$ of the step ϕ-currents on the interval [0,1] with discontinuities in the binary rational points (see Chapter 5), G is a completion of the group G_0 with respect to the projective limit topology.

7.2. G_0-quasi-invariant measures on G.

In what follows we give a construction of G_0-quasi-invariant measures on G that generalizes the construction of $\mathbb{R}_0^\infty$-quasi-invariant product measures on $\mathbb{R}^\infty$.

Consider the product $\prod\limits_{k=1}^{\infty} G^{(k)}$ $(G^{(1)} = G_1)$, and define the mappings $\phi_r, \phi_l : \prod\limits_{k=1}^{\infty} G^{(k)} \to G$ by

$$\phi_r(g_1, g_2, \ldots, g_n, \ldots) = (g_1, g_2 g_1, \ldots, g_n \cdots g_1, \ldots),$$

$$\phi_l(g_1, g_2, \ldots, g_n, \ldots) = (g_1, g_1 g_2, \ldots, g_1 \cdots g_n, \ldots).$$

LEMMA 1. The mappings ϕ_r and ϕ_l are homeomorphisms between $\prod_{k=1}^{\infty} G^{(k)}$ and G.

Proof. It is obvious that ϕ_r maps $\prod_{k=1}^{\infty} G^{(k)}$ into G. We show that it has an inverse. Indeed, it follows from (1) that any element $g \in G$ can be represented as

$$g = (g_1, g_2 g_1, \ldots, g_n \cdots g_1, \ldots)$$

so the mapping ϕ_r is invertible.

Under the mapping $\phi_r^{-1} : G \to \prod_{k=1}^{\infty} G^{(k)}$, the projections $p_n : G \to G_n$ transform into the projections

$$\tilde{p}_n : \prod_{k=1}^{\infty} G^{(k)} \ni (g_1, g_2 \cdots) \mapsto (g_1, g_2, \ldots, g_n) \in \prod_{k=1}^{n} G^{(k)}.$$

Because the weakest topology in $\prod_{k=1}^{\infty} G^{(k)}$ in which the projections $\tilde{p}_n : \prod_{k=1}^{\infty} G^{(k)} \to \prod_{k=1}^{\infty} G^{(k)}$ are continuous is the product topology, ϕ_r is a homeomorphism.

To prove that ϕ_l is a homeomorphism, it is sufficient to note that

$$\phi_l(g_1, g_2, \ldots, g_n, \ldots) = (g_1, g_1 g_2, \ldots, g_1 \cdots g_n, \ldots) =$$
$$= (g_1^{-1}, (g_1 g_2)^{-1}, \ldots, (g_1 \cdots g_n)^{-1}, \ldots)^{-1} =$$
$$= (g_1^{-1}, g_2^{-1} g_1^{-1}, \ldots, g_n^{-1} g_1^{-1}, \ldots)^{-1} = \phi_r(g_1^{-1}, g_2^{-1}, \ldots, g_n^{-1}, \ldots)^{-1}.$$

The lemma is proved. □

Let $d\mu_k(\cdot)$ be a $G^{(k)}$-quasi-invariant probability measure on the group $G^{(k)}$, $k \in \mathbb{N}$. On the measurable space $(\prod_{k=1}^{\infty} G^{(k)}, \mathrm{B}(\prod_{k=1}^{\infty} G^{(k)}))$ we introduce the product measure $d\mu(\cdot) = \bigotimes_{k=1}^{\infty} d\mu_k(\cdot)$. Denote the image of the measure $d\mu(\cdot)$ under the mapping ϕ_r (correspondingly ϕ_l) by $d\mu_r(\cdot)$ (correspondingly $d\mu_l(\cdot)$).

THEOREM 1. The measure $d\mu_r(\cdot)$ is quasi-invariant and ergodic with respect to the right action of the group $G_0 \subset G$. Similarly, the measure $d\mu_l(\cdot)$ is quasi-invariant and ergodic with respect to the left action of the group $G_0 \subset G$.

Proof. The action of the group G_0 in $\prod_{k=1}^{\infty} G^{(k)}$ is defined by the formula

$$\prod_{k=1}^{\infty} G^{(k)} \ni \tilde{g} \mapsto \phi_r^{-1}(\phi_r(\tilde{g}) g_0) = \tilde{g} g_0, \quad g_0 \in G_0.$$

For $g_0 \in G_0$, the element $\tilde{g} g_0$ is different from $\tilde{g}$ only by the projection $p_n(g)$. Indeed, for

$$\bar{g} = (g_1, g_2, \ldots, g_n, \ldots),\ \phi_r(\bar{g}) = (g_1, g_2 g_1, \ldots, g_n \cdots g_1, \ldots).$$

If $g_0 = (g_1^0, g_2^0 g_1^0, \ldots, g_n^0 \cdots g_1^0, g_n^0 \cdots g_1^0, \ldots)$ then

$$\phi_r(\bar{g})\, g_0 = (g_1 g_1^0, g_2 g_1 g_2^0 g_1^0, \ldots, g_n \cdots g_1 g_n^0 \cdots g_1^0, g_{n+1}(g_n \cdots g_1 g_n^0 \cdots g_1^0), \ldots$$

so one can see that the action by $g_0 \in G_0$ changes only $g_1, \ldots, g_n$. As $d\mu(\cdot) = (\bigotimes_{k=1}^{n} d\mu_k(\cdot) \bigotimes_{k=n+1}^{\infty} d\mu_k(\cdot))$ the measure $d\mu(\cdot)$ is G_0-quasi-invariant because the measure $\bigotimes_{k=1}^{n} d\mu_k(\cdot)$ is G_n-quasi-invariant.

To prove that the measure $d\mu_r$ is ergodic we state the following:

LEMMA 2. Let $A \subset \prod_{k=1}^{\infty} G^{(k)}$ be a measurable set which is invariant with respect to the right action of the group G_0. For any $n \in \mathbb{N}$, the set A can be described as the direct product, $A = \prod_{k=1}^{n} G^{(k)} \times A_n$.

Proof. Let

$$\bar{g} = (g_1, g_2, \ldots, g_n, \ldots) \in \prod_{k=1}^{\infty} G^{(k)},\ \bar{g}' = (g_1', g_2', \ldots, g_n', g_{n+1}, \ldots) \in \prod_{k=1}^{\infty} G^{(k)}.$$

For some $g_0 \in G_0 (\phi_r^{-1}(g_0) = (g_1^0, g_2^0, \ldots, g_n^0, e, \ldots))$

$$\bar{g}^{-1} = \bar{g}\, g_0.$$

Indeed,

$$\bar{g}\, g_0 = (g_1 g_1^0, g_2 g_1 g_2^0 g_1^{-1}, \ldots, g_n \cdots g_1 g_n^0 (g_{n-1} \cdots g_1)^{-1}, g_{n+1}, \ldots)$$

so

$$g_1^0 = g_1^{-1} g_1', \ldots, g_n^0 = (g_{n-1} \cdots g_1)^{-1} g_n^{-1} g_n' (g_{n-1} \cdots g_1).$$

Thus, because the set A is invariant, it follows that together with the element $\bar{g}$ it contains all the elements $\bar{g}'$ which differ from $\bar{g}$ only by the projection $\bar{p}_n(\bar{g}')$. The lemma is proved. []

The remaining part of the proof, viz. that the measure $d\mu_r(\cdot)$ is ergodic, is similar to the proof that the product-measure on $\mathbb{R}^\infty$ is ergodic with respect to finite shifts (zero-one property).

Because A is measurable, there exists a sequence of cylindrical sets A_n, $n \in \mathbb{N}$ such that $\mu(A_n \Delta A) \to 0$. It is obvious that here, $\mu(A_n) \to \mu(A)$ and $\mu(A \cap A_n) \to \mu(A)$. On the other hand, $\mu(A_n \cap A) = \mu(A_n)\,\mu(A) \to \mu(A)^2$, so $\mu(A) = \mu(A)^2$. This is possible only if $\mu(A) = 0$ or $\mu(A) = 1$.

We can similarly prove that the measure $d\mu_l(\cdot)$ is ergodic with respect to the left shifts. []

7.3. Irreducible representations of the group $G = \lim\limits_{\leftarrow} G_n$.

In Chapter 6 we described representations of the group $B_0(\mathbb{N}, \mathbb{R})$ than can be continuously extended to representations of the group $B(\mathbb{N}, \mathbb{R}) = \lim\limits_{\leftarrow} B(n, \mathbb{R})$. Such representations have a kernel of finite codimension and are determined by irreducible representations of some prelimiting group $B(n, \mathbb{R})$.

For arbitrary groups of the form $G = \lim\limits_{\leftarrow} G_n$ the situation is similar.

Let $G_n \ni g \mapsto T_n(g)$ be an irreducible unitary representation of the group G_n. Because $G_n = G_m / G^{(m,n)}$, $m > n$, the representation T_n can be extended to a representation of each group G_m, $m > n$ (which is trivial on $G^{(m,n)}$) and so it can be extended to a representation T_0 of the group G_0. On the other hand, setting

$$T(g) = T_n(p_n(g)), \quad g \in G \tag{2}$$

we get an irreducible representation of the group G whose restriction to G_0 coincides with T_0. So we have constructed a family of irreducible representations of the group G_0 than can be continuously extended to representations of the group G.

THEOREM 2. The representations (2) form a complete collection of irreducible unitary representations of the group G.

Proof. Let T be an irreducible unitary representation of the group G on a separable Hilbert space H. Choose an arbitrary nonzero vector $f \in H$ and let O_f be a convex neighbourhood of f whose closure does not contain zero. The mapping $S_g : g \mapsto T(g)f$ is continuous and so $S_g^{-1}(O_f)$ is open in G. For any $n \in \mathbb{N}$ the group G can be decomposed into the semi-direct product $G = G^{(\infty,n)} \wedge G_n$ with projective limit topology generated by the sets $G^{(\infty,n)} \times O$, where O is open in G_n. Therefore, for some $n \in \mathbb{N}$, $G^{(\infty,n)} \subset S_g^{-1}(O_f)$. Let $f \in H$ be the closed convex span of the vectors $\{T(g_0)f \mid g_0 \in G^{(\infty,n)}\}$. It is evident that the set F is contained in the closure of O_f. It follows from the Alaoglu-Birkoff theorem that there exists a vector $f_0 \in F$ such that $T(g_0)f_0 = f_0$, $g_0 \in G^{(\infty,n)}$. For any $g \in G$ the vector $\tilde{f} = T(g)f_0$ also has the property that $T(g_0)\tilde{f} = \tilde{f}$, $g_0 \in G^{(\infty,n)}$. Indeed, because $G^{(\infty,n)}$ is a normal subgroup of G,

$$T(g_0)\tilde{f} = T(g_0)\,T(g)f_0 = T(g)\,T(g^{-1})\,T(g_0)\,T(g)f_0 =$$
$$= T(g)\,T(g^{-1}\,g_0\,g)f_0 = T(g)f_0 = \tilde{f}.$$

Because the representation T is irreducible, the nonzero vector $f_0 \in H$ is cyclic in H and the restriction of the representation T to the subgroup $G^{(\infty,n)}$ is trivial. There, the representation T can be obtained from the restriction $T \upharpoonright G_n$ using the irreducible representation T_n of the group G_n. □

Comments to Chapter 7.

1. The class of groups $G_0 = \lim\limits_{\rightarrow} G^{(n)} \wedge G_{n-1}$ for which there can be naturally defined the maximal group $G = \lim\limits_{\leftarrow} G^{(n)} \wedge G_{n-1}$ as an everywhere dense subgroup containing G_0 was introduced by V.L. Ostrovskiĭ and the author.

2 Exposition of the article V.L. Ostrovskiĭ [3].

3. The invariant vectors technique used to describe a complete collection of irreducible unitary representation of the group G is due to G.I. Olshanskiĭ [2].

PART III
COLLECTIONS OF UNBOUNDED SELF-ADJOINT OPERATORS SATISFYING GENERAL RELATIONS

If unbounded operators satisfy some algebraic relations (commutative, anti-commutative or others), then it is necessary to specify the exact meaning of these relations. In this chapter we give a definition of anti-commutation of self-adjoint operators (it is analogous to the definition of commutation of self-adjoint operators in terms of commutation of their spectral projections), and study the spectral properties of the operators (Chapter 8). The same questions will be solved for finite and certain countable collections of mutually commuting or anti-commuting (commuting with respect to the grading) self-adjoint operators (Chapter 9). In Chapter 10 these questions are solved both for pairs of unbounded self-adjoint operators A and B such that

$$AB = B\,F(A)$$

where $F(\cdot): \mathbb{R}^1 \to \mathbb{R}^1$ is a measurable function and, correspondingly, for finite and countable collections. For countable collections of operators satisfying relations such that the structure problems are difficult (in particular, for representations of the algebra of locally observable spin systems with countable degree of freedom) we construct commutative models.

Chapter 8.
ANTICOMMUTING SELF-ADJOINT OPERATORS

For non-degenerate bounded self-adjoint operators A and B the relation

$$AB - \alpha BA = 0 \quad (\alpha \in \mathbb{C}^1)$$

implies that $\alpha = 1$ or $\alpha = -1$, i.e. the operators commute or anticommute. In this chapter we study spectral questions for unbounded anticommuting self-adjoint operators.

8.1. Anticommuting self-adjoint operators. The joint domain. The rigging.

Two bounded operators A and B anticommute if

$$\{A, B\} = AB + BA = 0 \ .$$

DEFINITION 1. The unbounded self-adjoint operators

$$A = \int_{\mathbb{R}^1} \lambda_1 \, dE_A(\lambda_1) \text{ and } B = \int_{\mathbb{R}^1} \lambda_2 dE_B(\lambda_2)$$

are said to be anticommuting if for all $0 \le l, m < \infty$ the bounded operators

$$A_l = \int_{-l}^{l} \lambda_1 \, dE_A(\lambda_1) \text{ and } B_m = \int_{-m}^{m} \lambda_2 \, dE_B(\lambda_2)$$

anticommute.

If the operators are bounded, then Definition 1 is equivalent to their anticommutativity.

PROPOSITION. For the bounded self-adjoint operators

$$A = \int_{-\|A\|}^{\|A\|} \lambda_1 \, dA_A(\lambda_1) \text{ and } B = \int_{-\|B\|}^{\|B\|} \lambda_2 \, dE_B(\lambda_2)$$

to be anticommuting it is necessary and sufficient that

$$\{A_l, B_m\} = A_l B_m + B_m A_l = 0 \tag{1}$$

for all non-negative l and m.

Proof. Because for $l \ge \|A\|$ and $m \ge \|B\|$, $A_l = A$ and $B_m = B$, (1) is automatically sufficient for the operators A and B to anticommute

We show the necessity. Let $\{A, B\} = AB + BA = 0$ and $\|A\| < \infty$, $\|B\| < \infty$. Then $\|A^2\| = \|A\|^2 < \infty$ and $\|B^2\| = \|B\|^2 < \infty$ and $[A, B^2] = [A^2, B] = 0$. The spectral projections of the operators $A^2 = \int_0^{\|A\|^2} \lambda_1 d\mathrm{E}_A(\lambda_1)$ and $B^2 = \int_0^{\|B\|^2} \lambda_2 \, d\mathrm{E}_B(\lambda_2)$ commute with both A and B and so

$$\{A_l, B_m\} = A_l B_m + B_m A_l =$$
$$= \mathbf{E}_A([0, \sqrt{l}])\, \mathbf{E}_B([0, \sqrt{m}])\, \{A, B\} = 0 \ . \qquad \Box$$

Also note that the bounded self-adjoint operators A and B anticommute if and only if the square of the operator $A + iB$ is self-adjoint.

Now we show that Definition 1 means that the unbounded self-adjoint operators A and B anticommute on a dense set Φ in H invariant relatively to A and B and consisting of joint entire vectors for A and B.

LEMMA 1. The self-adjoint operators A and B anticommute if and only if they anticommute on a dense set Φ in H invariant relatively to A and B and consisting of joint entire vectors of A and B.

Proof. If the operators A and B anticommute, then $[A, B^2] = 0$ (recall that commutativity of self-adjoint operators means commutativity of the spectral projections of the self-adjoint operators A and B^2). Indeed, to prove that the spectral projections commute it is sufficient to show that the bounded operators $\int_{-l}^{l} \lambda_1\, dE_A(\lambda_1)$ and $\int_{-m}^{m} \lambda_2^2\, dE_B(\lambda_2)$ commute for all $0 \le l, m < \infty$. Because the operators $\int_{-l}^{l} \lambda_1\, dE_A(\lambda_1)$ and $\int_{-m}^{m} \lambda_2\, dE_B(\lambda_2)$ anticommute, $[A_l, B_m^2] = 0$. Similarly, we can show that $[A^2, B] = 0$.

Now, because $[A^2, B^2] = 0$ we have the equality $\overline{H^b(A^2) \cap H^b(B^2)} = H$ (here $H^b(A)$ is the set of bounded vectors of the operator A (see § 1.6)). But because $H^b(A) = H^b(A^2)$ and $H^b(B) = H^b(B^2)$, $\overline{H^b(A) \cap H^b(B)} = H$.

So, the set $H^b(A) \cap H^b(B) \subset H^c(A) \cap H^c(B)$ which is invariant relatively to A and B, is dense in H and thus necessity part is proved.

Now suppose that the operators $A = \int_{-\infty}^{\infty} \lambda_1\, dE_A(\lambda_1)$ and $B = \int_{-\infty}^{\infty} \lambda_2\, dE_B(\lambda_2)$ are anticommuting on a dense set Φ consisting of entire vectors of A and B. Then, for any $x_1, x_2 \in \mathbb{R}^1$, the bounded self-adjoint operators

$$\sin x_1 A = \int_{-\infty}^{\infty} \sin x_1 \lambda_1\, dE_A(\lambda_1)$$

and

$$\sin x_2 B = \int_{-\infty}^{\infty} \sin x_2 \lambda_2\, E_B(\lambda_2)$$

anticommute. Indeed, for any vector $f \in \Phi$ the series $x_1 A f + \frac{x_1^3}{3!} A^3 f + \cdots$ and

$x_2 Bf + \frac{x_2^3}{3!} B^3 f + \cdots$ absolutely converge. Now, it follows from Fubini's theorem that for any odd functions $u(\cdot)$ and $v(\cdot)$ belonging to the Schwartz space $S(\mathbb{R}^1)$, $\{u(A), v(B)\} = 0$ so choosing a uniformly bounded sequence of odd functions $u_n(\cdot) \in S(\mathbb{R}^1)$ pointwise convergent to the odd function

$$\chi_{[-l,l]}(\lambda_1) = \begin{cases} \lambda_1 \ , \ |\lambda_1| \le l \\ 0 \ , \ |\lambda_1| > l \end{cases}$$

(then $u_n(A)$ will strongly converge to $\chi_{[-l,l]}(\lambda_1)(A) = A_l$) and a uniformly bounded sequence of odd functions $v_n(\cdot) \in S(\mathbb{R}^1)$ pointwise convergent to

$$\chi_{[-m,m]}(\lambda_2) = \begin{cases} \lambda_2 \ , \ |\lambda_2| \le m \\ 0 \ , \ |\lambda_2| > m \end{cases}$$

(then $v_n(B)$ strongly converges to $\chi_{[-m,m]}(\lambda_2)(B) = B_m$), we find that $\{A_l, B_m\} = 0$ for all $l, m \ge 0$. □

Remark 1. Following E. Nelson, [1], we can replace in Definition 1 the invariant set Φ of joint entire vectors of A and B by the dense set of joint analytic vectors.

Example 1. Lemma 1 does not hold if Φ is replaced by an arbitrary core. To construct a counterexample take two operators which commute on a core which do not commute strongly, i.e. in the sense of the resolution of the identity: $\mathbf{A} = \mathbf{A}^*$, $\mathbf{B} = \mathbf{B}^*$ and $\mathbf{AB}f = \mathbf{BA}f$ for all $f \in D$, where D is a core of $\mathbf{A}$ and $\mathbf{B}$ invariant relatively to $\mathbf{A}$ and $\mathbf{B}$. The operators $A = \begin{bmatrix} 1 & 0 \\ 0 & -1 \end{bmatrix} \otimes \mathbf{A}$ and $B = \begin{bmatrix} 0 & 1 \\ 1 & 0 \end{bmatrix} \otimes \mathbf{B}$ on $\mathbf{H} = \mathbb{C}^2 \otimes H$ anticommute on the set $\mathbb{C}^2 \otimes D$ dense in $\mathbf{H}$, but they do not anticommute in the sense of Definition 1. □

For self-adjoint operators, commutativity of the spectral projections is equivalent to commutativity of the operators on a dense set of their joint analytic vectors, and also equivalent to commutativity of the one-parameter groups generated by these operators. Now we state a similar theorem for anticommuting self-adjoint operators.

THEOREM 1. Let A and B be self-adjoint operators on H.

The following are equivalent:

(i) A and B anticommute;

(ii) $(AB + BA)f = 0$ for all $f \in D$ where $D \subset H^{\omega}(A, B)$ is a dense set invariant relatively to A and B;

(iii) $\{\sin tA, \sin sB\} = 0 \quad \forall t, s \in \mathbb{R}^1$.

To prove that $(i) \Rightarrow (ii) \Rightarrow (iii) \Rightarrow (i)$ we essentially follow the proof of Lemma 1 using Remark 1. □

For two anticommuting self-adjoint operators A and B the set $H^b(A) \cap H^b(B)$ consisting of bounded vectors for the operators A and B is dense in H.

THEOREM 2. For the operators A and B there exists a standard nuclear rigging, $\Phi \subset H^b(A) \cap H^b(B)$ i.e. a nuclear linear topological space, Φ, dense in H and invariant with respect to the operators A and B such that $A\restriction\Phi, B\restriction\Phi \in L(\Phi)$.

Proof. Denote

$$K_n = \{(\lambda_1, \lambda_2) \in \mathbb{R}^2 \mid n-1 \le \max(|\lambda_1|, |\lambda_2|) < n\} \ ,$$

$$H_n = (\mathbf{E}_A \otimes \mathbf{E}_B)(K_n) H \, (n = 1, 2, \cdots) \ .$$

Because $\bigcup_{n=1}^{\infty} K_n = \mathbb{R}^2$, it follows that $H = \oplus \sum_{n=1}^{\infty} H_n$.

The subspaces H_n are invariant with respect to the operators A and B, and $A_n = A\restriction H_n$ and $B_n = B\restriction H_n$ are bounded, $H_n \subset H^B(A) \cap H^B(B)$. The rest of the proof is based on the following lemma.

LEMMA 2. For any countable collection of bounded operators $C_1, \cdots, C_n, \cdots$, there exists a nuclear rigging, $\Phi \subset H$, related in a standard way to the operators $C_j (j = 1, 2, \cdots)$.

Proof. Choose a basis $e_1, e_2, \cdots$ in H and denote

$$C_0 = I, \ \mathbf{M}_n = LS\{C^{\#}_{i_1} \cdots C^{\#}_{i_n} \Omega_1\} \ , \ i_k \le n.$$

(The # means taking or not taking the adjoint, $\Omega_1 = e_1$). If $\Phi_1 = \bigcup_{n=0}^{\infty} \mathbf{M}_n$ is dense in H, then endowing Φ_1 with the nuclear topology of the limit of the finite-dimensional spaces $\mathbf{M}_n$ (this topology is the same as the topology of the projective limit of the Hilbert spaces

$$H((p_k)_{k=1}^{\infty}) = \{f = \sum_{k=1}^{\infty} c_k \hat{e}_k \mid \|f\|^2_{H((p_k)_{k=1}^{\infty})} = \sum_{k=1}^{\infty} |c_k|^2 p_k < \infty\}$$

where $p_k \ge 1$, and $\hat{e}_k \in \bigcup_{n=0}^{\infty} \mathbf{M}_n$, $k = 1, 2, \cdots$ is an orthonormal basis in H with the limit taken relatively to all possible weights $p = (p_k)_{k=1}^{\infty})$, we can construct the needed rigging

$$\Phi_1' = p - \varinjlim H\left[\left(\frac{1}{p_k}\right)_{k=1}^{\infty}\right] \supset H \supset \Phi_1 = p - \varprojlim H((p_k)_{k=1}^{\infty}) \ .$$

If $\Phi_1 = \bigcup_{n=0}^{\infty} \mathbf{M}_n$ is not dense in H, then choosing the first basis vector $e_{m_1} \notin \bigcup_{n=0}^{\infty} \mathbf{M}_n = H_1$ we set $\Omega_2 = e_{m_1} - P_{H_1} e_{m_1}$ (P_{H_1} is the orthogonal projection on H_1) and introduce the corresponding nuclear space Φ_2. Continuing in this manner we construct a nuclear linear topological space $\Phi = \oplus \sum_{k=1}^{\infty} \Phi_k$ which is dense in H and endowed with an inductive limit (or, equivalently, with a projective limit) topology. $\square$

For the bounded operators A_n and B_n following the proof of Lemma 2 we construct a nuclear rigging $\Phi_n \subset H_n$.

Now, the nuclear linear topological space $\Phi = \oplus \sum_{n=1}^{\infty} \Phi_n$ is dense in H and invariant with respect to the operators A and B which continuously map Φ into Φ with $\Phi \subset H^b(A) \cap H^b(B)$.

For two anticommuting operators A and B we give one more construction of a rigging Φ standardly related to the operators A and B.

First we construct a rigging if the pair of anticommuting operators has a simple joint spectrum.

DEFINITION 2. We say that the pair A, B has simple joint spectrum if there exists a vector Ω (a vacuum) such that

$$\text{CLS}\,\{A_l^i B_m^j \Omega\} = H\,, \quad i, j = 0, 1, \cdots; 0 \le l, m < \infty.$$

DEFINITION 3. A vector $\Omega_1 \in H^{\infty}(A, B)$ is called a joint cyclic vector for the pair A, B if $\Omega \subset \bigcap_{i=1}^{\infty} D(A^i) \cap \bigcap_{j=1}^{\infty} D(B^j)$ and $CLS\,\{A^i B^j \Omega\} = H, i, j = 0, 1, \cdots$.

PROPOSITION 2. A cyclic vector Ω_1 for the pair A, B is also a vacuum.

Suppose the contrary. Assume that Ω_1 is not a vacuum for A, B, i.e. there exists a vector $h \ne 0$ such that

$$(h, A_l^i B_m^j \Omega_1) = 0 \quad (i, j = 0, 1, \cdots; 0 \le l, m < \infty).$$

But then $(h, A^i B^j \Omega_1) = 0$ $(i, j = 0, 1, \cdots)$ and so it is not a cyclic vector. $\square$

It is clear that the converse does not hold in general, i.e. a vacuum for the pair A, B is not necessarily a cyclic vector (for example, not all the powers $A^i B^j$ may be defined on a vacuum). However, we have

PROPOSITION 3. If the joint spectrum of the pair A, B is simple (there exists a vacuum Ω), then there exists a cyclic vector Ω_1.

Proof. We show that the vector $\Omega_1 = e^{-A^2} e^{-B^2} \Omega \in H^c(A, B) = H^c(A) \cap H^c(B)$ is a cyclic vector for the pair A, B. Indeed, $\Omega_1 \in H^{\infty}(A, B) = \bigcap_{i,j=0}^{\infty} D(A^i B^j)$ because the operators $A^i B^j e^{-A^2} e^{-B^2}$ are bounded on H. Now,

$$\Omega_1 \in H^c(A) \cap H^c(B) \text{ because } \Omega_1 \in (\bigcap_{\varepsilon>0} D(e^{\varepsilon |A|})) \cap (\bigcap_{\varepsilon>0} D(e^{\varepsilon |B|}))\ .$$

Let $(h, A^i B^j e^{-A^2} e^{-B^2} \Omega) = 0$ $(i, j = 0, 1, \cdots)$. Then $h = 0$. Indeed, for compact sets Δ and δ $(h, E_A(\Delta) E_B(\delta) \Omega) = 0$ because $E_A(\Delta) = \chi_\Delta(A)$, $E_B(\delta = \chi_\delta(B)$ and because the functions $\chi_\Delta(\cdot)$ and $\chi_\delta(\cdot)$ can be represented as Fourier-Hermite series. $\square$

The space $LS\{A^i B^j \Omega_1\}$ with the inductive limit topology of finite-dimensional spaces yields the rigging in question.

If the spectrum of the operators A and B is not simple, then decomposing H into a direct sum of invariant spaces H_n such that $A \restriction H_n$ and $B \restriction H_n$ have a simple spectrum, and rigging them with the spaces Φ_n, we get the needed rigging $\Phi = \oplus \sum_{n=1}^{\infty} \Phi_n$. □

8.2. The Spectral theorem for a pair of anticommuting self-adjoint operators.

First we describe up to unitary equivalence the irreducible pairs of anticommuting self-adjoint operators.

Because A^2 and B^2 commute with A and B, $A^2 = \lambda_1^2 I$, $B^2 = \lambda_2^2 I$ $(\lambda_1, \lambda_2 \in \mathbb{R}^1)$, and so the operators A and B are bounded.

If $\lambda_1 = 0$ or $\lambda_2 = 0$, then the operators A and B are operators on $\mathbb{C}^1 : A = \lambda_1 I, B = \lambda_2 I$ and the set of such unequivalent representations can be indexed by the points of the set

$$\mathbb{R}^2_+ = \{(\lambda_1, \lambda_2) \in \mathbb{R}^2 \mid \lambda_1 = 0, \lambda_2 \in \mathbb{R}^1\} \cup \{(\lambda_1, \lambda_2) \in \mathbb{R}^2 \mid \lambda_1 \in \mathbb{R}^1, \lambda_2 = 0\}.$$

If the operators A and B are not degenerate, then

$$H = \mathbb{C}^2 \otimes H_+, A = \begin{bmatrix} \mathbf{A}_1 & 0 \\ 0 & -\mathbf{A}_1 \end{bmatrix}, B = \begin{bmatrix} 0 & \mathbf{B}_1 \\ \mathbf{B}_1 & 0 \end{bmatrix}$$

where $\mathbf{A}_1, \mathbf{B}_1$ are commuting positive self-adjoint operators on H_+. The irreducible pairs of anticommuting self-adjoint operators act on $\mathbb{C}^2$ and, using unitary transformations, can be reduced to the form: $A = \begin{bmatrix} \lambda_1 & 0 \\ 0 & -\lambda_1 \end{bmatrix}$ and $B = \begin{bmatrix} 0 & \lambda_2 \\ \lambda_2 & 0 \end{bmatrix}$ $(\lambda_1 > 0, \lambda_2 > 0\}$. Different pairs of points $(\lambda_1, \lambda_2) \in \mathbb{R}^2_{++} = \{\lambda_1 > 0, \lambda_2 > 0\}$ correspond to unitarily unequivalent irreducible representations of the pair A and B.

For an arbitrary pair of anticommuting operators, decompose now the representation space H into two invariant subspaces: the subspace H_1, in which the operators A^2 and B^2 are not degenerate, and its orthogonal complement, H_0 :

$$H = H_0 \oplus H_1 \ .$$

The operators A and B commute on the invariant subspace H_0 and it follows from the spectral theorem for commuting operators that they can be represented in the form:

$$A \restriction H_0 = \int_{\mathbb{R}^2_+} \lambda_1 \, dE_0(\lambda_1, \lambda_2), \; B \restriction H_0 = \int_{\mathbb{R}^2_+} \lambda_2 \, dE_0(\lambda_1, \lambda_2) \ .$$

Here $dE_0(\lambda_1, \lambda_2)$ is an orthogonal operator-valued measure defined on $\mathbb{R}^2_+$ with values in the set of projections onto the subspaces of H_0.

As the operators $A \upharpoonright H_1$ and $B \upharpoonright H_1$ are not degenerate, using unitary transformations one can bring them into the form: $\begin{bmatrix} \mathbf{A} & 0 \\ 0 & -\mathbf{A} \end{bmatrix}$ and $\begin{bmatrix} 0 & \mathbf{B} \\ \mathbf{B} & 0 \end{bmatrix}$ where $\mathbf{A}$ and $\mathbf{B}$ are commuting positive self-adjoint operators on $H_+ (H_1 = \mathbb{C}^2 \otimes H_+)$. Representing $A \upharpoonright H_1$, $B \upharpoonright H_1$ in the form

$$A \upharpoonright H_1 = \begin{bmatrix} 1 & 0 \\ 0 & -1 \end{bmatrix} \otimes \mathbf{A}, \; B \upharpoonright H_1 = \begin{bmatrix} 0 & 1 \\ 1 & 0 \end{bmatrix} \otimes \mathbf{B}$$

and using the spectral theorem for the positive CSO $\mathbf{A}$ and $\mathbf{B}$ we get the following theorem:

THEOREM 3. (Spectral theorem in terms of the resolution of the indentity for a pair of anticommuting self-adjoint operators).
To any representation of a pair of anticommuting self-adjoint operators on a separable Hilbert space H, there correspond an orthogonal decomposition, $H = H_0 \oplus H_1 = H_0 \oplus (\mathbb{C}^2 \otimes H_+)$ and orthogonal measures $dE_0(\lambda_1, \lambda_2)$ on $\mathbb{R}^2_+ = \{\lambda_1 = 0, \lambda_2 \in \mathbb{R}^1\} \cup \{\lambda_1 \in \mathbb{R}^1, \lambda_2 = 0\}$ with values in the set of projections onto the subspaces of H_0 and $dE_1(\lambda_1, \lambda_2)$ on $\mathbb{R}^2_{++} = \{\lambda_1 > 0, \lambda_2 > 0\}$ with values in the set of projections onto the subspaces of H_+ such that

$$A = \int_{\mathbb{R}^2_+} \lambda_1 \, dE_0(\lambda_1, \lambda_2) + \begin{bmatrix} 1 & 0 \\ 0 & -1 \end{bmatrix} \otimes \int_{\mathbb{R}^2_{++}} \lambda_1 \, dE_1(\lambda_1, \lambda_2)$$

$$B = \int_{\mathbb{R}^2_+} \lambda_2 \, dE_0(\lambda_1, \lambda_2) + \begin{bmatrix} 0 & 1 \\ 1 & 0 \end{bmatrix} \otimes \int_{\mathbb{R}^2_{++}} \lambda_2 \, dE_1(\lambda_1, \lambda_2) \; .$$

It should be noted that to the operator-valued measures $dE_0(\lambda_1, \lambda_2)$ on the set $\mathbb{R}^2_+ \subset \mathbb{R}^2$ and $dE_1(\lambda_1, \lambda_2)$ on the set $\mathbb{R}^2_{++} \subset \mathbb{R}^2$ there correspond spectral types and dimension functions. They form a complete system of unitary invariants of the pair of anticommuting self-adjoint operators A and B.

In a similar way, we can obtain the spectral theorem in the multiplication operators form. We state it for the case of a representation with a simple joint spectrum, i.e. when every irreducible representation is contained in the decomposition of the pair not more than once.

THEOREM 3′. (Spectral theorem in multiplication operator form for a pair of anticommuting self-adjoint operators with a simple joint spectrum).
For any pair of anticommuting self-adjoint operators A and B on H with a simple joint spectrum, there exists a finite measure $\rho(\cdot)$ on $\mathbb{R}^2_+ \cup \mathbb{R}^2_{++}$ such that the pair A, B on H is unitarily equivalent to the operators $\tilde{A} = \lambda_1 \oplus \begin{bmatrix} 1 & 0 \\ 0 & -1 \end{bmatrix} \otimes \lambda_1$, and $\tilde{B} = \lambda_2 \oplus \begin{bmatrix} 0 & 1 \\ 1 & 0 \end{bmatrix} \otimes \lambda_2$ on

$$L_2(\mathbb{R}^2_+, d\rho(\lambda_1,\lambda_2)) \oplus \mathbb{C}^2 \otimes L_2(\mathbb{R}^2_{++}, d\rho(\lambda_1,\lambda_2)) \ .$$

The measure $\rho(\cdot)$ is determined up to equivalence.

8.3. The $*$-algebra and C*-algebra of a pair of anticommuting self-adjoint operators.

Let U be the complex $*$-algebra generated by two self-adjoint generators $a_1 = a_1^*, a_2 = a_2^*$ satisfying the relation $a_1 a_2 + a_2 a_1 = 0$. As a linear space it is isomorphic to a set of polynomials with complex coefficients of two variables λ_1 and λ_2:

$$P(\lambda_1,\lambda_2) = \sum_{j=0}^{n} \sum_{k=0}^{m} c_{jk} \lambda_1^j \lambda_2^k$$

where

a) involution is given by

$$P^*(\lambda_1,\lambda_2) = \sum_{j=0}^{n} \sum_{k=0}^{m} (-1)^{jk} \bar{c}_{jk} \lambda_1^j \lambda_2^k$$

b) product is given by

$$P_1(\lambda_1,\lambda_2) P_2(\lambda_1,\lambda_2) = (\sum_{j_1=0}^{n_1} \sum_{k_1=0}^{m_1} c_{j_1k_1} \lambda_1^{j_1} \lambda_2^{k_1}) \circ$$

$$\circ (\sum_{j_2=0}^{n_2} \sum_{k_2=0}^{m_2} c_{j_2k_2} \lambda_1^{j_2} \lambda_2^{k_2}) =$$

$$= \sum_{j=0}^{n_1+n_2} \sum_{k=0}^{m_1+m_2} (\sum_{j_1+j_2=j} \sum_{k_1+k_2=k} (-1)^{k_1 j_2} c_{j_2k_2}) \lambda_1^j \lambda_2^k \ .$$

There is a one-to-one correspondence between bounded self-adjoint anticommuting operators A, B on H and $*$-homomorphisms ϕ of the algebra U into the $*$-algebra $L(H)$ with the usual involution, norm and multiplication operation.

Every such pair of operators we put into correspondence with a representation of a C^*-algebra $C_a([-\|A\|, \|A\|] \times [-\|B\|, \|B\|])$ containing U as an everywhere dense $*$-subalgebra.

As a linear space $C_a([-\|A\|, \|A\|] \times [-\|B\|, \|B\|])$ equals $C([-\|A\|, \|A\|] \times [-\|B\|, \|B\|])$, so its elements are continuous functions on the compact set $[-\|A\|, \|A\|] \times [-\|B\|, \|B\|]$.

Set

$$\hat{\phi} : C([-\|A\|, \|A\|] \times [-\|B\|, \|B\|]) \ni f(\lambda_1,\lambda_2) \mapsto M_f(\lambda_1,\lambda_2)$$

$$\text{with } M_f(\lambda_1,\lambda_2)=\begin{bmatrix}(f_{ee}+f_{oe})(\lambda_1,\lambda_2) & (f_{eo}+f_{oo})(\lambda_1,\lambda_2)\\ (f_{eo}-f_{oo})(\lambda_1,\lambda_2) & (f_{ee}-f_{oe})(\lambda_1,\lambda_2)\end{bmatrix}.$$

Here

$$f_{ee}(\lambda_1,\lambda_2)=\frac{1}{4}\left[f(\lambda_1,\lambda_2)+f(\lambda_1,-\lambda_2)+f(-\lambda_1,-\lambda_2)\right],$$

$$f_{oe}(\lambda_1,\lambda_2)=\frac{1}{4}\left[f(\lambda_1,\lambda_2)-f(-\lambda_1,\lambda_2)+f(\lambda_1,-\lambda_2)-f(-\lambda_1,-\lambda_2)\right],$$

$$f_{eo}(\lambda_1,\lambda_2)=\frac{1}{4}\left[f(\lambda_1,\lambda_2)+f(-\lambda_1,\lambda_2)-f(\lambda_1,-\lambda_2)-f(-\lambda_2,-\lambda_2)\right],$$

$$f_{oo}(\lambda_1,\lambda_2)=\frac{1}{4}\left[f(\lambda_1,\lambda_2)-f(-\lambda_1,\lambda_2)-f(\lambda_1,-\lambda_2)+f(-\lambda_1,-\lambda_2)\right]$$

are correspondingly the even and odd parts of the function $f(\lambda_1,\lambda_2)$ taken relatively to each variable separately on $[0,\|A\|]\times[0,\|B\|]\ni(\lambda_1,\lambda_2)$.

Note that

$$\hat{\phi}(\lambda_1)=\begin{bmatrix}\lambda_1 & 0\\ 0 & -\lambda_1\end{bmatrix} \text{ and } \hat{\phi}(\lambda_2)=\begin{bmatrix}0 & \lambda_2\\ \lambda_2 & 0\end{bmatrix} \quad (\lambda_1\geq 0,\ \lambda_2\geq 0).$$

The image of $C([-\|A\|,\|A\|]\times[-\|B\|,\|B\|])$ under this linear mapping $\hat{\phi}$ is the closed $*$-subalgebra $\hat{\mathbf{U}}$ in the C^*-algebra of continuous 2×2 matrix functions

$$M(\lambda_1,\lambda_2)=\begin{bmatrix}f_{11}(\lambda_1,\lambda_2) & f_{12}(\lambda_1,\lambda_2)\\ f_{21}(\lambda_1,\lambda_2) & f_{22}(\lambda_1,\lambda_2)\end{bmatrix}$$

defined on $[0,\|A\|]\times[0,\|B\|]$ with

$$\|M(\cdot)\|=\max_{(\lambda_1,\lambda_2)\in[0,\|A\|]\times[0,\|B\|]}\|M(\lambda_1,\lambda_2)\|$$

and pointwise involution and product.

PROPOSITION 4. For $M(\cdot,\cdot)$ to belong to $\hat{\mathbf{U}}$ it is necessary and sufficient that

$$\begin{aligned}&1)\ f_{11}(0,\lambda_2)=f_{22}(0,\lambda_2)\\ &2)\ f_{12}(0,\lambda_2)=f_{21}(0,\lambda_2)\\ &3)\ f_{12}(\lambda_1,0)=f_{21}(\lambda_1,0)=0,\quad \lambda_1\in[0,\|A\|],\ \lambda_2\in[0,\|B\|].\end{aligned}\tag{2}$$

The proof of the necessity is obvious. Conversely, we show how to reconstruct the function $f(\cdot,\cdot)$ on $[-\|A\|,\|A\|]\times[-\|B\|,\|B\|]$ from the continuous matrix function

$$M(\cdot,\cdot)=\begin{bmatrix}f_{11}(\cdot,\cdot) & f_{12}(\cdot,\cdot)\\ f_{21}(\cdot,\cdot) & f_{22}(\cdot,\cdot)\end{bmatrix}$$

on $[0,\|A\|]\times[0,\|B\|]$ which satisfy the conditions 1) - 3):

$$f(\lambda_1, \lambda_2) = f_{11}(\lambda_1, \lambda_2) + f_{12}(\lambda_1, \lambda_2) \qquad (\lambda_1 \geq 0, \lambda_2 \leq 0)$$

$$f(\lambda_1, \lambda_2) = f_{11}(\lambda_1, |\lambda_2|) - f_{12}(\lambda_1, |\lambda_2|) \qquad (\lambda_1 \geq 0, \lambda_2 < 0)$$

$$f(\lambda_1, \lambda_2) = f_{21}(|\lambda_1|, \lambda_2) + f_{22}(|\lambda_1|, \lambda_2) \qquad (\lambda_1 < 0, \lambda_2 \geq 0)$$

$$f(\lambda_1, \lambda_2) = f_{22}(|\lambda_1|, |\lambda_2|) - f_{21}(|\lambda_1|, |\lambda_2) \quad (\lambda_1 < 0, \lambda_2 < 0) \ .$$

Continuity of the function $f(\cdot,\cdot)$ on $[-\|A\|, \|A\|] \times [-\|B\|, \|B\|]$ follows from 1) - 3). □

Because the mapping $\hat{\phi} : C([-\|A\|, \|A\|] \times [-\|B\|, \|B\|]) \to \hat{\mathrm{U}}$ is bijective this allows to endow $C([-\|A\|, \|A\|] \times [-\|B\|, \|B\|])$ with the structure of a C^*-algebra $C_a([-\|A\|, \|A\|] \times [-\|B\|, \|B\|])$ by setting for $f(\cdot,\cdot) \in C([-\|A\|, \|A\|] \times [-\|B\|, \|B\|])$

$$f^*(\cdot,\cdot) = \hat{\phi}^{-1}(\hat{\phi}(f(\cdot,\cdot))^*) \ ,$$

$$\|f(\cdot,\cdot)\|_{C_a([-\|A\|,\|A\|]\times[-\|B\|,\|B\|])} = \|M_{f(\cdot,\cdot)}\|_{\hat{\mathrm{U}}}$$

and for $f_1(\cdot,\cdot), f_2(\cdot,\cdot) \in C([-\|A\|, \|A\|] \times [-\|B\|, \|B\|])$

$$(f_1(\cdot,\cdot) \cdot f_2(\cdot,\cdot)) = \hat{\phi}^{-1}(M_{f_1(\cdot,\cdot)} \cdot M_{f_2(\cdot,\cdot)}) \ .$$

PROPOSITION 5. The $*$-algebra U is everywhere dense in $C_a([-\|A\|, \|A\|] \times [-\|B\|, \|B\|])$.

Proof. $\hat{\phi}(\mathrm{U})$ consists of all polynomial matrix functions P on $[0, \|A\|] \times [0, \|B\|]$

$$P(\lambda_1, \lambda_2) = \begin{bmatrix} P_{11}(\lambda_1, \lambda_2) & P_{12}(\lambda_1, \lambda_2) \\ P_{21}(\lambda_1, \lambda_2) & P_{22}(\lambda_1, \lambda_2) \end{bmatrix}$$

such that

$$\begin{aligned} P_{12}(0, \lambda_2) &= P_{21}(0, \lambda_2) \quad (\lambda_2 \geq 0) \\ P_{11}(0, \lambda_2) &= P_{22}(0, \lambda_2) \quad (\lambda_2 \geq 0) \\ P_{12}(\lambda_1, 0) &= P_{21}(\lambda_1, 0) \quad (\lambda_1 \geq 0). \end{aligned} \tag{3}$$

The set of such polynomial matrix function is dense in $\hat{\mathrm{U}}$

Indeed, let

$$M_{f(\cdot,\cdot)} = \begin{bmatrix} f_{11}(\cdot,\cdot) & f_{12}(\cdot,\cdot) \\ f_{21}(\cdot,\cdot) & f_{22}(\cdot,\cdot) \end{bmatrix}$$

satisfy (2). We can approximate $M_{f(\cdot,\cdot)}$ in $\hat{\mathrm{U}}$ by polynomial functions $P^{(n)}(\cdot,\cdot)$ which satisfy (3). To do this we use the $f_{ij}(\lambda_1, \lambda_2)$ $(\lambda_1 \geq 0, \lambda_2 \geq 0; i, j = 1, 2)$ to reconstruct the function

$$f(\lambda_1, \lambda_2), \quad (\lambda_1, \lambda_2) \in [-\|A\|, \|A\|] \times [-\|B\|, \|B\|]$$

and uniformly approximate this function on the compact set $[-\|A\|, \|A\|] \times [-\|B\|, \|B\|]$ by polynomials $P^{(n)}(\lambda_1, \lambda_2)$. Consider the polynomial matrix function

$$\hat{\phi}(P^{(n)}(\cdot,\cdot)) = \begin{bmatrix} P_{11}^{(n)}(\cdot,\cdot) & P_{12}^{(n)}(\cdot,\cdot) \\ P_{21}^{(n)}(\cdot,\cdot) & P_{22}(\cdot,\cdot) \end{bmatrix}$$

which satisfy condition (3). Because

$$\|M_{f(\cdot,\cdot)}\|_{\hat{U}}^2 = \max_{(\lambda_1,\lambda_2)\in[0,\|A\|]\times[0,\|B\|]} \|M_f(\lambda_1,\lambda_2)\|^2 \le$$

$$\le 2 \max_{(\lambda_1,\lambda_2)\in[0,\|A\|]\times[0,\|B\|]} (|f_{ee}(\lambda_1,\lambda_2)|^2 + |f_{eo}(\lambda_1,\lambda_2)|^2 +$$

$$+ |f_{oe}(\lambda_1,\lambda_2)|^2 + |f_{oo}(\lambda_1,\lambda_2)|^2) \le$$

$$\le 8 \max_{(\lambda_1,\lambda_2)\in[-\|A\|,\|A\|]\times[-\|B\|,\|B\|]} |f(\lambda_1,\lambda_2)|^2 = 8\,\|f(\cdot,\cdot)\|^2_{C([-\|A\|,\|A\|]\times[-\|B\|,\|B\|]}$$

we have

$$\|M_f(\cdot,\cdot) - \hat{\phi}(P^{(n)}(\cdot,\cdot))\|_{\hat{U}} \le \text{const}\cdot \|f(\cdot,\cdot) - P^{(n)}(\cdot,\cdot)\|_{C([-\|A\|,\|A\|]\times[-\|B\|,\|B\|])}$$

and so $\hat{\phi}(U)$ is dense everywhere in $\hat{U}$. □

PROPOSITION 6. There is a one-to-one correspondence between the pairs of bounded self-adjoint anticommuting operators A, B on H and $*$- representations of the C^*-algebra $\hat{U}$.

Indeed, to a pair of bounded anticommuting self-adjoint operators there corresponds a $*$-homomorphism of U, but any $*$-homomorphism of the dense $*$-algebra U into the C^*-algebra $L(H)$ can be continuously extended to a $*$-homomorphism of the C^*-algebra $C_a([-\|A\|,\|A\|]\times[-\|B\|,\|B\|])$. Conversely, a representation of $C_a([-\|A\|,\|A\|]\times[-\|B\|,\|B\|])$ on the generators a_1 and a_2 gives a pair of bounded self-adjoint anticommuting operators. □

For unbounded anticommuting self-adjoint operators, we have the following

PROPOSITION 7. There is a one-to-one correspondence between pairs of anticommuting unbounded self-adjoint operators A, B and $*$-representations of the C^*-algebra $\hat{U}_C$ of continuous matrix functions

$$M(\lambda_1,\lambda_2) = \begin{bmatrix} f_{11}(\lambda_1,\lambda_2) & f_{12}(\lambda_1,\lambda_2) \\ f_{21}(\lambda_1,\lambda_2) & f_{22}(\lambda_1,\lambda_2) \end{bmatrix}$$

on $[0,\infty)\times[0,\infty)$ satisfying the conditions 1) - 3) and

4) $$\lim_{\lambda_1+\lambda_2\to\infty} M(\lambda_1,\lambda_2) = 0.$$

8.4. Functions of anticommuting operators.

The spectral theorem for commuting self-adjoint operators allows us to construct a functional calculus for a wide class of functions which are measureable and finite almost everywhere relatively to the joint spectral measure. We use the structure Theorem 3 to construct ordered functions $f(\lambda_1, \lambda_2)$ $((\lambda_1, \lambda_2) \in \mathbb{R}^2)$ of a pair of self-adjoint anticommuting operators A, B.

For the pair of bounded self-adjoint anticommuting operators A, B on H define a $*$-homomorphism ϕ of the algebra U into the C^*-algebra $L(H)$ by:

$$\mathrm{U} \ni P(\lambda_1, \lambda_2) = \sum_{j=0}^{n} \sum_{k=0}^{m} c_{jk} \lambda_1^j \lambda_2^k \overset{\phi}{\mapsto}$$

$$\overset{\phi}{\mapsto} P(A, B) = \sum_{j=0}^{n} \sum_{k=0}^{m} c_{jk} A^k B^k \in L(H) \ .$$

Because the operators A and B are bounded, the $*$-homomorphism ϕ can be continuously extended to a $*$-homomorphism of the C^*-algebra $C_a([-\|A\|, \|A\|] \times [-\|B\|, \|B\|])$. For every continous function $f(\cdot,\cdot) \in C([-\|A\|, \|A\|] \times [-\|B\|, \|B\|])$ it yields a bounded operator $f(A, B) = \phi(f(\cdot,\cdot)) \in L(H)$.

If the operators A and B are unbounded, then $f(A, B)$ is defined for all $f(\cdot,\cdot) \in C(\mathbb{R}^2)$ such that $f(\lambda_1, \lambda_2) \to 0$, $|\lambda_1| + |\lambda_2| \to \infty$.

However, as in the case of two commuting self-adjoint operators, the structure theorem allows to extend the ordered functional calculus to functions of two variables, $f(\lambda_1, \lambda_2)$ measurable and finite almost everywhere with respect to $dE_0(\cdot)$ and $dE_1(\cdot)$:

$$f(A, B) \overset{def}{=} \int_{\mathbb{R}_+^2} f(\lambda_1, \lambda_2)\, dE_0(\lambda_1, \lambda_2) \oplus$$

$$\oplus [\sigma_0 \otimes \int_{\mathbb{R}_{++}^2} f_{ee}(\lambda_1, \lambda_2)\, dE_1(\lambda_1, \lambda_2) +$$

$$+ \sigma_z \otimes \int_{\mathbb{R}_{++}^2} f_{oe}(\lambda_1, \lambda_2)\, dE_1(\lambda_1, \lambda_2) +$$

$$+ \sigma_x \otimes \int_{\mathbb{R}_{++}^2} f_{eo}(\lambda_1, \lambda_2)\, dE_1(\lambda_1, \lambda_2) -$$

$$- i\sigma_y \otimes \int_{\mathbb{R}_{++}^2} f_{oo}(\lambda_1, \lambda_2)\, dE_1(\lambda_1, \lambda_2)]$$

where $\sigma_0 = \begin{bmatrix} 1 & 0 \\ 0 & 1 \end{bmatrix}$, $\sigma_x = \begin{bmatrix} 0 & 1 \\ 1 & 0 \end{bmatrix}$, $\sigma_y = \begin{bmatrix} 0 & i \\ -i & 0 \end{bmatrix}$, $\sigma_z = \begin{bmatrix} 1 & 0 \\ 0 & -1 \end{bmatrix}$ are the Pauli matrices, and $f_{ee}(\cdot,\cdot)$, $f_{eo}(\cdot,\cdot)$, $f_{oe}(\cdot,\cdot)$ and $f_{oo}(\cdot,\cdot)$ are correspondingly even and odd parts of the function $f(\cdot,\cdot)$ taken with respect to each variable.

Comments to Chapter 8.

The broadened exposition of the author's outline [5].

For anticommuting self-adjoint unbounded operators see also F.-H. Vasilescu [1, 2].

Chapter 9.
FINITE AND COUNTABLE COLLECTIONS OF GRADED-COMMUTING SELF-ADJOINT OPERATORS (GCSO)

In this chapter we study both finite and countable collections of self-adjoint operators which are mutually connected by relations of commutation or anticommutation (graded commutation). To describe which operators of the collection commute or anticommute, we can consider the operators of the collection as operators of the self-adjoint representation of an unordered simple graph Γ. There the vertices of the graph Γ correspond to the operators of the collection (A_k). If there is an edge between the vertices a_i and a_j, it means that the operators A_i and A_j anticommute. If there is no edge, it means that they commute. For finite and "tame" countable collections of graded-commuting operators we can prove a spectral theorem, for "wild" countable collections we study particular classes of representations and the objects which arise: measures and cocycles.

9.1. Finite collections of graded-commuting self-adjoint operators.

We start the study of finite collections of mutually commuting or anticommuting self-adjoint operators by considering representations of the self-adjoint generators of the algebra of observables of a one-dimensional quantum spin system with finite degrees of freedom. This is a collection of unitary self-adjoint operators $(Z_k, X_k)_{k=1}^{n}$ where all the operators commute except for the pairs Z_k, X_k $(k=1,\dots,n)$ which anticommute. Then we study the graded-commuting collections of self-adjoint operators $(A_k)_{k=1}^{\infty}$ with the condition $A_k^2 = I$ $(k=1,\dots,n)$. For these operators, the structure questions can be solved rather easily because there are only a finite number of unitarily unequivalent irreducible representations. In the next step we will study the general case of self-adjoint representations of an unoriented graph Γ without loops and multiple arcs. In this case the operators are supposed to satisfy the condition $A_k^2 = I$ $(k=1,\dots,n)$. Representations of such a graph can be regarded as self-adjoint representations of the basis of a commutative $\times \mathbb{Z}_2^m$-graded Lie algebra.

First, we consider representations of the generators of the algebra of observables of a one-dimensional spin system with finite degrees of freedom. An algebra of observables of a quantum system with n degrees of freedom is a $*$-algebra S_n, generated by self-adjoint unitary generators Z_k and X_k $(k=1,\dots,n)$ which satisfy the relations

$$Z_k Z_j - Z_j Z_k = X_k X_j - X_j X_k = 0 \quad (k,j=1,\dots,n)$$

$$Z_k X_j - X_j Z_k = 0 \quad (k \neq j)$$

$$Z_k X_k + X_k Z_k = 0 \quad (k=1, \ldots, n).$$

Maintaining X_k, Z_k $(k=1, \ldots, n)$ as notations for the representation operators corresponding to the generators, consider a $*$-representation of the algebra S_n. The operators $(Z_k, X_k)_{k=1}^n$ establish a collection of self-adjoint operators $(Z_k)_{k=1}^n$ and $(X_k)_{k=1}^n$ on a complex Hilbert space H, satisfying the relations

$$Z_k^2 = X_k^2 = I \quad (k=1, \ldots, n)$$

$$[X_k, X_j] = [Z_k, Z_j] = 0 \quad (k,j=1, \ldots, n)$$

$$[Z_k, X_j] = 0 \quad (k \neq j)$$

$$\{Z_k, X_k\} = Z_k X_k + X_k Z_k = 0 \quad (k=1, \ldots, n). \tag{1}$$

On the space $\mathbb{C}^2$, the pair of anticommuting unitary self-adjoint operators Z, X is unique up to unitary equivalence and can be realized by the Pauli matrices

$$\sigma_z = \begin{pmatrix} 1 & 0 \\ 0 & -1 \end{pmatrix}, \quad \sigma_x = \begin{pmatrix} 0 & 1 \\ 1 & 0 \end{pmatrix}.$$

An irreducible representation of a finite collection of such pairs is in the same way unique. We have the following theorem.

THEOREM 1. There exists a unique up to unitary equivalence irreducible collection of self-adjoint operators $(Z_k, X_k)_{k=1}^n$ which satisfy the relations (1). Simplicity of the spectrum of the system of commuting operators $(Z_k)_{k=1}^n$ is equivalent to simplicity of the spectrum of the system of commuting operators $(X_k)_{k=1}^n$ and is equivalent to irreducibility of the whole collection.

Proof. We begin the study of the irreducible representations of the algebra of observables of a quantum spin system with finite degrees of freedom by diagonalizing in $H = H_1 \oplus H_2$ the self-adjoint operator Z_1 satisfying the condition $Z_1^2 = I$

$$Z_1 = \begin{pmatrix} I_1 & 0 \\ 0 & -I_2 \end{pmatrix}$$

where I_1 is the identity on H_1 and I_2 the identity on H_2.

Because the operators X_1 and Z_1 anticommute, we can bring X_1 into the form

$$X_1 = \begin{pmatrix} 0 & C^* \\ C & 0 \end{pmatrix}.$$

Because the operator X_1 is non-degenerate, the dimensions m_1 and m_2 of the spaces H_1 and H_2 are equal, so that $H_1 = H_2 = \hat{H}$ and $I_1 = I_2 = \hat{I}$. By choosing an appropriate unitary operator U on H

$$U = \begin{bmatrix} U_1 & 0 \\ 0 & U_2 \end{bmatrix}$$

we can then transform the operator X_1 into the form

$$X_1 = \begin{bmatrix} 0 & \hat{I} \\ \hat{I} & 0 \end{bmatrix} .$$

Because the operators $(Z_k)_{k=2}^n$ and $(X_k)_{k=2}^n$ together with Z_1 and X_1 satisfy the commutation relations (1), we deduce that they have the form

$$Z_k = \begin{bmatrix} \hat{Z}_k & 0 \\ 0 & \hat{Z}_k \end{bmatrix} , \quad X_k = \begin{bmatrix} \hat{X}_k & 0 \\ 0 & \hat{X}_k \end{bmatrix} \quad (k=2, \ldots , n)$$

where the systems of self-adjoint operators $(\hat{Z}_k)_{k=2}^n$ and $(\hat{X}_k)_{k=2}^n$ satisfy the initial conditions (1).

Representing the space H as $\mathbf{C}^2 \otimes \hat{H}$ we can rewrite the systems of operators in the form

$$Z_1 = \sigma_z \otimes \hat{I} , \quad Z_k = \sigma_0 \otimes \hat{Z}_k \quad (k=2, \ldots , n) ,$$

$$X_1 = \sigma_x \otimes \hat{I} , \quad X_k = \sigma_0 \otimes \hat{X}_k$$

where the Pauli matrices

$$\sigma_0 = \begin{bmatrix} 1 & 0 \\ 0 & 1 \end{bmatrix} , \quad \sigma_x = \begin{bmatrix} 0 & 1 \\ 1 & 0 \end{bmatrix} , \quad \sigma_z = \begin{bmatrix} 1 & 0 \\ 0 & -1 \end{bmatrix} .$$

Repeating the described procedure for the system of operators $(Z_k, X_k)_{k=2}^n$ we finally get

$$H = \underbrace{\mathbf{C}^2 \otimes \cdots \otimes \mathbf{C}^2}_{n \text{ times}} \otimes H' ,$$

$$Z_k = \sigma_0 \otimes \cdots \otimes \overbrace{\sigma_0 \otimes \sigma_z \otimes \sigma_0}^{k\text{-th place}} \otimes \cdots \otimes \sigma_0 \otimes I' ,$$

$$Z_k = \sigma_0 \otimes \cdots \otimes \overbrace{\sigma_0 \otimes \sigma_x \otimes \sigma_0}^{k\text{-th place}} \otimes \cdots \otimes \sigma_0 \otimes I'$$

$$(k = 1, \ldots , n).$$

From this it follows that the irreducible representation is unique. The space H' is necessarily one-dimensional and the operators on $H = \underbrace{\mathbf{C}^2 \otimes \cdots \otimes \mathbf{C}^2}_{n \text{ times}}$ have the form

$$Z_k = \sigma_0 \otimes \cdots \otimes \sigma_0 \otimes \underbrace{\sigma_z}_{k\text{-th place}} \otimes \sigma_0 \otimes \cdots \otimes \sigma_0$$

$$X_k = \sigma_0 \otimes \cdots \otimes \sigma_0 \otimes \underbrace{\sigma_x}_{k\text{-th place}} \otimes \sigma_0 \otimes \cdots \otimes \sigma_0 \ , \quad k = 1, \ldots, n. \tag{2}$$

The dimension of the irreducible representation equals 2^n.

To prove that the systems of operators $(Z_k)_{k=1}^n$ and $(X_k)_{k=1}^n$ of the irreducible representation have a simple spectrum it is sufficient to give a cyclic vector for both systems of operators. Such a vector can be chosen to be

$$f = f_1 \otimes f_2 \otimes \cdots \otimes f_k \otimes \cdots \otimes f_n$$

where $f_k = C_1^k e_1 + C_{-1}^k e_{-1}$ $(k = 1, \ldots, n)$, where $e_{\pm 1}$ is the orthonormal basis consisting of the eigenvectors of the operator σ_z and where the coefficients $C_{\pm 1}^k$ satisfy the following conditions:

1) $C_{\pm 1}^k \neq 1$ which guarantees that every f_k is cyclic for σ_z;

2) $C_1^k \neq \pm C_{-1}^k$, which implies that f_k is cyclic for σ_x;

3) $|C_1^k|^2 + |C_{-1}^k|^2 = \|f_k\|^2 = 1.$ □

In the sequel we will also need the collection of self-adjoint operators $(Y_k)_{k=1}^n = (i Z_k X_k)_{k=1}^n$. In the unique irreducible representation on $H = \bigotimes_{k=1}^n \mathbb{C}^2$ of the algebra of observables, the operators Y_k can be written as:

$$Y_k = \sigma_0 \otimes \cdots \otimes \sigma_0 \otimes \underbrace{\sigma_y}_{k\text{-th place}} \otimes \sigma_0 \otimes \cdots \otimes \sigma_0 \quad (k = 1, \ldots, n)$$

where the Pauli matrix $\sigma_y = \begin{pmatrix} 0 & i \\ -i & 0 \end{pmatrix}$.

The algebra F_n of observables of a Fermi system with finite degrees of freedom is called the $*$-algebra generated by the operators of creation and annihilation a_k and a_k^* $(k = 1, \ldots, n)$ which satisfy the Fermi anticommutation relations:

$$\{a_k, a_j\} = \{a_k^*, a_j^*\} = 0 \quad (k, j = 1, \ldots, n) \tag{3}$$

$$\{a_k^*, a_j\} = \delta_{kj} I.$$

where $\{a, b\} = ab + ba$.

Consider the self-adjoint generators $A_k = a_k + a_k^*$ and $B_k = \frac{1}{i}(a_k - a_k^*)$. We have the following:

$$A_k^2 = B_k^2 = I\ , \quad \{A_k, B_l\} = 0\ , \quad (k, l = 1, \ldots, n) \tag{4}$$

$$\{A_k, A_l\} = \{B_k, B_l\} = 0 \quad (k \neq l = 1, \ldots, n).$$

The $*$-algebra, K_{2n} generated by the self-adjoint generators $(A_k, B_k)_{k=1}^n$ is a Clifford algebra with an even number of generators. The following proposition can be checked by direct calculations.

PROPOSITION 1. The algebras S_n of observables of spin systems and F_n of Fermi systems with n degrees of freedom are each $*$-isomorphic to a Clifford algebra K_{2n} with $2n$ generators. The isomorphism is defined for the generators (1) and (4) as:

$$A_k = X_1 \cdots X_{k-1} Z_k, \quad B_k = X_1 \cdots X_{k-1} Y_k \quad (k = 1, \ldots, n).$$

Under this isomorphism, the operators of creation and annihilation correspond to the operators:

$$a_k = \frac{1}{2}(A_k + i B_k) = X_1 \cdots X_{k-1} \frac{Z_k + i Y_k}{2}$$

$$a_k^* = \frac{1}{2}(A_k - i B_k) = X_1 \cdots X_{k-1} \frac{Z_k - i Y_k}{2}.$$

As irreducibility of the collection $(A_k, B_k)_{k=1}^n$ implies irreducibility of the collection $(Z_k, X_k)_{k=1}^n$ and vice versa, and unitary equivalence of the collections $(A_k, B_k)_{k=1}^n$ and $(\tilde{A}_k, \tilde{B}_k)_{k=1}^n$ implies unitary equivalence of the corresponding collections $(Z_k, X_k)_{k=1}^n$ and $(\tilde{Z}_k, \tilde{X}_k)_{k=1}^n$. From Theorem 1 it follows

THEOREM 2. There exists a unique up to unitary equivalence irreducible collection $(A_k, B_k)_{k=1}^n$ of self-adjoint operators which satisfy the relations (4).

This is the representation on $H = \bigotimes_{k=1}^{n} \mathbb{C}^2$ where

$$A_k = \underbrace{\sigma_x \otimes \cdots \otimes \sigma_x}_{(k-1) \text{ times}} \otimes \sigma_z \otimes \sigma_0 \otimes \cdots \otimes \sigma_0 \qquad (k = 1, \ldots, n)$$

$$B_k = \underbrace{\sigma_x \otimes \cdots \otimes \sigma_x}_{(k-1) \text{ times}} \otimes \sigma_y \otimes \sigma_0 \otimes \cdots \otimes \sigma_0$$

Here the creation and annihilation operators $(a_k, a_k^*)_{k=1}^n$ act through the formulas:

$$a_k = \underbrace{\sigma_x \otimes \cdots \otimes \sigma_x}_{(k-1) \text{ times}} \otimes s \otimes \sigma_0 \otimes \cdots \otimes \sigma_0$$

$$a_k^* = \underbrace{\sigma_x \otimes \cdots \otimes \sigma_x}_{(k-1) \text{ times}} \otimes s^* \otimes \sigma_0 \otimes \cdots \otimes \sigma_0$$

where

$$s = \frac{1}{2}\begin{bmatrix} 1 & -1 \\ 1 & -1 \end{bmatrix}.$$

Now, consider a finite collection of self-adjoint unitary operators $(A_k)_{k=1}^n$ ($A_k^2 = I$, $k = 1, \ldots, n$), where some pairs commute and others anticommute. Relate to such a collection a simple graph $\Gamma = (S, R)$ (without loops and multiple edges). Here, S is a set of vertices $(a_k)_{k=1}^n$ and R is a set of two element subsets of S corresponding to the edges. The vertices a_k and a_j of the graph Γ are connected with an edge if $\{A_k, A_j\} = 0$ and there is no edge if the operators commute.

In what follows we will regard such collections as self-adjoint and unitary representations of the graph Γ.

Our task will be to study the spectral questions for such collections of operators: to describe up to unitary equivalence the irreducible collections and to describe the decomposition of any collection into irreducible ones.

THEOREM 3. A simple graph Γ with n vertices have $2^{r(\Gamma)}$ ($0 \le r(\Gamma) \le n$) unitarily unequivalent irreducible unitary self-adjoint representations of equal dimension $2^{m(\Gamma)}$ where $2m(\Gamma) + r(\Gamma) = n$.

Proof. We enumerate the irreducible collections of unitary CSO $(A_k)_{k=1}^n$ which form representations of a graph consisting of n isolated points: $\Gamma = (\overset{a_1}{\cdot}\ \overset{a_2}{\cdot}\ \cdots\ \overset{a_n}{\cdot})$. The collection of unitary CSO $A_1, \ldots, A_n$ has 2^n unitarily unequivalent irreducible one-dimensional representations $A_k = \pm 1$ $(k = 1, \ldots, n)$. The subspaces $H_{i_n, \ldots, i_n}$ in

$$H = \oplus \sum_{i_1, \ldots, i_n \in \{-1, 1\}} H_{i_1, \ldots, i_n}$$

are stable under the representation operators $(A_k)_{k=1}^n$ and operate in them as operators of multiplication by $(i_k)_{k=1}^n$. In this case $r(\Gamma) = n$, $m(\Gamma) = 0$.

Suppose now that at least two of the vertices (without loss of generality we can assume that these are a_1 and a_2) are connected by an edge. Then the operators A_1 and A_2 anticommute and $A_1^2 = A_2^2 = I$. So we can decompose $H = \mathbb{C}^2 \otimes H_1$ in such a way that A_1 and A_2 will have the following block structure in H:

$$A_1 = \begin{bmatrix} I_1 & 0 \\ 0 & -I_2 \end{bmatrix} = \sigma_z \otimes I, \quad A_2 = \begin{bmatrix} 0 & I_1 \\ I_1 & 0 \end{bmatrix} = \sigma_x \otimes I$$

(I_1 is the identity operators on H_1). So

if A_k commutes with A_1 and A_2 then $A_k = \sigma_0 \otimes B_k$;

if A_k commutes with A_1 and anticommutes with A_2 then $A_k = \sigma_z \otimes B_k$;

if A_k anticommutes with A_1 and commutes with A_2 then $A_k = \sigma_x \otimes B_k$;

if A_k anticommutes with A_1 and A_2, then $A_k = \sigma_y \otimes B_k$, where the B_k are self-adjoint unitary operators on H_1, which also either commute or anticommute. One can check directly the following construction rule for the graph $\Gamma_1 = (S_1, R_1)$ which corresponds to the collection of self-adjoint unitary operators $(B_k)_{k=3}^n$ from the graph $\Gamma = (S, R)$ corresponding to the collection $(A_k)_{k=1}^n$ where the vertices a_1 and a_2 are joined by an edge:

a) the graph Γ_1 contains the vertices $(b_k)_{k=3}^n$;

b) if in the graph Γ the vertex a_k is contained in the star of the vertex a_1 (i.e. it is connected with a_1 by an edge), and a_j is contained in the star of a_2 but at least one of them is not in both of the stars (i.e. it is not connected with both vertices), then the edge $(b_k, b_j) \in R_1$ if $(a_k, a_j) \notin R$ and conversely, $(b_k, b_j) \notin R_1$ if $(a_k, a_j) \in R$;

c) in all other cases $(b_k, b_j) \in R_1$ if $(a_k, a_j) \in R$ and $(b_k, b_j) \notin R_1$, if $(a_k, a_j) \notin R$.

So the unitary self-adjoint representation problem for the graph Γ is reduced to the unitary self-adjoint representation problem for the graph Γ_1, which contains two vertices less than Γ. If in Γ_1 all the vertices are isolated, then $m(\Gamma) = 1$ and $r(\Gamma) = n-2$. If in Γ_1 at least two vertices (suppose b_3 and b_4) are connected by an edge, then we construct Γ_2 from Γ_1, etc.

At the end, we find that either in the graph $\Gamma_{m(\Gamma)}$ all vertices will become isolated and so the graph $\Gamma_{m(\Gamma)}$ and so the graph Γ will have $2^{n-2m(\Gamma)} = 2^{r(\Gamma)}$ unitarily unequivalent irreducible representations (the dimension of each irreducible representations of the graph Γ is equal to $2^{m(\Gamma)}$, or $m(\Gamma) = \frac{n}{2}$. In this case we have the theorem of uniqueness up to unitary equivalence of the irreducible unitary self-adjoint representation of the graph Γ. In particular, it follows from Theorem 2 that for the graph $(\mathrm{I}\,\mathrm{I}\cdots\mathrm{I})$, the number $r(\mathrm{I}\,\mathrm{I}\cdots\mathrm{I}) = 0$. □

Now, the spectral decomposition of the CSO leads to the spectral decomposition of the unitary self-adjoint representations of the graph Γ. Indeed, any unitary self-adjoint representation of the graph Γ on the separable Hilbert space $H = \underbrace{\mathbb{C}^2 \otimes \cdots \otimes \mathbb{C}^2}_{m(\Gamma)} \otimes \mathbf{H}$ has the form

$$A_k = \sigma_{k1} \otimes \cdots \otimes \sigma_{km(\Gamma)} \otimes \mathbf{A}_k \quad (k = 1, \ldots, n) \tag{5}$$

($\mathbf{A}_1 = \cdots = \mathbf{A}_{2m(\Gamma)} = I$, I is the identity on $\mathbf{H}$, σ_{kj} is the Pauli matrix contained as the j-th factor in A_k and $\mathbf{A}_{2m(\Gamma)+1}, \ldots, \mathbf{A}_n$ are unitary CSO). Identify the points of the spectrum of the collection $(A_k)_{k=1}^n$ with the points of the spectrum of the collection of the unitary CSO $(\mathbf{A}_k)_{k=1}^n \ni (1, \ldots, 1, i_{2m(\Gamma)+1}, \ldots, i_n)$, where $i_k \in \{-1, 1\}$. The space $\mathbf{H}$ can be decomposed into the direct sum $\mathbf{H} = \oplus \sum_{i_{2m(\Gamma)+1}, \ldots, i_n} \mathbf{H}_{i_{2m(\Gamma)+1}, \ldots, i_n}$ of subspaces which are invariant for $(\mathbf{A}_k)_{k=1}^n$ and in which these operators are operators of multiplication by $(1, \ldots, 1, i_{2m(\Gamma)+1}, \ldots, i_n)$ correspondingly. But then the operators A_k on

$$H = \underbrace{\mathbb{C}^2 \otimes \cdots \otimes \mathbb{C}^2}_{m(\Gamma)} \otimes \mathrm{H} =$$

$$\oplus \sum_{i_{2m(\Gamma)+1}, \ldots, i_n} \underbrace{\mathbb{C}^2 \otimes \cdots \otimes \mathbb{C}^2}_{m(\Gamma)} \otimes \mathrm{H}_{i_{2m(\Gamma)+1}, \ldots, i_n} =$$

$$\bigoplus_{i_{2m(\Gamma)+1}, \ldots, i_n} H_{i_{2m(\Gamma)+1}, \ldots, i_n}$$

reduce the invariant subspaces $H_{i_{2m(\Gamma)}, \ldots, i_n}$ and have the form

$$A_k \restriction H_{i_{2m(\Gamma)+1}, \ldots, i_n} = \sigma_{k1} \otimes \cdots \otimes \sigma_{km(\Gamma)} \otimes i_k \quad (k = 1, \ldots, n). \tag{6}$$

If the CSO $(\mathbf{A}_k)_{k=1}^n$ have a simple joint spectrum, then the spaces $\mathrm{H}_{i_{2m(\Gamma)+1}, \ldots, i_n}$ have dimension ≤ 1, and the decomposition of the space $H = \oplus \sum_{i_{2m(\Gamma)+1}, \ldots, i_n} H_{i_{2m(\Gamma)+1}, \ldots, i_n}$ is the decomposition into relatively to $(A_k)_{k=1}^n$ invariant subspaces, in which the irreducible representations of the collection are realized. If $(\mathbf{A}_k)_{k=1}^n$ has a multiple spectrum, then we get the decomposition of the unitary self-adjoint representation of the graph Γ into irreducible ones by additionally choosing a basis $e^{(k)}_{i_{2m(\Gamma)+1}, \ldots, i_n}$ in every $\mathrm{H}_{i_{2m(\Gamma)+1}, \ldots, i_n}$

$$\mathrm{H}_{i_{2m(\Gamma)+1}, \ldots, i_n} = \oplus \sum_k \mathrm{H}^{(k)}_{i_{2m(\Gamma)+1}, \ldots, i_n},$$

$$H_{i_{2m(\Gamma)+1}, \ldots, i_n} = \oplus \sum_k \underbrace{\mathbb{C}^2 \otimes \cdots \otimes \mathbb{C}^2}_{m(\Gamma)} \otimes \mathrm{H}_{i_{2m(\Gamma)+1}, \ldots, i_n} =$$

$$= \oplus \sum_k H_{i_{2m(\Gamma)+1}, \ldots, i_n}.$$

Here, the equivalent irreducible unitary self-adjoint representations of Γ are realized on every $H^{(k)}_{i_{2m(\Gamma)+1}, \ldots, i_n}$. So, we have the theorem:

THEOREM 4. Unitarily unequivalent irreducible unitary self-adjoint representations of the graph Γ are realized on $H_{i_{2m(\Gamma)+1}, \ldots, i_n}$ $(i_k = \pm 1)$ according to formulas (6). Any unitary self-adjoint representation of the graph Γ is unitarily equivalent to the direct sum (possible with multiples) of irreducible representations on $H_{i_{2m(\Gamma)+1}, \ldots, i_n}$.

Now consider an arbitrary collection $(A_k)_{k=1}^n$ of self-adjoint operators which establish a self-adjoint representation of a graph $\Gamma = (S, R)$, i.e. the operators that anticommute if the vertices a_i and a_j in Γ are joined by an edge and commute if the edge is absent.

PROPOSITION 2. $(A_i^2)_{i=1}^n$ are nonnegative CSO which commute with all A_k $(k = 1, \ldots, n)$.

The proof is similar to the proof of Lemma 1.8. □

Next we describe, up to unitary equivalence, the irreducible non-degenerated self-adjoint representations of Γ. Because the operators $(A_i^2)_{i=1}^n$ commute with all the operators of the representation, assuming that the representation is irreducible, we get that $A_k^2 = \lambda_k I$ $(\lambda_k > 0;\ k = 1, \ldots, n)$.

THEOREM 5. There is a one-to-one correspondence between the irreducible unequivalent self-adjoint representations of the graph Γ and the points of the set

$$\{\lambda_1 > 0\} \times \cdots \times \{\lambda_{2m(\Gamma)} > 0\} \times [\{\lambda_{2m(\Gamma)+1} > 0\} \cup \{\lambda_{2m(\Gamma)+1} < 0\}] \times \cdots \times [\{\lambda_n > 0\} \cup \{\lambda_n < 0\}].$$

Proof. Because the operators $(A_i)_{i=1}^n$ are non-degenerate, following the proof of Theorem 3, all of them using unitary transformations of the space $\underbrace{\mathbb{C}^2 \otimes \cdots \otimes \mathbb{C}^2}_{m(\Gamma)}$, can be brought into the form

$$A_k = \lambda_k \sigma_{k1} \otimes \cdots \otimes \sigma_{km(\Gamma)} \quad (k = 1, \ldots, n)$$

(here σ_{kj} correspond to the operator A_k in the irreducible self-adjoint unitary representation of Γ (see (5)),

$$(\lambda_1, \ldots, \lambda_{2m(\Gamma)}) \in \{\lambda_1 > 0\} \times \cdots \{\lambda_{2m(\Gamma)} > 0\}$$

and $(\lambda_{2m(\Gamma)+1}, \ldots, \lambda_n)$ is a point of the spectrum of the non-degenerate CSO $(\mathbf{A}_k)_{k=2m(\Gamma)+1}^n$. So, to any non-degenerate irreducible self-adjoint representation of Γ there corresponds a point of the set

$$\{\lambda_1 > 0\} \times \cdots \times \{\lambda_{2m(\Gamma)} > 0\} \times [\{\lambda_{2m(\Gamma)+1} > 0\} \cup \{\lambda_{2m(\Gamma)+1} < 0\}] \times \cdots \times [\{\lambda_n > 0\} \cup \{\lambda_n < 0\}].$$

If the collections $(\lambda_1, \ldots, \lambda_n)$ and $(\tilde{\lambda}_1, \ldots, \tilde{\lambda}_n)$ are different, then the spectrum of at least one of the operators $(A_k)_{k=1}^n$ differs from the spectrum of the corresponding operator from $(\tilde{A}_k)_{k=1}^n$. □

Now, the nondegenerate operators $(A_k)_{k=1}^n$ of a representation of the graph Γ using unitary transformations, can be brought on $H = \underbrace{\mathbb{C}^2 \otimes \cdots \otimes \mathbb{C}^2}_{m(\Gamma)} \otimes \mathbf{H}$ into the form

$$A_k = \sigma_{k1} \otimes \cdots \otimes \sigma_{km(\Gamma)} \otimes \mathbf{A}_k \quad (k = 1, \ldots, n),$$

where $\mathbf{A}_1, \ldots, \mathbf{A}_{2m(\Gamma)}$ are positive self-adjoint operators, and $\mathbf{A}_{2m(\Gamma)+1}, \ldots, \mathbf{A}_n$ are non-degenerate self-adjoint operators on $\mathbf{H}$. Applying the spectral theorem to the collection $(\mathbf{A}_k)_{k=1}^n$ of CSO leads to the following theorem.

THEOREM 6. (Spectral theorem for non-degenerate self-adjoint representations of a graph Γ in the form of a resolution of the identity).

To any non-degenerate self-adjoint representation of the graph Γ on a separable Hilbert space, H, there corresponds a decomposition, $H = \mathbb{C}^2 \otimes \cdots \otimes \mathbb{C}^2 \otimes \mathbf{H}$ and an orthogonal operator-valued measure $E(\cdot)$ on

$$\sigma_+(\Gamma) = \{\lambda_1 > 0\} \times \cdots \times \{\lambda_{2m(\Gamma)} > 0\} \times$$

$$\times [\{\lambda_{2m(\Gamma)+1} > 0\} \cup \{\lambda_{2m(\Gamma)+1} < 0\}] \times \cdots \times [\{\lambda_n > 0\} \cup \{\lambda_n < 0\}]$$

taking as its values projections onto the subspaces of $\mathbf{H}$ such that

$$A_k = \sigma_{k1} \otimes \cdots \otimes \sigma_{km(\Gamma)} \otimes \int_{\sigma_+(\Gamma)} \lambda_k \, dE(\lambda_1, \ldots, \lambda_n) \quad (k = 1, \ldots, n).$$

Note that to an orthogonal operator-valued measure $E(\cdot)$ on $\sigma_+(\Gamma)$ there correspond the spectral type and the dimension function both of which form a complete system of unitary invariants of the non-degenerate representations of the graph Γ.

We recall now the definition of (non-graded) representation of a graded real Lie algebra.

Let $L = L_0 \oplus L_1$ be a $\mathbb{Z}_2$-graded real vector space, $\dim L_0 = p$, $\dim L_1 = q$. The elements of L_0 are called even, the elements of L_1 odd, and both even and odd elements are called homogeneous. The grading $\delta(\cdot)$ is defined for the homogeneous elements: $\delta(x) = 0$ for $x \in L_0$, $\delta(x) = 1$ for $x \in L_1$.

DEFINITION 1. A real (p,q)-dimensional $\mathbb{Z}_2$-graded Lie algebra (Lie superalgebra) is a real vector space $L = L_0 \oplus L_1$ with a bilinear operation $<\cdot,\cdot>$ such that

a) $\delta(<x,y>) = \delta(x) + \delta(y)$;

b) $<x,y> = -(-1)^{\delta(x)\delta(y)} <y,x>$;

c) $(-1)^{\delta(x)\delta(z)} <x, <y,z>> + (-1)^{\delta(z)\delta(y)} <z, <x,y>> + (-1)^{\delta(y)\delta(x)} <y, <z,x>> = 0$ for homogeneous $x,y,z \in L$.

In what follows we will consider only the commutative Lie superalgebras, i.e. the Lie superalgebras such that $<x,y> = 0 \ \forall x, y \in L$.

DEFINITION 2. A (non-graded) representation $\pi(\cdot)$ is a homomorphism of L into a set of linear closed operators on a Hilbert space H such that for all homogeneous $x,y \in L$

$$\pi(x)\,\pi(y) f - (-1)^{\delta(x)\delta(y)}\,\pi(y)\,\pi(x) f = \pi(<x,y>) f = 0$$

on a dense domain $\Phi \ni f$ invariant with respect to the operators of the representation.

Note, that because there are odd elements in L, we cannot consider a non-trivial representation of L_1 to be self-adjoint, for if $\delta(x) = 1$ then $(\pi(x))^2 = 0$ which is impossible for a non-trivial self-adjoint operator $\pi(x)$. Thus, if we need to consider the graded-commuting collections of self-adjoint operators as the representation operators for a basis in a graded Lie algebra, we need to use the $\mathbb{Z}_2 \times \cdots \times \mathbb{Z}_2$-grading.

Let now

$$L = \oplus \sum L_{(i_1,\dots,i_n)} \quad (i_k \in \mathbb{Z}_2\,,\ \dim L_{(i_1,\dots,i_n)} = p_{(i_1,\dots,i_n)}).$$

As before, the elements $x \in L_{(i_1,\dots,i_n)} = ((i_1,\dots,i_n) \in \mathbb{Z}_2 \times \cdots \times \mathbb{Z}_2)$ will be called homogeneous. Set

$$\delta(x) = (\delta_1(x),\dots,\delta_n(x)) = (i_1,\dots,i_n) \in \mathbb{Z}_2 \times \cdots \times \mathbb{Z}_2.$$

DEFINITION 3. A real $(p_{(i_1,\dots,i_n)})$-dimensional $\mathbb{Z}_2 \times \cdots \times \mathbb{Z}_2$-graded Lie algebra is a real vector space $L = \oplus \sum L_{(i_1,\dots,i_n)}$ with a bilinear operation $<\cdot,\cdot>$ such that

a) $\delta_k(<x,y>) = \delta_k(x) + \delta_k(y) \quad (k = 1,\dots,n)$;

b) $<x,y> = -(-1)^{\sum \delta_k(x)\delta_k(y)} <y,x>$;

c) $(-1)^{\sum \delta_k(x)\delta_k(z)} <x, <y,z>> + ((-1)^{\sum \delta_k(z)\delta_k(y)} <z, <x,y>> +$
$+ (-1)^{\sum \delta_k(y)\delta_k(x)} <y, <z,x>> = 0$ for homogeneous $x,y,z \in L$.

We will consider only the commutative $\mathbb{Z}_2 \times \cdots \times \mathbb{Z}_2$-graded Lie algebras, i.e. $<x,y> = 0 \ \forall x,y \in L$.

DEFINITION 4. A (non-graded) representation $\pi(\cdot)$ is a homomorphism of L into the closed operators of H such that for homogeneous $x,y \in L$

$$\pi(x)\,\pi(y) f - (-1)^{\sum \delta_k(x)\delta_k(y)}\, \pi(y)\,\pi(x) f = \pi(<x,y>) f = 0$$

on a dense domain $\Phi \ni f$ invariant with respect to the representation operators.

The essentially self-adjoint representation operators $\pi(\cdot)$ defined on Φ are trivial on the subspaces whose grading contains an odd number of ones.

It is natural to say that two self-adjoint representations, $\pi_1(\cdot)$ on H_1 and $\pi_2(\cdot)$ on H_2 are equivalent if there exists a unitary operator $U : H_1 \to H_2$ such that $\pi_1(x) = U^* \pi_2(x)\, U$.

We will consider only self-adjoint representations $\pi(\cdot)$ for which $\Phi \subset H^{\omega}(\pi(x))$ for all $x \in L$.

The description problem for such self-adjoint representations of a commutative $\mathbb{Z}_2 \times \cdots \times \mathbb{Z}_2$-graded Lie algebra is equivalent to the description problem for all self-adjoint representations of a graph.

PROPOSITION 3. For every commutative $\mathbb{Z}_2 \times \cdots \times \mathbb{Z}_2$-graded Lie algebra L there is a graph Γ such that there is a one-to-one correspondence between the self-adjoint representations of Γ and the self-adjoint representations of a homogeneous basis in L. Conversely, for any graph Γ there exists a commutative $\mathbb{Z}_2 \times \cdots \times \mathbb{Z}_2$-graded Lie algebra such that there exists a one-to-one correspondence between the self-adjoint representations of its generators and the self-adjoint representations of the graph Γ.

Proof. Let $(e_k)_{k=1}^{n}$ be a homogeneous basis of a commutative $\mathbb{Z}_2 \times \cdots \times \mathbb{Z}_2$-graded Lie algebra. Then the graph $\Gamma = (S, R)$ which contains the vertices $(a_k)_{k=1}^{n}$ and the edges

(a_i, a_j) has the same self-adjoint representations as the basis $(e_i)_{i=1}^n$ in L, if $\sum \delta_k(e_i)\,\delta_k(e_j) = 1$.

Conversely, for every graph Γ given an n-dimensional commutative $\mathbb{Z}_2 \times \cdots \times \mathbb{Z}_2$-graded Lie algebra, we choose the grading as follows: if the vertices a_1 and a_2 are joined by an edge, grade the vector e_1 by the collection $(110\,000 \cdots 000) \in \mathbb{Z}_2^{3n}$ and the vector e_2 by $(101\,110 \cdots 0) \in \mathbb{Z}_2^{3n}$, the vector e_k will be graded by the collection $(\ldots \cdots 110\,0 \cdots 0)$, choosing the first $3(k-1)$ numbers depending on whether the vertex a_k is joined with the vertices $a_1, \ldots, a_{k-1}$ or it is not. Any self-adjoint representation of the thus graded commutative Lie algebra is a representation of the graph Γ. $\square$

9.2. Countable collections of GCSO. The Spectral theorem for "tame" countable collections of GCSO.

Consider a finite collection $A_1, \ldots, A_n$ of unbounded self-adjoint operators satisfying

$$A_i A_j - (-1)^{g(i,j)} A_j A_i = 0. \tag{7}$$

Here, $g(\cdot,\cdot)$ is a function of $i,j = 1, \ldots, n$ taking the values zero or one such that

a) $g(i,i) = 0, \quad i = 1, \ldots, n;$

b) $g(i,j) = g(j,i) \quad i,j = 1, \ldots, n$

(we will call such functions grading).

For such families of operators:

1) $\bigcap_{k=1}^{n} D(A_k)$ is dense in H and there exists a nuclear linear topological space Φ which is dense in H and invariant with respect to the operators of the family;

2) one can give a description of the irreducible families of self-adjoint operators satisfying (7) and prove a theorem similar to the spectral theorem on the decomposition of an arbitrary family into irreducible ones.

Let us prove that for a countable family of unbounded self-adjoint operators $A_1, \ldots, A_n, \ldots$ satisfying the relations

$$A_i A_j - (-1)^{g(i,j)} A_j A_i = 0 \tag{8}$$

(here $g(\cdot,\cdot)$ is a grading function of $i,j = 1, \ldots, n, \ldots$ taking the values zero or one such that

a) $g(i,i) = 0 \quad i = 1, \ldots, n, \ldots;$

b) $g(i,j)=g(j,i) \quad i,j=1,\ldots,n,\ldots;$)

the set $\bigcap_{k=1}^{\infty} D(A_k)$ is still dense in H and let us prove that there exists a nuclear linear topological space Φ which is dense in H and invariant with respect to the operators of the family. However, in case of an arbitrary family $A_1,\ldots,A_n,\ldots$ of self-adjoint operators satisfying relations (8), it depends on the choice of the function $g(\cdot,\cdot)$ whether or not we can describe all such irreducible families. Relations (8) can be splitted into two classes: the class of those, for which it is possible to prove a structural theorem, and the class of the rest, for which this problem contains, for example, a yet unsolved problem of describing the representations of the canonical anticommutation relations of systems with infinite degrees of freedom (see Section 9.3.).

Consider a countable family of unbounded self-adjoint operators $A_1,\ldots,A_n,\ldots$ on H which pairwise commute or anticommute depending on the given grading function $g(\cdot,\cdot)$

$$A_i A_j - (-1)^{g(i,j)} A_j A_i = 0 \quad (i,j=1,2,\ldots).$$

LEMMA 1. The set $\bigcap_{k=1}^{\infty} D(A_k)$ is dense in H.

Proof. Consider the set

$$H^b(A_k) = \{f \in \bigcap_{n=0}^{\infty} D(A_k^n) \mid \|A_k^n f\| \le C_f^n \quad (n=0,1,\ldots)\}$$

of bounded vectors for the operator A_k. As the operators $A_1^2,\ldots,A_n^2,\ldots$ commute, $H^b(A_k)=H^b(A_k^2)$ $(k=1,2,\ldots)$ and the set $\bigcap_{k=1}^{\infty} H^b(A_k)=\bigcap_{k=1}^{\infty} H^b(A_k^2)$ is dense so the set $\bigcap_{k=1}^{\infty} D(A_k) \supset \bigcap_{k=1}^{\infty} H^b(A_k)$ is also dense in H. □

Now we construct a nuclear linear topological space Φ topologically dense in H, $\Phi \subset \bigcap_{k=1}^{\infty}\bigcap_{n=0}^{\infty} D(A_k^n)$ with $A_k : \Phi \to \Phi$ being continuous for all $k=1,2,\ldots$.

First, suppose that the family $A_1,\ldots,A_n,\ldots$ has a cyclic vector, i.e. a vector $\Omega \in \bigcap_{k=1}^{\infty}\bigcap_{n=0}^{\infty} D(A_k^n)$ such that $\operatorname{CLS}_{\substack{k=1,2,\ldots \\ n=0,1,\ldots}} \{A_k^n \Omega\} = H$. Write

$$H_{i_1\cdots i_n\cdots} = \operatorname{LS}\{A_1^{k_1}\cdots A_n^{k_n}\cdots\Omega \mid 0\le k_1\le i_1,\ldots,0\le k_n\le i_n,\ldots\}$$

$$(i_1,\ldots,i_n,\ldots)\in \mathbb{N}_0^{\infty}$$

and introduce on $\Phi = \bigcup_{i_1=0}^{\infty}\cdots\bigcup_{i_n=0}^{\infty}\cdots H_{i_1\cdots i_n\cdots}$ the inductive limit topology induced by the finite-dimensional spaces $H_{i_1\cdots i_n\cdots}$. The space Φ is a nuclear topological space, topologically dense in H, $\Phi\subset\bigcap_{k=1}^{\infty}\bigcap_{n=0}^{\infty} D(A_k^n)$ and $A_k : \Phi\to\Phi$ is continuous for all $k=1,2,\ldots$.

Choosing in Φ and orthonormal basis $(e_n)_{k\in I\!N}$ we have

$$\Phi = \bigcap_{(p_k)} l_2((p_k))$$

where

$$l_2((p_k)) = \{\sum_{k=1}^{\infty} c_k e_n \mid \sum_{k=1}^{\infty} |c_k|^2 p_n < \infty\},$$

and (p_k) denotes any sequence of positive numbers. So on Φ we can introduce also the projective limit topology brought about by the Hilbert space $l_2((p_k))$. In this case the projective limit topology, and the inductive limit topology are the same.

Now, if the family of the operators $A_1, \ldots, A_n, \ldots$ does not have a cyclic vector, then choosing the vectors $\Omega_1, \ldots, \Omega_n, \ldots$ ($\Omega_j \in \bigcap_{k=1}^{\infty} \bigcap_{n=0}^{\infty} D(A_k^n)$) such that $\underset{j=1,2,\ldots}{\mathrm{CLS}} \{\Omega_j\} = H$ and denoting

$$H_{i_1 \cdots i_n \cdots}^{(j)} = \mathrm{LS}\,\{A_1^{k_1} \cdots A_n^{k_n} \cdots \Omega_i \mid 0 \le i \le j,\ 0 \le k_1 \le i_1, \ldots,\ 0 \le k_n \le i_n, \ldots\}$$

and

$$\Phi = \bigcup_{j=1}^{\infty} \bigcup_{i_1=0}^{\infty} \cdots \bigcup_{i_n=0}^{\infty} \cdots H_{i_1 \cdots i_n \cdots}^{(j)}$$

with the inductive (= projective) limit topology, we get the following theorem.

THEOREM 7. There exists a nuclear linear topological space, $\Phi \subset \bigcap_{k=1}^{\infty} \bigcap_{n=0}^{\infty} D(A_k^n)$ topological dense in H such that $A_k : \Phi \to \Phi$ is continuous ($k = 1, 2, \ldots$).

The possibility to describe all such irreducible collections up to unitary equivalence depends on the choice of the grading function $g(\cdot,\cdot)$.

We give a description of all irreducible families of self-adjoint operators $A_1, \ldots, A_n, \ldots$ which satisfy conditions (8) in case that this family can be splitted into a finite number of subfamilies

$$A_1^{(1)}, \ldots, A_n^{(1)}, \ldots;$$

$$A_1^{(2)}, \ldots, A_n^{(2)}, \ldots;$$

....

$$A_1^{(m)}, \ldots, A_n^{(m)}, \ldots,$$

such that each subfamily contains a finite countable number of operators, and a grading function $\tilde{g}(\cdot,\cdot)$ exists that $\forall k, j$

$$A_k^{(p)} A_j^{(l)} - (-1)^{\tilde{g}(p,l)} A_j^{(l)} A_k^{(p)} = 0 \quad (p, l = 1, \ldots, m). \tag{9}$$

Suppose first that the operators $A_1^{(1)}, \ldots, A_1^{(m)}$ are not degenerate. The description up to unitary equivalence of the irreducible representations of families of non degenerate self-adjoint operators $\mathbf{A}_1^{(1)}, \ldots, \mathbf{A}_1^{(m)}$ with

$$\mathbf{A}_1^{(0)} \mathbf{A}_1^{(l)} - (-1)^{g(p,l)} \mathbf{A}_1^{(l)} \mathbf{A}_1^{(p)} = 0 \quad (p, l = 1, \ldots, m)., \tag{10}$$

is given in Section 9.1. Let $(\mathbf{A}_1^{(l)})_{l=1}^m$ be the operators of such a representations on H.

THEOREM 8. There is a one-to-one correspondence between the set of irreducible unitary non-equivalent representations of the family (9) such that the operators $A_1^{(1)}, \ldots, A_1^{(m)}$ are non-degenerate, the direct product of the set of non-degenerate none-quivalent representations of the family (10), and the set

$$\mathbb{R}^1 \times \mathbb{R}^1 \times \cdots \ni (\lambda_2^{(1)}, \lambda_3^{(1)}, \ldots; \lambda_2^{(2)}, \lambda_3^{(2)}, \ldots; \lambda_2^{(m)}, \lambda_3^{(m)} \cdots).$$

The representation operators are the operators on H

$$A_1^{(l)} = \mathbf{A}_1^{(l)} \quad \text{and} \quad A_k^{(l)} = \lambda_k^{(l)} \mathbf{A}_1^{(l)} \quad (k \geq 2;\, l = 1, \ldots, m). \tag{11}$$

Proof. Any irreducible representation of the family (9) such that the operators $(A_1^{(l)})_{l=1}^m$ are non degenerate has the form (11) because the representation space H can be represented as $H = \mathbf{H} \otimes H_1$ and the operators

$$A_1^{(l)} = \mathbf{A}_1^{(l)} \otimes I_1, \quad A_k^{(l)} = \mathbf{A}^{(l)} \otimes \mathbf{B}_k^{(l)} \quad (k \geq 2;\, l = 1, \ldots, m)$$

where $\mathbf{B}_k^{(l)}$ $(k \geq 2;\, l = 1, \ldots, m)$ are commuting self-adjoint operators on H_1, I_1 is the identity operator on H_1.

Conversely, the given representation of the family (9) is irreducible and since the operators $\mathbf{A}_1^{(l)}$ $(l = 1, \ldots, m)$ are non-degenerate, any two such representations $(A_k^{(l)})_{l=1}^m$, $k = 1,2,\ldots$ on $H = \mathbf{H}$ and $(\tilde{A}_k^{(l)})_{l=1}^m$, $k = 1,2,\ldots$ on $\tilde{H} = \tilde{\mathbf{H}}$ are unitarily equivalent if and only if the families $(\mathbf{A}_1^{(l)})_{l=1}^m$ on $\mathbf{H}$ and $(\tilde{\mathbf{A}}_1^{(l)})_{l=1}^m$ on $\tilde{\mathbf{H}}$ are unitarily equivalent and the sequence $(\lambda_2^{(1)}, \ldots; \cdots; \lambda_2^{(m)}, \ldots)$ is the same as the sequence $(\tilde{\lambda}_2^{(1)}, \ldots; \cdots; \tilde{\lambda}_2^{(m)}, \ldots)$. ▯

If any of the operators $A_1^{(l)}$ $(l = 1, \ldots, n)$ are degenerate then choosing among the operators $(A_k^{(l)})$ $(k = 1,2,\ldots)$ the first non-degenerate ones $A_{k_l}^{(l)}$, construct the representations on H

$$A_1^{(l)} = \cdots = A_{k_l-1}^{(l)} = 0, \quad A_{k_l}^{(l)} = \mathbf{A}_1^{(l)}, \quad A_k^{(l)} = \lambda_k^{(l)} A_1^{(l)},$$

$$(l = 1, \ldots, m;\, k > k_l)$$

which give all such irreducible unequivalent representations of the family (9).

If all the operators $A_1^{(p)} = A_2^{(p)} = \cdots = 0$, then the problem of describing the representations of (9) is reduced to the problem of describing the representations of the family (A_k^l) $(l = 1, \ldots, p-1, p+1, \ldots, m;\, k = 1,2,\ldots)$.

9.3. Representations of the algebra of local observables of a spin system with countable degrees of freedom.

In this chapter we will study unitary self-adjoint representations of the graph

$$(\mathbf{I}\,\mathbf{I}\cdots\mathbf{I}\cdots)=(\mathbf{I})^{\infty}$$

i.e. we will study the collections of unitary self-adjoint operators on a complex Hilbert space H, $(Z_k)_{k=1}^{\infty}$ and $(X_k)_{k=1}^{\infty}$ satisfying the relations

$$Z_k^2=X_k^2=I \quad (k=1,2,\ldots),$$

$$[Z_k,Z_j]=[X_k,X_j]=0 \quad (k,j=1,2,,\ldots),$$

$$[Z_k,X_j]=0 \quad (k\neq j=1,2,\ldots),$$

$$\{Z_k,X_k\}=Z_kX_k+X_kZ_k=0 \quad (k=1,2,\ldots).$$

Consider the $*$-algebra M_2 of all operators on $\mathbb{C}^2$ with the usual operations of addition, multiplication and conjugation. Let $M_{2^n}=\bigotimes_1^n M_2$ be the $*$-algebra of the operators on the 2^n-dimensional space $\mathbb{C}^{2n}$.

M_{2^n} is a complex Banach algebra, i.e. it is a Banach space with norm $\|\cdot\|_n$ (the norm of the operators on $\mathbb{C}^{2^n}$) for which the multiplication satisfies the inequality $\|xy\|_n\le\|x\|_n\,\|y\|_n$ $(x,y\in M_{2^n})$.

M_{2^n} is a $*$-algebra (involutory Banach algebra); the adjoint induces a mapping $*$: $M_{2^n}\to M_{2^n}$ with the following properties:

1) $(x^*)^*=x$;

2) $(x+y)^*=x^*+y^*$;

3) $(\alpha x)^*=\bar{\alpha}x^*$;

4) $(xy)^*=y^*x^*$;

5) $\|x^*\|=\|x\|$

for all $x,y\in M_{2^n}$ and $\alpha\in\mathbb{C}^1$; moreover, M_{2^n} is a C^*-algebra because

6) $\|xx^*\|=\|x^*\|\,\|x\|$.

Define the imbeddings $J_{n+1}^n: M_{2^n}\hookrightarrow M_{2^{n+1}}$ induced by the tensor multiplication by σ_0 which is the identity matrix in $\mathbb{C}^2$: $J_{n+1}^n\,M_{2^n}=M_{2^n}\otimes\sigma_0\subset M_{2^{n+1}}$. Thus M_{2^n} can be considered as a C^*-subalgebra of $M_{2^{n+1}}$.

DEFINITION 5. A $*$-algebra M of local observables of a spin system is called the union of the increasing sequence of the imbedded C^*-algebras M_{2^n} $(n=1,2,\ldots)$ if $M=J-\lim\limits_{\rightarrow}M_{2^n}$ $(J=(J_{n+1}^n))$.

Because the imbeddings J_{n+1}^n $(n=1,2,\ldots)$ preserve the norm of the operators, there is a natural norm $\|\cdot\|$ on M.

DEFINITION 6. The completion of M with respect to the norm $\|\cdot\|$ is called a C^*-algebra of quasi-local observables of a one-dimensional spin system.

Denote $J_n : M_{2^n} \hookrightarrow M$ $(n=1,2,...)$.

Similar to Chapter 2 one can choose the generators of the $*$-algebra M in different ways. Let, for example,

$$a_k^{(n)} = \underbrace{\sigma_x \otimes \cdots \otimes \sigma_x}_{k-1} \otimes s \otimes \sigma_0 \otimes \cdots \otimes \sigma_0 \qquad (k=1,\ldots,n)$$

$$a_k^{(n)*} = \underbrace{\sigma_x \otimes \cdots \otimes \sigma_x}_{k-1} \otimes s^* \otimes \sigma_0 \otimes \cdots \otimes \sigma_0$$

be generators of M_{2^n} satisfying the Fermi commutation relations. Then $J_n(a_k^{(n)}) = a_k$ and $J_n(a_k^{(n)*}) = a_k^*$ $(k=1,2,...)$ are generators of the algebra M satisfying the Fermi commutation relations for systems with countable degrees of freedom:

$$\{a_k, a_j\} = \{a_k^*, a_j^*\} = 0, \quad \{a_k^*, a_j\} = \delta_{kj} I \quad (k,j=1,2,...).$$

A $*$-algebra with generators satisfying the Fermi commutation relations is called an algebra of local observables of a Fermi system.

Choosing in M_{2^n} the generators

$$A_k^{(n)} = \underbrace{\sigma_x \otimes \cdots \otimes \sigma_x}_{k-1} \otimes \sigma_z \otimes \sigma_0 \otimes \cdots \otimes \sigma_0 \qquad (k=1,\ldots,n)$$

$$B_k^{(n)} = \underbrace{\sigma_x \otimes \cdots \otimes \sigma_x}_{k-1} \otimes \sigma_y \otimes \sigma_0 \otimes \cdots \otimes \sigma_0$$

and setting $A_k = J_n(A_k^{(n)})$ and $B_k = J_n(B_k^{(n)})$, we get a collection of self-adjoint unitary generators in M, satisfying the relations

$$A_k^* = A_k, \quad B_k^* = B_k, \quad A_k^2 = B_k^2 = I \quad (k=1,2,...)$$

$$\{A_k, A_l\} = \{B_k, B_l\} = 0 \quad (k \neq l = 1,2,...)$$

$$\{A_k, B_l\} = 0 \quad (k,l = 1,2,...).$$

The generated $*$-algebra is called a Clifford algebra with a countable number of generators. Finally, if we choose the following generators in M_{2^n}

$$Z_k^{(n)} = \sigma_0 \otimes \cdots \otimes \sigma_0 \otimes \overbrace{\sigma_z}^{k\text{-th place}} \otimes \sigma_0 \otimes \cdots \otimes \sigma_0$$

$(k=1,\ldots,n)$

$$X_k^{(n)} = \sigma_0 \otimes \cdots \otimes \underbrace{\sigma_0 \otimes \sigma_x \otimes \sigma_0}_{k\text{-th place}} \otimes \cdots \otimes \sigma_0$$

and set $Z_k = J_n(Z_k^{(n)})$ and $X_k = J_n(X_k^{(n)})$ $(k=1,2,\ldots)$, we get a collection of self-adjoint unitary generators in M which satisfy the following relations:

$$Z_k^* = Z_k\,,\ X_k^* = X_k\,,\ Z_k^2 = X_k^2 = I \quad (k=1,2,\ldots)\,,$$

$$[Z_k, Z_j] = [X_k, X_j] = 0 \quad (k,j = 1,2,\ldots)\,,$$

$$[Z_k, X_j] = 0 \quad (k \neq j = 1,2,\ldots)\,,$$

$$\{Z_k, X_k\} = Z_k X_k + X_k Z_k = 0 \quad (k=1,2,\ldots).$$

Thus we get the following

PROPOSITION 4. The $*$-algebras of local observables of Fermi systems with countable degrees of freedom, the Clifford algebra with a countable number of generators and the $*$-algebra M of local observables of a spin system are $*$-isomorphic.

This $*$-isomorphism is first defined for the generators of the $*$-algebras $(a_k, a_k^*)_{k=1}^{\infty}$, $(A_k, B_k)_{k=1}^{\infty}$ and $(Z_k, X_k)_{k=1}^{\infty}$ and then extended to a $*$-isomorphism of the algebras of quasi-local observables.

Because a representation of a $*$-algebra M i.e. a $*$-isomorphism of the algebra M into the algebra of linear bounded operators on H is completely determined by the representation of its generators, we will consider only the collections of self-adjoint unitary operators $(Z_k, X_k)_{k=1}^{\infty}$ satisfying relations (12).

We study the representations of the relations (12) by first diagonalizing the system of commuting operators and then using the relations (12).

Let $\mathbb{Z}_2^{\infty} = \prod_1^{\infty} \mathbb{Z}_2 = \{-1,1\}^{\infty}$ be a commutative group of sequences of +1 and −1, $\mathbb{Z}_{2,0}^{\infty} = \sum_1^{\infty} \mathbb{Z}_2$ be the subgroup in $\mathbb{Z}_2^{\infty}$ of the sequences that have only a finite number of −1, and set $s_k = (1, 1, \ldots, \underbrace{1, -1, 1}_{k-\text{th place}}, \ldots)$. We have the following theorem.

THEOREM 9. For the operators $(Z_k, X_k)_{k=1}^{\infty}$ to define a representation of the collection (12) on a Hilbert space H it is necessary and sufficient that the space H can be represented as

$$H = \oplus \int_{\mathbb{Z}_2^{\infty}} H_\lambda \, d\rho(\lambda)\,,$$

where $\rho(\cdot)$ is a $\mathbb{Z}_{2,0}^{\infty}$-quasi-invariant measure on $\mathbb{Z}_2^{\infty}$, and the operators Z_k and

X_k $(k=1,2,...)$ are defined as

$$(Z_k \vec{f})(\lambda)=\lambda_k \vec{f}(\lambda)$$
$$(X_k \vec{f})(\lambda)=C_k(\lambda)\sqrt{\frac{d\rho(s_k\lambda)}{d\rho(\lambda)}}\,\vec{f}(s_k\lambda), \tag{12_0}$$

where $C_k(\lambda): H_{s_k\lambda}\to H_\lambda$ are unitary measurable operator-functions satisfying the equations

$$C_k(\lambda)\,C_k(s_k\lambda)=I$$
$$C_k(\lambda)\,C_j(s_k\lambda)=C_j(\lambda)\,C_k(s_j\lambda). \tag{12_K}$$

Proof. First, suppose that the joint spectrum of the countable collection of unitary CSO $(Z_k)_{k=1}^{\infty}$ is simple. Then H is unitarily equivalent to $L_2(\mathbb{Z}_2^{\infty}, d\rho(\lambda))$. The operators $(Z_k)_{k=1}^{\infty}$ are realized on

$$L_2(\mathbb{Z}_2^{\infty}, d\rho(\lambda))=\oplus\int_{\mathbb{Z}_2^{\infty}} H_\lambda\, d\rho(\lambda)\quad(\dim H_\lambda=1)$$

as operators of multiplication by the independent variables $(\lambda_k)_{k=1}^{\infty}$. Let us prove that the measure $\rho(\cdot)$ on $(\mathbb{Z}_2^{\infty}, \mathbf{B}(\mathbb{Z}_2^{\infty}))$ is quasi-invariant with respect to multiplication by the elements $t\in\mathbb{Z}_{2,0}^{\infty}$, i.e with respect to the measurable transformations of $(\mathbb{Z}_2^{\infty}, \mathbf{B}(\mathbb{Z}_2^{\infty}))$ which change a finite fixed number of variables. For any cylindrical function

$$f(\lambda)=\sum c_{i_1\cdots i_n}\lambda_1^{i_1}\cdots\lambda_n^{i_n}\quad(i_1\cdots i_n\in\{0,1\}^n,\ n=1,2,...)$$
$$(X_k f)(\lambda)=\sum c_{i_1\cdots i_n} X_k\lambda_1^{i_1}\cdots\lambda_n^{i_n}=$$
$$=(\sum c_{i_1\cdots i_n}\lambda_1^{i_1}\cdots\lambda_{k-1}^{i_{k-1}}(s_k\lambda_k)^{i_k}\lambda_{k+1}^{i_{k+1}}\cdots\lambda_n^{i_n})\,a_k(\lambda)=$$
$$=f(s_k\lambda)\,a_k(\lambda)\quad(k=1,2,...),$$

where $a_k(\lambda)=X_k 1$. Here, as $X_k^2 1=I1=1$,

$$a_k(\lambda)\,a_k(s_k\lambda)=1,$$

and since $X_kX_j1=X_jX_k1$,

$$a_k(\lambda)\,a_j(s_k\lambda)=a_j(\lambda)\,a_k(s_j\lambda).$$

Consequently, because the operator X_k is unitary, for any function $f(\cdot)\in L_2(\mathbb{Z}_2^{\infty}, d\rho(\lambda))$

$$\int_{\mathbb{Z}_2^{\infty}}|f(\lambda)|^2\,d\rho(\lambda)=\int_{\mathbb{Z}_2^{\infty}}|f(s_k\lambda)|^2\,|a_k(\lambda)|^2\,d\rho(\lambda)$$

and

$$\frac{d\rho(s_k\lambda)}{d\rho(\lambda)} = |a_k(\lambda)|^2 \quad (|a_k(\lambda)| > 0 \ (\operatorname{mod} \rho(\cdot);\ k = 1,2,\ldots).$$

This means that the measure $\rho(\cdot)$ is $\mathbb{Z}_{2,0}^{\infty}$-quasi-invariant.

Introduce the $\mathbf{B}(\mathbb{Z}_2^{\infty})$-measurable functions $\alpha_k(\lambda) = \dfrac{a_k(\lambda)}{|a_k(\lambda)|}$ $(k=1,2,\ldots)$ such that

a) $|\alpha_k(\lambda)| = 1$;

b) $\alpha_k(\lambda)\, \alpha_k(s_k\lambda) = 1$;

c) $\alpha_k(\lambda)\, \alpha_j(s_k\lambda) = \alpha_j(\lambda)\, \alpha_k(s_j\lambda)$.

Then the operators will be given on $L_2(\mathbb{Z}_2^{\infty}, d\rho(\lambda))$ by the formulas

$$(Z_k f)(\lambda) = \lambda_k f(\lambda), \quad (X_k f) = \alpha_k(\lambda)\sqrt{\frac{d\rho(s_k\lambda)}{d\rho(\lambda)}}\, f(s_k\lambda) \tag{$12_0'$}$$

and in the case of a simple joint spectrum of the CSO $(Z_k)_{k=1}^{\infty}$ the theorem is proved.

Now, consider the case that the joint spectrum of the CSO $(Z_k)_{k=1}^{\infty}$ is multiple. Then H is unitarily equivalent to

$$\oplus \int_{\mathbb{Z}_2^{\infty}} H_\lambda\, d\rho_1(\lambda) = \oplus \sum_k L_2(\mathbb{Z}_2^{\infty}, d\rho_k(\lambda)) \quad (\rho_1(\cdot) \gg \rho_2(\cdot) \gg \cdots)$$

and the operators $(Z_k)_{k=1}^{\infty}$ are unitarily equivalent to the operators of multiplication by the independent variables $(\lambda_k)_{k=1}^{\infty}$.

We show that all the measures $\rho_k(\cdot)$ $(k=1,2,\ldots)$ are $\mathbb{Z}_{2,0}^{\infty}$-quasi-invariant. Let $\Omega_k = 1 \ (\operatorname{mod} \rho_k(\cdot))$

$$(\Omega_k \in L_2(\mathbb{Z}_2^{\infty}, d\rho_k(\lambda)) \ \text{and} \ X_j\Omega_k = \vec{a}_j^{(k)}(\cdot) \in \oplus \sum_k L_2(\mathbb{Z}_2^{\infty}, d\rho_k(\lambda))).$$

Because for any complex-valued function $f(\lambda) \in L_2(\mathbb{Z}_2^{\infty}, d\rho_k(\lambda))$

$$\int_{\mathbb{Z}_2^{\infty}} |f(\lambda)|^2\, d\rho_k(\lambda) = \int_{\mathbb{Z}_2^{\infty}} |f(s_j\lambda)|^2\, \|\vec{a}_j^{(k)}(\lambda)\|_{H_\lambda}^2\, d\rho_1(\lambda)$$

$$(j = 1,2,\ldots),$$

we can use induction on k to show that all the measures $\rho_k(\cdot)$ $(k=1,2,\ldots)$ are $\mathbb{Z}_{2,0}^{\infty}$-quasi-invariant and that

$$\frac{d\rho_k(s_j\lambda)}{d\rho_k(\lambda)} = \|a_j^{(k)}(\lambda)\|_{H_\lambda}^2 \quad (k = 1,2,\ldots).$$

The representation space can be written as

$$H = \oplus \sum_k L_2(\operatorname{supp}\rho_k(\cdot), d\rho_1(\lambda)) = \oplus \int_{\mathbb{Z}_2^\infty} H_\lambda \, d\rho_1(\lambda),$$

because $\rho_1(\cdot) \gg \rho_2(\cdot) \gg \cdots$, $\operatorname{supp}\rho_1(\cdot) \supset \operatorname{supp}\rho_2(\cdot) \supset \cdots$ and $\rho_1(\cdot)\restriction \operatorname{supp}\rho_k(\cdot) = \rho_k(\cdot)$. The quasi-invariance of all the measures $\rho_k(\cdot)$ with respect to the shifts by $t \in \mathbb{Z}_{2,0}^\infty$ implies the $\mathbb{Z}_{2,0}^\infty$-invariance of the function $\nu(\lambda) = \dim H_\lambda$.

Introduce the unitary operators V_k on H:

$$(V_k \vec{f})(\lambda) = ((V_k \vec{f})_j(\lambda)) = (\|a_j^{(k)}(\lambda)\|_{H_\lambda} f_j(s_k \lambda))$$

$$(j = 1,2,\ldots;\ \vec{f}(\cdot) = (f_j(\cdot)) \in \oplus \int_{\mathbb{Z}_2^\infty} H_\lambda \, d\rho_1(\lambda)).$$

The unitary operators $C_k = X_k V_k$ on H commute with all the operators of multiplication by essentially bounded measurable functions, and so, they are of the form $(C_k \vec{f})(\lambda) = C_k(\lambda) \vec{f}(\lambda)$ for $\rho_1(\cdot)$-almost all $\lambda \in \mathbb{Z}_2^\infty$ where $C_k(\lambda)$ are unitary operators on H_λ. Hence

$$(X_k \vec{f})(\lambda) = C_k(\lambda) \sqrt{\frac{d\rho_1(s_k \lambda)}{d\rho_1(\lambda)}}\, \vec{f}(s_k \lambda).$$

Next we show that the operator-functions $C_k(\lambda) : H_{s_k\lambda} = H_\lambda \to H_\lambda$ satisfy equations (12_K). Because $X_k^2 = I$ $(k = 1,2,\ldots)$ and $V_k^2 = I$, $C_k(\lambda)\, C_k(s_k \lambda) = I$. The commutation of X_k and X_j for $k \neq j$ leads to

$$C_k(\lambda)\, C_j(s_k \lambda) = C_j(\lambda)\, C_k(s_j \lambda).$$

□

So we see that the study of representations of the algebra of local observables of a spin system can be reduced to the study of the following:

a) $\mathbb{Z}_2^\infty$-quasi-invariant probability measures on $(\mathbb{Z}_2^\infty, \mathbf{B}(\mathbb{Z}_2^\infty))$;

b) measurable vector functions $(C_k(\lambda))_{k=1}^\infty$ with their values unitary operators of $H_{s_k\lambda} = H_\lambda \to H_\lambda$ and satisfying the equations (12_K) up to unitary equivalence.

Suppose, first, that the joint spectrum of CSO $(Z_k)_{k=1}^\infty$ is simple. Let $\rho(\cdot)$ be the spectral measure of this collection on $(\mathbb{Z}_2^\infty, \mathbf{B}(\mathbb{Z}_2^\infty))$.

By Theorem 9, the measure $\rho(\cdot)$ is $\mathbb{Z}_{2,0}^\infty$-quasi-invariant, i.e. for all $t \in \mathbb{Z}_{2,0}^\infty$, the measures $\rho(\cdot)$ and $\rho(\cdot\, t)$ are equivalent. The irreducibility of a representation of the algebra M of the CSO $(Z_k)_{k=1}^\infty$ with simple joint spectrum depends on the ergodic properties of the measure $\rho(\cdot)$.

PROPOSITION 5. The representation is irreducible if and only if the measure $\rho(\cdot)$ on $(\mathbb{Z}_2^\infty, \mathbf{B}(\mathbb{Z}_2^\infty))$ is ergodic with respect to all $\mathbb{Z}_{2,0}^\infty$-transformations.

Proof. If a bounded operator on $L_2(\mathbb{Z}_2^\infty, d\rho(\lambda))$ commutes with all the operators $(\lambda_k)_{k=1}^\infty$, then it is an operator of multiplication by an essentially bounded measurable function in $L_2(\mathbb{Z}_2^\infty, d\rho(\lambda))$, and if the operator commutes with all the operators X_k $(k=1,2,...)$, then this function is $\mathbb{Z}_{2,0}^\infty$-invariant. In this case, such an operator has to be a multiple of the identity because the measure is $\mathbb{Z}_{2,0}^\infty$-ergodic. So we have proved that a sufficient condition for the representation to be irreducible is that the measure is $\mathbb{Z}_{2,0}^\infty$-ergodic. To prove the necessity, we assume the contrary, and consider an operator of multiplication by a $\mathbb{Z}_{2,0}^\infty$-invariant function which is not constant. Such an operator commutes with the representation and is not a multiple of the identity. This contradicts the fact that the representation is irreducible. □

Now, $C_k(\lambda) = \alpha_k(\lambda)$ is a complex-valued $\mathbf{B}(\mathbb{Z}_2^\infty)$-measurable function on $\mathbb{Z}_2^\infty$ such that

$$\begin{aligned} &\text{a)}\quad |\alpha_k(\lambda)| = 1\,; \\ &\text{b)}\quad \alpha_k(\lambda)\,\alpha_k(s_k\lambda) = 1\,; \\ &\text{c)}\quad \alpha_k(\lambda)\,\alpha_j(s_k\lambda) = \alpha_j(\lambda)\,\alpha_k(s_j\lambda) \end{aligned} \qquad (12_K{}')$$

for $\rho(\cdot)$-almost all $\lambda \in \mathbb{Z}_2^\infty$.

The properties $(12_K{}')$ allow to construct a function $\alpha_t(\lambda)$ dependent on the parameter $t \in \mathbb{Z}_{2,0}^\infty$ and $\mathbf{B}(\mathbb{Z}_2^\infty)$-measurable for all t such that

$$\begin{aligned} &\text{a)}\quad |\alpha_t(\lambda)| = 1\,; \\ &\text{b)}\quad \alpha_{t^{(1)}t^{(2)}}(\lambda) = \alpha_{t^{(1)}}(\lambda)\,\alpha_{t^{(2)}}(t^{(1)}\lambda)\,; \\ &\text{c)}\quad \alpha_{s_k}(\lambda) = \alpha_k(\lambda)\,, \end{aligned} \qquad (13)$$

for $\rho(\cdot)$ almost all $\lambda \in \mathbb{Z}_2^\infty$.

Indeed, for $t = s_{k_1} \cdots s_{k_n}$, set

$$\alpha_t(\lambda) = \alpha_{s_{k_1}\cdots s_{k_n}}(\lambda) = \alpha_{s_{k_1}}(\lambda)\,\alpha_{s_{k_2}}(s_{k_1}\lambda)\cdots\alpha_{s_{k_n}}(s_{k_1}\cdots s_{k_{n-1}}\lambda).$$

So defined, it follows from $(12_K{}'$ c)) that the function $\alpha_t(\lambda)$ does not depend on the order of multiplication in the product $s_{k_1} \cdots s_{k_n}$. The properties (13) of the function $\alpha_t(\lambda)$ follow directly from (14). The function $\alpha_t(\lambda)$ will be called a cocycle (1-cocycle).

Note that by (13) there is a one-to-one correspondence between the cocycles $\alpha_t(\lambda)$ and the collections of functions $(\alpha_k(\lambda))_{k=1}^\infty$ which satisfy $(12_K{}')$.

The pair $(\rho(\cdot), \alpha_t(\cdot))$ is the simplest "non-commutative" measure on $\mathbb{Z}_2^\infty$.

For representations of M for which the joint spectrum of the CSO $(Z_k)_{k=1}^\infty$ is simple, we can give a $\alpha_t(\cdot)$ criterion for the unitary equivalence of the representations in terms of the pairs $(\rho(\cdot), \alpha_t(\lambda))$.

DEFINITION 7. A cocycle $\alpha_t(\lambda)$ is called trivial if there exists a measurable function $a(\lambda)$ ($|a(\lambda)| = 1$ for $\rho(\cdot)$-almost all λ) such that

$$\alpha_t(\lambda) = \frac{a(\lambda t)}{a(\lambda)}.$$

Example 1. The simplest example of a, generally speaking, non-trivial cocycle is a product-cocycle

$$\delta_t(\lambda) = \frac{a_1(\lambda_1 t_1)}{a_1(\lambda_1)} \; \frac{a_2(\lambda_2 t_2)}{a_2(\lambda_2)} \cdots$$

($a_k(\lambda_k)$ is a function with absolute value equal to one on $\mathbb{Z}_2 = \{-1,1\}$, $k = 1,2,\ldots$). □

The ratio of two cocycles $\dfrac{\alpha_t(\cdot)}{\beta_t(\cdot)}$ is a cocycle because for all $t \in \mathbb{Z}_{2,0}^{\infty}$

$$\left|\frac{\alpha_t(\lambda)}{\beta_t(\lambda)}\right| = 1 \quad (\text{mod}\, \rho(\cdot))$$

and

$$\frac{\alpha_{t^{(1)}t^{(2)}}(\lambda)}{\beta_{t^{(1)}t^{(2)}}(\lambda)} = \frac{\alpha_{t^{(1)}}(\lambda)}{\beta_{t^{(1)}}(\lambda)} \; \frac{\alpha_{t^{(2)}}(\lambda t^{(1)})}{\beta_{t^{(2)}}(\lambda t^{(1)})}$$

for $\rho(\cdot)$-almost all $\lambda \in \mathbb{Z}_2^{\infty}$.

DEFINITION 8. The cocycles $\alpha_t(\cdot)$ and $\beta_t(\cdot)$ are called equivalent if the cocycle $\gamma_t(\cdot) = \dfrac{\alpha_t(\cdot)}{\beta_t(\cdot)}$ is trivial.

PROPOSITION 6. For a representation on $L_2(\mathbb{Z}_2^{\infty}, d\rho(\lambda))$ of the algebra M of local observables determined by a cocycle $\alpha_t(\cdot)$ using ($12_0'$), to be unitarily equivalent to a representation of M on $L_2(\mathbb{Z}_2^{\infty}, d\mu(\lambda))$ determined by a cocycle $\beta_t(\cdot)$, it is necessary and sufficient that the measures $\rho(\cdot)$ and $\mu(\cdot)$ as well as the cocycles $\alpha_t(\cdot)$ and $\beta_t(\cdot)$ be equivalent.

Proof. If two representations of M on $L_2(\mathbb{Z}_2^{\infty}, d\rho(\lambda))$ and on $L_2(\mathbb{Z}_2^{\infty}, d\mu(\lambda))$ are equivalent, then the spectral measures of the unitarily equivalent collections of the CSO $(Z_k)_{k=1}^{\infty}$ and $(\tilde{Z}_k)_{k=1}^{\infty}$ are equivalent and therefore can be chosen to be equal, $\rho(\cdot) = \mu(\cdot)$. The operator on $L_2(\mathbb{Z}_2^{\infty}, d\rho(\lambda))$ establishing the unitary equivalence commutes with the operators $(\lambda_k)_{k=1}^{\infty}$ and so, it is the operator of multiplication by a measurable function $a(\lambda)$, with the absolute values $|a(\lambda)| = 1$. Now, from the equality

$$a(\lambda)\, X_k = \tilde{X}_k\, a(\lambda)$$

using the relations

$$(X_k f)(\lambda) = \alpha_{s_k}(\lambda) \sqrt{\frac{d\rho(s_k\lambda)}{d\rho(\lambda)}}\, f(s_k\lambda)\,,$$

$$(\tilde{X}_k f)(\lambda) = \beta_{s_k}(\lambda) \sqrt{\frac{d\rho(s_k\lambda)}{d\rho(\lambda)}}\, f(s_k\lambda)\,,$$

$$(f(\cdot) \in L_2(\mathbb{Z}_2^\infty, d\rho(\lambda)))$$

we get that $\dfrac{a(s_k\lambda)}{a(\lambda)} = \dfrac{\alpha_{s_k}(\lambda)}{\beta_{s_k}(\lambda)}$, i.e. the cocycles $\alpha_t(\cdot)$ and $\beta_t(\cdot)$ are equivalent.

Conversely, if the cocycles $\alpha_t(\cdot)$, $\beta_t(\cdot)$ and the measures $\rho(\cdot)$, $\mu(\cdot)$ are equivalent, then this yields a straightforward construction of an operator which settles the unitary equivalence of the two representations of the collections $(Z_k, X_k)_{k=1}^\infty$ on $L_2(\mathbb{Z}_2^\infty, d\rho(\lambda))$, and $(\tilde{Z}_k, \tilde{X}_k)_{k=1}^\infty$ on $L_2(\mathbb{Z}_2^\infty, d\mu(\lambda))$ using the cocycles $\alpha_t(\cdot)$ and $\beta_t(\cdot)$. □

Example 2. The simplest irreducible representations of the collections $(Z_k)_{k=1}^\infty$ and $(X_k)_{k=1}^\infty$ satisfying relations (12) are the independent ones. These representations are realized on the space $H = \underset{\{f_k\}}{\otimes} \mathbb{C}^2$, where $\{f_k\}_{k=1}^\infty$ is a stabilizing sequence. The operators $(Z_k)_{k=1}^\infty$ and $(X_k)_{k=1}^\infty$ have the form:

$$Z_k = \sigma_0 \otimes \cdots \otimes \sigma_0 \otimes \underbrace{\sigma_z}_{k-\text{th place}} \otimes\, \sigma_0 \otimes \cdots\,,$$

$$X_k = \sigma_0 \otimes \cdots \otimes \sigma_0 \otimes \underbrace{\sigma_x}_{k-\text{th place}} \otimes\, \sigma_0 \otimes \cdots\,.$$

We have the following theorem:

THEOREM 10. Independent representations of the algebra have the following properties:

1) each system of operators $(Z_k)_{k=1}^\infty$, $(X_k)_{k=1}^\infty$, $(Y_k = i\,Z_k X_k)_{k=1}^\infty$ has a simple joint spectrum;

2) in the realization of these representations in the form given in Theorem 9, the measure $\mathbb{Z}_2^\infty$ is a product-measure, and the cocycle is a product-cocycle;

3) the independent representations are irreducible;

4) the independent representations on the spaces $H = \underset{\{f_k\}}{\otimes} \mathbb{C}^2$ and $H' = \underset{\{f_k'\}}{\otimes} \mathbb{C}^2$ are unitarily equivalent if and only if the stabilizing sequences $\{f_k\}_{k=1}^\infty$ and $\{f_k'\}_{k=1}^\infty$ are weakly equivalent.

Proof. 1) To show that the joint spectrum of the three systems is simple we can give a common cyclic vector, Ω in the representation space H. To do this, first take a sequence $\{g_k\}_{k=1}^\infty$ which is weakly equivalent to a given stabilizing sequence $\{f_k\}_{k=1}^\infty$ such that the vectors g_k are cyclic for the operators $\sigma_z, \sigma_x, \sigma_y$ for all k. Then the vector Ω can be written as

$$\Omega = \lim_{n \to \infty} \Omega_n \otimes f_{n+1} \otimes f_{n+2} \otimes \cdots ,$$

where Ω_n is the common cyclic vector constructed in Theorem 1 for the finite system of operators $(Z_k)_{k=1}^n, (X_k)_{k=1}^n, (Y_k)_{k=1}^n$.

2) For the system of operators $(Z_k)_{k=1}^\infty$, we can construct a σ-additive joint resolution of the identity on the space $\mathbb{Z}_2^\infty$ by defining the spectral operator-valued function $E(\Delta)$ on the cylindrical sets $\Delta = \Delta_1 \times \Delta_2 \times \cdots \times \{-1,1\} \times \{-1,1\} \times \cdots$ as

$$E(\Delta) = E_1(\Delta_1) \otimes E_2(\Delta_2) \otimes \cdots \otimes I \otimes I \otimes \cdots$$

and then extending it to the σ-algebra $\mathbf{B}(\mathbb{Z}_2^\infty)$ generated by these sets. Because the spectrum of the system of operators $(\mathbb{Z}_k)_{k=1}^\infty$ is simple, we have $\overline{\text{LS}\{E(\Delta)\Omega\}_\Delta} = \underset{\{f_k\}}{\otimes} \mathbb{C}^2$. The isometric correspondence $E(\Delta)\Omega \mapsto h(\Delta)$, where $h(\Delta)$ is the characteristic function of the set $\Delta \in \mathbf{B}(\mathbb{Z}_2^\infty)$ generates a unitary mapping of the space $H = \underset{\{f_k\}}{\otimes} \mathbb{C}^2$ into $L_2(\mathbb{Z}_2^\infty, d\rho(\lambda))$. Because the vector Ω is a product-vector, $\Omega = g_1 \otimes g_2 \otimes \cdots$ where $g_k = g_k(1)e_1 + g_k(-1)e_{-1}, e_{\pm 1}$ are the eigenvectors of the matrix σ_z, the measure $d\rho(\lambda) = \otimes\, d\rho_k(\lambda_k)$, where $\rho_k(\lambda) = |g_k(\lambda_k)|^2$ is $\mathbb{Z}_{2,0}^\infty$-quasi-invariant. Using the form of the operators Z_k in the space $\underset{\{f_k\}}{\otimes} \mathbb{C}^2$ and the form of their images under the unitary mapping $\underset{\{f_k\}}{\otimes} \mathbb{C}^2 \to L_2(\mathbb{Z}_2^\infty, d\rho(\lambda))$, define the form of the function $\alpha_k(\lambda) = \dfrac{\arg g_k(s_k \lambda)}{\arg g_k(\lambda)}$.

3) The constructed representation is irreducible because the product-measure $\rho(\cdot)$ is $\mathbb{Z}_{2,0}^\infty$-ergodic.

4) The weak equivalence of the two stabilizing sequences $\{f_k\}_{k=1}^\infty$ and $\{f_k'\}_{k=1}^\infty$ is necessary and sufficient for the corresponding measures and cocycles to be equivalent, and so for the unitary equivalence of the corresponding representations of the algebra M.▯

Example 3. We give one more method to construct the cocycle $\alpha_t(\lambda)$. It is similar to the method of constructing a Markov measure on $\mathbb{Z}_2^\infty$. Give a matrix

$$\begin{bmatrix} p_{1,1} & p_{1,-1} \\ p_{-1,1} & p_{-1,-1} \end{bmatrix}$$

consisting of the numbers $p_{i,j}$, $|p_{i,j}| = 1$, $i,j = \pm 1$. and set

$$\alpha_{s_k}(\lambda) = \alpha_k(\lambda) = \frac{p_{\lambda_{k-1},-\lambda_k}\, p_{-\lambda_k,\lambda_{k+1}}}{p_{\lambda_{k-1},\lambda_k}\, p_{\lambda_k,\lambda_{k+1}}}.$$

▯

Example 4. As a generalization of Example 3, we give a method to construct a cocycle $\alpha_t(\cdot)$ using a given sequence of functions $(f_k(\lambda_k, \ldots, \lambda_k))_{k=1}^\infty$ with absolute value equal to one:

$$\alpha_{s_k}(\lambda) = \alpha_k(\lambda) = \frac{f_1(\ldots, \lambda_{k-1}, s_k\lambda_k, \lambda_{k+1}, \ldots) \cdots f_k(s_k\lambda_k, \lambda_{k+1}, \ldots)}{f_1(\ldots, \lambda_{k-1}, \lambda_k, \lambda_{k+1}, \ldots) \cdots f_k(\lambda_k, \lambda_{k+1}, \ldots)}.$$

□

However, unlike the representations of a spin system with finite degrees of freedom, the representations of relations (12) can be irreducible even for an n-multiple or infinite-multiple spectrum of the CSO $(Z_k)_{k=1}^{\infty}$ (see A. Gårding, A. Wightman [1], V.Ya. Golodets [1,3]). It is a difficult ("wild") problem to describe such irreducible representations up to unitary equivalence.

9.4. On dividing countable collections of GCSO into "tame" and "wild".

In Section 9.2 we described all irreducible representations for "tame" countable collections of GCSO. There is a corresponding structure theorem that gives a decomposition of a "tame" collection of GCSO into irreducible ones. We show that the description problem for irreducible representations of a countable collection of GCSO which is not "tame" contains the so far unsolved problem of describing all irreducible representations of the algebra of local observables.

THEOREM 11. If a collection of GCSO $A_1, \ldots, A_n, \ldots$ is not "tame", then there exists a countable subcollection of GCSO of generators of the algebra of local observables.

Proof. Let the collection of GCSO $A_1, \ldots, A_n, \ldots$ be not "tame", i.e. it cannot divided into a finite number of commutative subfamilies:

$$A_1^{(1)}, \ldots, A_k^{(1)}, \ldots$$

$$A_1^{(2)}, \ldots, A_k^{(2)}, \ldots$$

....

$$A_1^{(m)}, \ldots, A_k^{(m)}, \ldots$$

so that the graded commutation relations have the form

$$A_k^{(p)} A_j^{(l)} - (-1)^{\tilde{g}(p,l)} A_j^{(l)} A_k^{(p)} = 0 \quad (p, l = 1, \ldots, m)$$

$$(k, j = 1, 2, \ldots)$$

for some grading function $\tilde{g}(\cdot, \cdot)$.

Divide the graph Γ of the collection $(A_k)_{k=1}^{\infty}$ into connected components $\Gamma_1, \ldots, \Gamma_n, \ldots$. The property of being "tame" does not depend on the one-point subsets of the graph Γ. So we can suppose that all the connected components of Γ contain more than one vertex. If the number of graphs $(\Gamma_j)_j$ is countable, then the representations of Γ

contain the representations $(\mathbf{1})^\infty$ and the theorem is proved. If the number of graphs is finite, then divide the set of graphs $\Gamma_1, \ldots, \Gamma_m$ into the graphs $(\Gamma_{i_1}^{(f)})$ that contain a finite number of vertices, and the graphs $(\Gamma_{j_1}^{(c)})$ in which the number of vertices is infinite. In every $\Gamma_{j_1}^{(c)}$ choose a pair of anticommuting operators A_{j_1}, B_{j_1} and construct, using $\Gamma_{j_1}^{(c)}$ as in Section 9.1, a new graph, which is then divided into connected components (as before we suppose that all the graphs $\Gamma_{j_1 j_2}$ contain more than one vertex). If the number of graphs $\Gamma_{j_1 j_2}$ is countable, then the representations of Γ_{j_1} contain the representations of $(\mathbf{1})^\infty$ and the theorem is proved. If for every j_1 the number of graphs $\Gamma_{j_1 j_2}$ is finite, then again divide $\Gamma_{j_1 j_2}$ into $\Gamma_{j_1 j_2}^{(f)}$ and $\Gamma_{j_1 j_2}^{(c)}$. There will be at least one graph containing a countable number of vertices. Continuing this process to the n-th step we get the connected graphs $\Gamma_{j_1 j_2 \cdots j_n}$ which have more than one vertex. Either the number of graphs is countable and so their representations contain the representations of the graph $(\mathbf{1})^\infty$ or the number of graphs $\Gamma_{j_1 j_2 \cdots j_n}$ is finite. But then there are countable graphs $\Gamma_{j_1 \cdots j_n}^{(c)}$, and, choosing a pair of anti-commuting operators $A_{j_1 \cdots j_n}$, the process can be continued. But if at every step the number of graphs $\Gamma_{j_1 \cdots j_n}$ is finite, then the representations of the collection of operators

$$A_j, B_j, A_{j_1 j_2}, B_{j_1 j_2}, \ldots, A_{j_1 \cdots j_n}, B_{j_1 \cdots j_n}, \ldots$$

(under an additional restriction, $A_{j_1 \cdots j_n}^2 = B_{j_1 \cdots j_n}^2 = I$ $(n = 1, 2, \ldots)$) are the representations of the generators of the algebra of local observables. □

So all collections of GCSO can be divided into two classes: the "tame" (for which the structure theorem is given in Section 9.2) and the "wild" (i.e. which contain a subcollection of generators of the algebra of local observables).

Comments to Chapter 9.

1. The classic theorem of P. Jordan, E. Wigner [1] on the uniqueness of an irreducible representation of the anticommutation relations for systems with finite degree of freedom and its reformulation are given. The theory of self-adjoint representations of the graphs Γ is an extended version of the article Yu.S. Samoĭlenko [4]. Definitions of graded Lie algebras are given, for example, in F.A. Berezin, G.I. Kats [1], F.A. Berezin [7]. For determining their (non-graded) representations see M. Scheunert [1]. Proposition 3 is due to the author.

2. The material follows Yu.S. Samoĭlenko [7].

3. The choice of different systems of generators in the algebra of local observables is well known and was used in a series of works on statistical physics (see, for example, D. Shale, W.F. Stinespring [1], F.A. Berezin [3]. S.A. Pirogov [1] and others). The classic results of L. Gårding and A. Wightman [1] are given (see also V.Ya. Golodets [1,2]). The construction of cocycles follow the article V.I. Kolomytsev, Yu.S. Samoĭlenko [3].

4. Theorem 11 on dividing collections of GCSO into "tame" and "wild" is due to the author.

Chapter 10.
COLLECTIONS OF UNBOUNDED CSO (A_k) AND CSO (B_k) SATISFYING GENERAL COMMUTATION RELATIONS

In this chapter we study self-adjoint operators, $A = \int_{\mathbb{R}^1} \lambda E_A(\lambda)$ and $B = \int_{\mathbb{R}^1} \mu\, dE_B(\mu)$, satisfying relations more general than commutation or anti-commutation relations i.e.

$$AB = BF(A). \tag{1}$$

Here, $\mathbb{R}^1 \ni \lambda \mapsto F(\lambda) \in \mathbb{R}^1$ is a fixed measurable function.

For the unbounded operators A and B, we point out and study the exact meaning of relations (1). We prove a structure theorem which describes such pairs of operators in the form of an integral of elementary pairs (Section 10.2). For $F(\lambda) = \lambda$, this yields the classic Spectral Theorem for the CSO A and B, for $F(\lambda) = -\lambda$, this yields the structure theorem 3.8 for a pair of anti-commuting self-adjoint operators, and for $F(\lambda) = -\lambda + 1$, this becomes the structure theorem for a pair of self-adjoint operators, A and B, satisfying $AB + BA = B$. We give (Section 10.3) generalizations of the structure theorem for families of CSO $\mathbf{A} = (A_k)_{k=1}^{\infty}$ and $\mathbf{B} = (B_j)_{j=1}^{n}$ satisfying

$$A_k B_j = B_j F_{jk}(\mathbf{A}) \quad (k = 1, 2, \ldots; j = 1, 2, \ldots, n).$$

Here $\mathbb{R}^{\infty} \ni \lambda = (\lambda_1, \ldots, \lambda_k, \ldots) \mapsto F_j(\lambda) = (F_{j1}(\lambda), \ldots, F_{jk}(\lambda), \ldots) \in \mathbb{R}^{\infty}$ are fixed functions measurable with respect to the σ-algebra $\mathbf{B}(\mathbb{R}^{\infty})$ and $F_{jk}(\mathbf{A}) = \int_{\mathbb{R}^{\infty}} F_{jk}(\lambda)\, dE_{\mathbf{A}}(\lambda)$, where $E_{\mathbf{A}}(\cdot)$ is the joint resolution of the identity of the family $\mathbf{A}$.

To get the structure theorem, it is essential that the collection $(B_j)_{j=1}^{n}$ is finite. If the collection $(B_j)_{j=1}^{\infty}$ is infinite, then instead of a structure theorem we construct a commutative model for the families of CSO $(A_k)_{k=1}^{\infty}$ and $(B_j)_{j=1}^{\infty}$ (Section 10.4). This model describes the form of the operators $(B_j)_{j=1}^{\infty}$ in the space of Fourier images of the CSO $(A_k)_{k=1}^{\infty}$.

10.1. Representations of the relation $AB = BF(A)$ using bounded self-adjoint operators.

In this section we give and study a description up to unitary equivalence of a pair (A, B) where $A = \int_{-\|A\|}^{\|A\|} \lambda\, dE_A(\lambda)$ and $B = \int_{-\|B\|}^{\|B\|} b\, dE_B(b)$ are bounded self-adjoint operators satisfying the relation (1)

$$AB = BF(A)$$

(F is a continuous real-valued function on $\mathbb{R}^1$). The simplest types of operator pairs are bounded commuting self-adjoint operators ($F(\lambda) = \lambda$), or bounded anti-commuting operators ($F(\lambda) = -\lambda$).

In what follows, we establish some commutation relations for functions of A and B, and some facts about the spectra of these operators. Then we study irreducible pairs (A,B) and prove a structure theorem for the operator pairs (A,B) with the mentioned properties.

First we establish some commutation relations for functions of A and some facts about the structure of its spectrum $\sigma(A)$.

PROPOSITION 1. Let

$$f(\cdot) : [-\|A\|, \|A\|] \cup [-\|F(A)\|, \|F(A)\|] \to \mathbb{R}^1,$$

be measurable and bounded. Then we have the following:

$$f(A)B = B f(F(A)) \tag{2}$$

$$B f(A) = f(F(A)) B. \tag{3}$$

Proof. For any power n, $n = 1,2,\ldots$

$$A^n B = A^{n-1}(AB) = A^{n-1} BF(A) = \cdots = B(F(A))^n.$$

This proves (2) for all polynomials. From the functional calculus for bounded self-adjoint operators, it follows that (2) holds for all bounded measurable functions $f(\cdot)$. Under the given conditions the operators $f(A)$, $f(F(A))$ are bounded and self-adjoint, so the left and the right parts of (3) are equal being adjoint to the corresponding parts of (2).

Example 1. Let the function $f(\cdot)$ satisfy the conditions of Proposition 1 and $\forall \lambda \in \sigma(A)\ f(\lambda) = f(F(\lambda))$. Then the operator $f(A)$ commutes with A and B, i.e. $[f(A), A] = [f(A), B] = 0$. Indeed, because

$$f(A) = \int_{\mathbb{R}^1} f(\lambda)\, dE_A(\lambda)$$

and

$$f(F(A)) = \int_{\mathbb{R}^1} f(F(\lambda))\, dE_A(\lambda) = f(A),\ f(A)B = B f(F(A)) = B f(A).$$

$\square$

Example 2. The operator $f(A) = A + F(A)$ commutes with A and B because $AB = BF(A)$ and $F(A)B = BA$. From these relations it follows that

$$(A - F(A))B = -B(A - F(A))$$

i.e. the operators B and $A - F(A)$ anticommute. So the operator $(A - F(A))^2$ also commutes with A and B. □

PROPOSITION 2. Let A and B satisfy (1) and let B be invertible (i.e. there exists a not necessarily bounded operator B^{-1}).

Then

a) $F : \sigma(A) \to \sigma(A)$ is one–to–one ,

b) $A = F(F(A))$, (4)

c) $\forall \lambda \subset \sigma(A) \ : \ \lambda = F(F(\lambda))$. (5)

Proof. a) If $\lambda \in \sigma(A)$, then $\exists \, \{e_n\} \subset H$, $\|e_n\| = 1$ and $\|A\, e_n - \lambda\, e_n\| \underset{n \to \infty}{\to} 0$. But since the point 0 is not in the discrete spectrum of the operator B, the operator $\operatorname{sign} B = \int_{-\|B\|}^{\|B\|} \operatorname{sign} \mu \, dE_B(\mu)$ is unitary, and (see Proposition 5 below) $A \operatorname{sign} B = \operatorname{sign} B\, F(A)$. Then

$$\|A(\operatorname{sign} B\, e_n) - F(\lambda)\,(\operatorname{sign} B\, e_n)\| = \|\operatorname{sign} B\,(F(A)\, e_n - F(\lambda)\, e_n)\| \underset{n \to \infty}{\to} 0$$

and

$$F(\lambda) \in \sigma(A).$$

b) Because it follows from Proposition 1 that

$$BA = F(A)\, B = BF(F(A))$$

and $\forall x \in H \ \ B A x \in D(B^{-1})$ ($D(B^{-1})$ is the domain of B^{-1}), $Ax = F(F(A))\, x$ and $A = F(F(A))$.

c) (5) can be proved using the theory of decomposition with respect to the generalized eigen-functions (see Yu.M. Berezanskiĭ [9.12]). Here we give a direct proof. For any

$$\lambda \in \sigma(A), \ \exists \, \{e_n\}_{n=1}^{\infty}, \ \|e_n\| = 1 \ \ (n = 1, 2, \ldots)$$

and

$$\lim_{n \to \infty} \|A\, e_n - \lambda\, e_n\| = 0.$$

Then, since F is continuous, $\|F(F(A))\, e_n - F(F(\lambda))\, e_n\| \underset{n \to \infty}{\to} 0$. From (4) it follows that

$$\|A\, e_n - F(F(\lambda))\, e_n\| \underset{n \to \infty}{\to} 0, \ \ \|(\lambda - F(F(\lambda))\, e_n\| \underset{n \to \infty}{\to} 0$$

and so we get (5). □

If the operator B is invertible and λ is a point of the discrete spectrum, then $A\, e_\lambda = \lambda\, e_\lambda$ $(e_\lambda \neq 0)$ and

$$A(B\,e_\lambda) = BF(A)\,e_\lambda = F(\lambda)\,(B\,e_\lambda) \quad (B\,e_\lambda \neq 0)\,, \tag{6}$$

i.e. the operator B maps the eigenvectors of the operator A corresponding to the eigenvalue λ into the eigenvectors corresponding to the eigenvalue $F(\lambda)$.

In general, A may not have a discrete spectrum, however we have:

PROPOSITION 3. Let B be invertible, Δ be a Borel set, $\Delta \subset \sigma(A)$. Then

$$E_A(\Delta)\,B = B\,E_A(F(\Delta))\,,\ \ E_A(F(\Delta))\,B = B\,E_A(\Delta). \tag{7}$$

In particular, if $\Delta = F(\Delta)$, then

$$[E_A(\Delta), A] = [E_A(\Delta), B] = 0.$$

Proof. From Proposition 1 it follows that

$$E_A(\Delta)\,B = \chi_\Delta(A)\,B = B\,\chi_\Delta(F(A)) = BE(F^{-1}(\Delta)).$$

But from 2a) and 2b), $F^{-1}(\Delta) = F(\Delta)$, so we get the first equality in (7). We get the second equality by taking the adjoint. □

Let H_0 denote the kernel of the operator B,

$$S_1 = \{\lambda \in \mathbb{R}^1 \mid \lambda = F(\lambda)\}\,,\ S_2 = \{\lambda \in \mathbb{R}^1 \mid \lambda \neq F(\lambda)\,,\ \lambda = F(F(\lambda))\}\,,$$

$$S_{2,+} = \{\lambda \in S_2 \mid \lambda > F(\lambda)\}\,,\ S_{2,-} = \{\lambda \in S_2 \mid \lambda < F(\lambda)\}.$$

PROPOSITION 4. The subspaces

$$H_0 = \mathrm{Ker}B, H_1 = E_A(S_1)\,H_0^\perp\,,\ H_2 = E_A(S_2)\,H_0^\perp$$

are invariant with respect to the operators A and B. And

$$H = H_0 \oplus H_1 \oplus H_2. \tag{8}$$

Proof. For any $x_0 \in H_0$, $B(A\,x_0) = F(A)\,B\,x_0 = 0$. The invariance of the subspaces H_1 and H_2 with respect to the operator B follows from Proposition 3. The decomposition $H = H_0 + H_1 + H_2$ follows from Proposition 2c). □

If we denote $A = A_0 = A \restriction H_0$, $A_1 = A \restriction H_1$, $A_2 = A \restriction H_2$ then

$$\sigma(A) = \sigma(A_0) \cup \sigma(A_1) \cup \sigma(A_2) \subset \sigma(A_0) \cup S_1 \cup S_2. \tag{9}$$

(S_2 in general is not closed, but $S_1 \cup S_2$ is closed).

Now we will consider commutation relations for the functions of B and give some statements on the structure of the spectrum of this operator.

Let $g(\cdot) : [-\|B\|, \|B\|] \to \mathbb{R}^1$ be a bounded measurable function. Denote

$$g_e(x) = \tfrac{1}{2}\,(g(x) + g(-x)) \ \text{ and } \ g_o(x) = \tfrac{1}{2}\,(g(x) - g(-x)).$$

PROPOSITION 5.

$$A\, g(B) = g_e(B) A + g_o(B)\, F(A). \tag{10}$$

Proof. Because $AB^2 = BF(A)B = B^2 A$, $A\, g_e(B) = g_e(B) A$, and since $AB = BF(A)$, $A\, g_o(B) = g_o(B)\, F(A)$ and so (10) follows. □

Example 3. If $B = \int_{-\|B\|}^{\|B\|} \mu\, dE_B(\mu)$ is the spectral representation of the operator B, $[A, E_B(\Delta)] = 0$ for a Borel set $\Delta = -\Delta$ because $E_B(\Delta) = \chi_\Delta(B)$ is an even function of the operator B. □

Example 4. $[B \operatorname{sign} B, A] = 0$, where $\operatorname{sign} B = \int_{-\|B\|}^{\|B\|} \operatorname{sign} \mu\, dE_B(\mu)$. □

Denote $B_0 = B \restriction H_0$, $B_1 = B \restriction H_1$, $B_2 = B \restriction H_2$.

Then $\sigma(B) = \sigma(B_0) \cup \sigma(B_1) \cup \sigma(B_2) \subset \{0\} \cup \{\sigma(B_1) \cup \{\sigma(B_2)\}$.

PROPOSITION 6. $\sigma(B_2) = -\sigma(B_2)$.

Proof. If $\mu \in \sigma(B_2)$ $(b \neq 0)$, then $\exists\ \{e_n\} \subset H_2$ $\|e_n\| = 1$ and $\|B_2 e_n - \mu e_n\| \underset{n \to \infty}{\to} 0$. Then

$$\|B \operatorname{sign}(A - F(A))\, e_n + \mu \operatorname{sign}(A - F(A))\, e_n\| = \|\operatorname{sign}(A - F(A))\, (B_2 e_n - \mu e_n)\| \underset{n \to \infty}{\to} 0.$$

Since $\lambda \neq F(\lambda)$ for almost all $\lambda \in \sigma(A)$ with respect to $E_A(\cdot) \restriction H_2$, $\|\operatorname{sign}(A - F(A))\, e_n\| = 1$ and $-\mu \in \sigma(B_2)$. □

Let now the pair of bounded self-adjoint operators A and B be irreducible and satisfy relations (1), i.e. suppose that any bounded operator on H which commutes with A and B is a multiple of the identity.

THEOREM 1. Any irreducible pair of bounded self-adjoint operators A and B satisfying (1) equals up to unitary equivalence one of the following:

a) $A = \lambda$, $B = 0$ on $\mathbb{C}^1$, $\lambda \in \mathbb{R}^1$;

b) $A = \lambda$, $B = \mu$ on $\mathbb{C}^1$, $\lambda \in S^1$, $\mu \in \mathbb{R}^1 \setminus \{0\}$

c) $A = \begin{pmatrix} \lambda & 0 \\ 0 & F(\lambda) \end{pmatrix}$, $B = \begin{pmatrix} 0 & \mu \\ \mu & 0 \end{pmatrix}$ on $\mathbb{C}^2$, $\lambda \in S_{2,+}$, $\mu > 0$.

The above given pairs are not unitarily equivalent.

Proof. Since B^2 commutes with A and B, and $B^2 \geq 0$, $B^2 = \mu^2 I$, $\mu \geq 0$.

If $\mu = 0$, then $B = 0$, and since H is irreducible, it is one-dimensional, i.e. we have case a).

Let $\mu > 0$, then B is invertible, $B^{-1} = \dfrac{1}{\mu^2} B$ and the operator $\dfrac{1}{\mu} B$ is unitary.

As seen in Example 2, the operators $A + F(A)$ and $(A - F(A))^2$ commute with A and B so from the irreducibility it follows that $A + F(A) = c\, I$, $(A - F(A))^2 = d^2 I$, $d \geq 0$. From the spectral theorem it follows that

$$\forall\lambda \in \sigma(A)\ \lambda + F(\lambda) = C\,,\ (\lambda - F(\lambda))^2 = d^2 \ \text{ or } \ \lambda + F(\lambda) = C\,,\ \lambda - F(\lambda) = \pm d.$$

If $d = 0$, the spectrum of A can contain only one point $\lambda = F(\lambda) = \frac{c}{2}$, and so we are in case b). If $d \neq 0$, then for any point of the spectrum of A , $\lambda = \frac{c \pm d}{2}$, and $F(\lambda) = \frac{c \mp d}{2}$. So the spectrum of A is discrete and contains two points $\lambda = \frac{c+d}{2}$, $F(\lambda) = \frac{c-d}{2}$, with $\lambda > F(\lambda)$ and $F(F(\lambda)) = \lambda$. Let e_λ be an eigenvector of A with eigenvalue λ, $\|e_\lambda\| = 1$. Since $\frac{1}{\mu} B$ is unitary, it follows from (6) that $B\, e_\lambda = \mu\, e_{F(\lambda)}$, where $A\, e_{F(\lambda)} = F(\lambda)\, e_{F(\lambda)}$, $\|e_{F(\lambda)}\| = 1$, Similarly, $B\, e_{F(\lambda)} = \mu\, e_\lambda$. Taking into consideration the irreducibility, we get case c). $\square$

Remark 1. If for the relation $AB = BF(A)$ $(A = A^*, B = B^*)$ there exist irreducible representations of the type c), i.e. $S_2 \neq \varnothing$ then there also exist irreducible representations of type b), i.e. $S_1 \neq \varnothing$.

Proof. Because $F(\lambda) - \lambda$ is continuous and because there exist $\lambda_1, \lambda_2 \in \mathbb{R}^1$ such that $\lambda_2 < \lambda_1$ and

$$F(\lambda_1) - \lambda_1 = \lambda_2 - \lambda_1 < 0\,,\ F(\lambda_2) - \lambda_2 = \lambda_1 - \lambda_2 > 0\,,\ S_2 \neq \varnothing\,,$$

we see that there also exists $\lambda_0 (\lambda_2 < \lambda_0 < \lambda_1)$ such that $F(\lambda_0) = \lambda_0$ and $S_1 \neq \varnothing$.

This remark could be phrased as follows: if the equation $F(\lambda) = \lambda$ does not have real solutions, the relation $AB = BF(A)$ has irreducible representations only of the form a). $\square$

THEOREM 2. There is a one-to-one correspondence between pairs A,B and orthogonal decompositions

$$H = H_0 \oplus H_1 \oplus H_2 = H_0 \oplus H_1 \oplus \mathbb{C}^2 \otimes H_+$$

together with respective resolutions of the identity:

a) the support of $dE_0(\lambda)$ is in a compact subset of $\mathbb{R}^1 \ni \lambda$, and the values are projections onto subspaces of H_0;

b) $dE_1(\lambda,b)$ has its support in a compact subset of $M_1 = S_1 \times (\mathbb{R}^1 \setminus \{0\}) \subset \mathbb{R}^2$ with as its values projections onto subspaces of H_1;

c) $dE_2(\lambda,\mu)$ has its support in a subspace of $M_2 = S_{2,+} \times \{\mu > 0\} \subset \mathbb{R}^2$ with as its values projections onto the subspaces of H_+, such that

$$A = \int_{\mathbb{R}^1} \lambda\, dE_0(\lambda) + \int_{M_1} \lambda\, dE_1(\lambda,\mu) + \int_{M_2} \begin{bmatrix} \lambda & 0 \\ 0 & F(\lambda) \end{bmatrix} \otimes dE_2(\lambda,\mu)$$

$$B = \int_{M_1} \mu\, dE_1(\lambda,\mu) + \begin{bmatrix} 0 & 1 \\ 1 & 0 \end{bmatrix} \otimes \int_{M_2} \mu\, dE_2(\lambda,\mu).$$

Proof. Proposition 4 gives a decomposition of H, $H = H_0 \oplus H_1 \oplus H_2$ into subspaces relatively to A and B.

a) The operators A_0 and B_0 have the form

$$A_0 = A \restriction H_0 = \int_{\mathbb{R}^1} \lambda \, dE_0(\lambda), \quad B_0 = B \restriction H_0 = 0.$$

b) Since A_1 and B_1 ($B_1 = B \restriction H_1$ is non-degenerate) commute, there exists a resolution of the identity $dE_1(\lambda,\mu)$ defined on $M_1 = S_1 \times \{ \mathbb{R}^1 \setminus \{0\} \}$ with as its values projections onto the subspaces of H_1 such that

$$A_1 = A \restriction H_1 = \int_{M_1} \lambda \, dE_1(\lambda,\mu), \quad B_1 = B \restriction H_1 = \int_{M_1} \mu \, dE_1(\lambda,\mu).$$

c) Now, denote $E_A(S_{2,+}) H_0^\perp = H_+$ and $E_A(S_{2,-}) H_0^\perp = H_-$. Since the operator $B_2 = B \restriction H_2 = \int_{-\|B\|}^{\|B\|} \mu \, dE_{B,2}(\mu)$ is non-degenerate, it follows from Proposition 3 that $B_2 : H_+ \to H_-$ and $B_2 : H_- \to H_+$. The operator $\operatorname{sign} B_2 = \int_{-\|B\|}^{\|B\|} \operatorname{sign} \mu \, dE_{B,2}(\mu)$ is unitary on $H_2 = H_+ \oplus H_-$, $\operatorname{sign} B_2 \, H_+ = H_-$, $\operatorname{sign} B_2 \, H_- = H_+$ and since the function $\operatorname{sign} B$ is odd it follows from Proposition 5 that

$$A_2 \operatorname{sign} B_2 = \operatorname{sign} B_2 \, F(A_2)$$

$$E_{A,2}(\Delta) \operatorname{sign} B_2 = \operatorname{sign} B_2 \, E_{A,2}(F(\Delta))$$

for $A_2 = A \restriction H_2 = \int_{S_2} \lambda \, dE_{A,2}(\lambda)$. Introducing a unitary operator

$$U : H_+ \oplus H_- \to H_+ \oplus H_+ = \mathbb{C}^2 \otimes H_+$$

by $U = I \oplus \operatorname{sign} B_2$ we get that $U A_2 U^* : \mathbb{C}^2 \otimes H_+ \to \mathbb{C}^2 \otimes H_+$ and

$$U A_2 U^* = (I \oplus \operatorname{sign} B_2) \Big(\int_{S_{2,+}} \lambda \, dE_{A,2}(\lambda) \oplus \int_{S_{2,-}} \lambda \, dE_{A,2}(\lambda) \Big) (I \oplus \operatorname{sign} B_2) =$$

$$= \int_{S_{2,+}} \lambda \, dE_{A,2}(\lambda) \oplus \int_{S_{2,-}} F(\lambda) \operatorname{sign} B_2 \, dE_{A,2}(F(\lambda)) \operatorname{sign} B_2 =$$

$$= \int_{S_{2,+}} \lambda \, dE_{A,2}(\lambda) \oplus \int_{S_{2,+}} F(\lambda) \, dE_{A,2}(\lambda) =$$

$$= \int_{S_{2,+}} \begin{bmatrix} \lambda & 0 \\ 0 & F(\lambda) \end{bmatrix} \otimes dE_{A,2}(\lambda),$$

where $E_{A,2}(\lambda)$ is an orthogonal operator-valued measure on $S_{2,+}$ with as its values projections onto subspaces of H_+. Since the operator $U B_2 U^* : \mathbb{C}^2 \otimes H_+ \to \mathbb{C}^2 \otimes H_+$, $B_2 \operatorname{sign} B_2$ commutes with the operator $A_2 \restriction H_+$, introducing a resolution of the identity for the positive operator $(B_2 \operatorname{sign} B_2) \restriction H_+ = \int_0^{\|B\|} \mu \, dE_B(\mu)$ which takes as its values projections onto subspaces of H_+, we find that

$$U\,B\,U^* = (I \oplus \operatorname{sign} B_2)\, B_2 (I \oplus \operatorname{sign} B_2) =$$

$$= \begin{bmatrix} 0 & 1 \\ 1 & 0 \end{bmatrix} \otimes B_2 \operatorname{sign} B_2 \restriction H_+ = \begin{bmatrix} 0 & 1 \\ 1 & 0 \end{bmatrix} \otimes \int_0^{\|B\|} b\, d\mathbf{E}_B(\mu).$$

Because the operator-valued measures $dE_{A,2}(\lambda)$ and $\mathbf{E}_B(\mu)$ defined on $S_{2,+}$ and $(0, \|B\|]$, respectively, commute, using the measure $dE_2(\lambda,\mu) = dE_{A,2}(\lambda) \otimes d\mathbf{E}_B(\mu)$ on M we get the proof of part c). □

Remark 2. If the equation $F(\lambda) = \lambda$ does not have real solutions, then for a pair (A,B) such that $AB = BF(A)$, we have $B = 0$. □

10.2. Representations of the relation $AB = BF(A)$ using unbounded operators, Riggings. Structure theorems.

If the self-adjoint operators $A = \int_{\mathbb{R}^1} \lambda\, dE_A(\lambda)$ and $B = \int_{\mathbb{R}^1} b\, dE_B(b)$ are unbounded, then we make the following definition.

DEFINITION 1. We say that the unbounded operators A and B satisfy the relation
(1)

$$AB = BF(A),$$

if $\forall l \geq 0$ and for every Borel $\Delta \subset \mathbb{R}^1$

$$E_A(\Delta)\, B_l = B_l\, E_A(F^{-1}(\Delta))$$

($F^{-1}(\Delta)$ is the full pre-image of the set Δ, $B_l = \int_{-l}^{l} b\ dE_B(b)$).

PROPOSITION 7. For bounded A and B the following statements are equivalent:

(i) $AB = BF(A)$;

(ii) $\forall l \geq 0$, for every Borel $\Delta \subset \mathbb{R}^1$ $E_A(\Delta) B_l = B_l E_A(F^{-1}(\Delta))$;

(iii) for every bounded measurable

$$f(\cdot) : [-\|A\|, \|A\|] \cup [-\|F(A)\|, \|F(A)\|] \to \mathbb{C}^1$$

and $g(\cdot) : [-\|B\|, \|B\|] \to \mathbb{C}^1$

$$f(A)\, g(B) = g_e(B)\, f(A) + g_o(B)\, f(F(A)),$$

here

$$g_e(\mu) = \frac{1}{2}\,(g(\mu) + g(-\mu))$$

$$g_o(\mu) = \frac{1}{2}(g(\mu) - g(-\mu)).$$

Proof. (i) $\Rightarrow$ (iii). Let $AB = BF(A)$. Then

$$A^n B = A^{n-1} BF(A) = \cdots = B(F(A))^n.$$

So for any polynomial, $P(A)B = BP(F(A))$, and using the functional calculus we get $f(A)B = Bf(F(A))$ for all bounded measurable

$$f(\cdot) : [-\|A\|, \|A\|] \cup [-\|F(A)\|, \|F(A)\|] \to \mathbb{C}^1.$$

Now, since $f(A)B^2 = B f(F(A))B = B^2 f(A)$, for any polynomials, and thus for any measurable function $g(\cdot) : [-\|B\|, \|B\|] \to \mathbb{C}^1$

$$f(A)\, g(B) = g_e(B)\, f(A) + g_o(B)\, f(F(A))$$

i.e. we get (iii).

(iii) $\Rightarrow$ (ii). Choosing $f(\cdot)$ to be the function $\chi_\Delta(\lambda) = \begin{cases} 1 & \lambda \in \Delta \\ 0 & \lambda \notin \Delta \end{cases}$ and $g(\cdot)$ to be the odd function

$$\chi_l(\mu) = \begin{cases} \mu, & |\mu| \le l \\ 0, & |\mu| > l \end{cases}$$

we get

$$E_A(\Delta)\, B_l = B_l\, \chi_\Delta \Big(\int_{-\|A\|}^{\|A\|} F(\lambda)\, dE_A(\lambda) \Big) = B_l\, E_A(F^{-1}(\Delta))$$

where $F^{-1}(\Delta)$ is a full pre-image of the set Δ.

(ii) $\Rightarrow$ (i). If $l = \|B\|$, we have $B_l = B$ and $E_A(\Delta)B = B\, E_A(F^{-1}(\Delta))$. Then

$$AB = \Big(\int_{-\|A\|}^{\|A\|} \lambda\, dE_A(\lambda) \Big) B = B \Big(\int_{\mathbb{R}^1} \lambda\, E_A(F^{-1}(\lambda)) \Big) =$$

$$B \Big(\int_{\mathbb{R}^1} F(\lambda)\, dE_A(\lambda) \Big) = BF(A).$$

□

THEOREM 3. The following relations are equivalent:

(1) $E_A(\Delta)\, B_l = B_l\, E_A(F^{-1}(\Delta))$
$(\forall l \ge 0,\ \forall \text{ Borel } \Delta \subset \mathbb{R}^1)$;

(2) $E_A(\Delta) \sin tB = \sin tB\ E_A(F^{-1}(\Delta))$
$(\forall t \in \mathbb{R}^1,\ \forall \text{ Borel } \Delta \subset \mathbb{R}^1)$;

(3) $f(A)\, g(B) = g_e(B)\, f(A) + g_o(B)\, f(F(A))$
$(\forall \text{ bounded measurable } f(\cdot), g(\cdot) : \mathbb{R}^1 \to \mathbb{C}^1)$.

(here $g_e(\cdot)$ and $g_o(\cdot)$ are the corresponding even and odd parts of the function $g(\cdot)$).

Proof. (1) $\Rightarrow$ (2). As the operators B_l, $E_A(\Delta)$ and $E_A(F^{-1}(\Delta))$ are self-adjoint, using (1) we find

$$B_l\, E_A(\Delta) = E_A(F^{-1}(\Delta))\, B_l \quad (\forall l \geq 0,\ \Delta \in \mathbf{B}(\mathbb{R}^1)).$$

Then for any $k = 1,2,\ldots$ and $\Delta \in \mathbf{B}(\mathbb{R}^1)$

$$E_A(\Delta)\, B_l^{2k} = B_l\, E_A(F^{-1}(\Delta))\, B_l^{2k-1} = B_l^2\, E_A(\Delta)\, B_l^{2k-2} = \cdots = B_l^{2k}\, E_A(\Delta)$$

and

$$E_A(\Delta)\, B_l^{2k+1} = B_l\, E_A(F^{-1}(\Delta))\, B_l^{2k} = B_l^{2k+1}\, E_A(F^{-1}(\Delta)).$$

Since B_l, $E_A(\Delta)$ and $E_A(F^{-1}(\Delta))$ are bounded

$$\sin t B_l = \sum_{k=1}^{\infty} \frac{B_l^{2k+1}}{(2k+1)!}\, t^{2k+1}, \quad \cos t B_l = \sum_{k=1}^{\infty} \frac{B_l^{2k}}{(2k)!}\, t^{2k}.$$

Then

$$E_A(\Delta)\, \sin t B_l = E_A(\Delta)\, \Big(\sum_{k=1}^{\infty} \frac{B_l^{2k+1}}{(2\lambda+1)!}\, t^{2k+1}\Big) =$$

$$= \Big(\sum_{k=1}^{\infty} \frac{B_l^{2k+1}}{(2k+1)!}\, t^{2k+1}\Big)\, E_A(F^{-1}(\Delta)) = \sin t B_l\, E_A(F^{-1}(\Delta))$$

and

$$E_A(\Delta)\, \cos t B_l = \cos t B_l\, E_A(\Delta).$$

For $l \to \infty$, $\sin t B_l$ strongly converges to $\sin t B$, and $\cos t B_l$ strongly converges to $\cos t B$. So

$$E_A(\Delta)\, \sin t B = \sin t B\ E_A(F^{-1}(\Delta))$$

and

$$E_A(\Delta)\, \cos t B = \cos t B\ E_A(\Delta).$$

(2) $\Rightarrow$ (3). Using the limiting process from (2), we get for an even measurable bounded function, $g_e(\cdot)$

$$E_A(\Delta)\ g_e(B) = g_e(B)\, E_A(\Delta)$$

and for any odd measurable bounded function, $g_o(\cdot)$

$$E_A(\Delta)\ g_o(B) = g_o(B)\, E_A(F^{-1}(\Delta)).$$

Now, for a measurable bounded $f(\cdot)$,

$$f(A)\, g_o(B) = (\int_{\mathbb{R}^1} f(\lambda)\, dE_A(\lambda))\, g_o(B) = g_o(B)\, (\int_{\mathbb{R}^1} f(F(\lambda))\, dE_A(\lambda)) = g_o(B)\, f(F(A)),$$

$$f(A)\, g_e(B) = (\int_{\mathbb{R}^1} f(\lambda)\, dE_A(\lambda))\, g_e(B) = g_e(B) \int_{\mathbb{R}^1} f(\lambda) = g_e(B)\, f(A).$$

(3) $\Rightarrow$ (1). It is obvious since (1) is a particular case of (3). ▯

For $F(\lambda) = \lambda$, Definition 1 defines commutativity for the operators $E_A(\Delta)$ and B_l for all Borel $\Delta \subset \mathbb{R}^1$ and $l \geq 0$ i.e. it defined the commutativity of the spectral projections $E_A(\Delta_1)$ and $E_B(\Delta_2)$ $\forall$ Borel $\Delta_1, \Delta_2 \subset \mathbb{R}^1$.

For $F(\lambda) = -\lambda$, Definition 1 and Definition I.8 for anti-commutativity of self-adjoint operators A and B are equivalent. Indeed, if $\forall m,l > 0$ $\{A_m, B_l\} = 0$ then it follows from Proposition 7 that $\forall$ Borel $\Delta \subset \mathbb{R}^1$ and $\forall l \geq 0$

$$E_A(\Delta)\, B_l = B_l\, E_A(-\Delta) = B_l\, E_A(F^{-1}(\Delta))$$

i.e. we get Definition 1 for $F(\lambda) = -\lambda$. Conversely, if we use Definition 1 for A and B with $F(\lambda) = -\lambda$ then it follows from Theorem 3 that

$$A_l\, B_m = B_m(-A)_l = -B_m\, A_l, \text{ i.e. } \{A_l, B_m\} = 0.$$

It does not follow from the relation $A\,B\,u = BF(A)u$ $(u \in \Phi)$ where Φ is a dense set in H invariant relatively to A,B and $F(A)$, that the operators A and B satisfy $AB = BF(A)$ even if Φ is a core for these operators.

However commutation, $AB\,u = BA\,u$, on a dense invariant $\Phi \ni u$ is equivalent to commutation of the spectral projections if one requires that Φ consists of analytic, in particular, entire vectors for the operators A and B. The same goes for the operators A and B satisfying relation (1). Let $F(\cdot)$ map compact subsets of $\mathbb{R}^1$ into compact subsets.

THEOREM 4. For self-adjoint operators A and B to satisfy (1) it is necessary and sufficient that

$$AB\,u = BF(A)\,u$$

on a dense set Φ in H invariant relatively to A, $F(A)$, B, which consists of entire vectors for these operators.

LEMMA 1. The self-adjoint operators $A, F(A), B^2$ commute.

Proof. It follows from the definition of $F(A)$ that A and $F(A)$ commute. Since for every Borel $\Delta \subset \mathbb{R}^1$ and $l \geq 0$ using the adjoint to (1), we get

$$B_l\, E_A(\Delta) = E_A(F^{-1}(\Delta))\, B_l$$

part (3) of Theorem 3 yields

$$E_A(\Delta)\, B_l^2 = B_l(E_A(F^{-1}(\Delta))\, B_l) = B_l^2\, E_A(\Delta)$$

and so the spectral projections of the operators A and B^2 commute. Being a function of

A, $F(A)$ also commutes with B^2. □

Let now $E_{B^2}(\cdot)$ and $E_{F(A)}(\cdot)$ be the corresponding resolutions of the identity for the operators B^2 and $F(A)$. Set

$$\Phi = \bigcup_{\Delta_1, \Delta_2, \Delta_3} E_A(\Delta)\, E_{F(A)}(\Delta_2)\, E_{B^2}(\Delta_3) H$$

($\Delta_1, \Delta_2, \Delta_3$ are compact subsets of $\mathbb{R}^1$).

Invariance of Φ with respect to the operators A and $F(A)$ follows directly from its definition. Indeed $A : \Phi \to \Phi$ because

$$A\, E_A(\Delta_1)\, E_{F(A)}(\Delta_2)\, E_{B^2}(\Delta_3) H =$$

$$= A_{\Delta_1} E_A(\Delta_1)\, E_{F(A)}(\Delta_2)\, E_{B^2}(\Delta_3) H = E_A(\Delta_1)\, E_{F(A)}(\Delta_2)\, E_{B^2}(\Delta_3) A_{\Delta_1} H \subset$$

$$\subset E_A(\Delta_1)\, E_{F(A)}(\Delta_2)\, E_{B^2}(\Delta_3) H$$

(here $A_{\Delta_1} = \int_{\Delta_1} \lambda\, dE_A(\lambda)$). Similarly we find that $F(A) : \Phi \to \Phi$. It also follows from the definition of Φ that it consists of entire (even bounded) vectors for the operators A and $F(A)$ (since Δ_1 and Δ_2 are compact).

Now, since $\operatorname{Ker} B = \operatorname{Ker} B^2$ and since the subspace $H_0 - E_B(\{0\})$ is invariant with respect to the operators A and $F(A)$, setting

$$H = H_0 \oplus H_0^{\perp}, \quad \Phi_0 = \bigcup_{\Delta_1, \Delta_2} E_A(\Delta_1)\, E_{F(A)}(\Delta_2) H_0 \subset H_0$$

(Δ_1, Δ_2 are compact in $\mathbb{R}^1$) and

$$\Phi_1 = \bigcup_{\Delta_1, \Delta_2, \Delta_3} E_A(\Delta_1)\, E_{F(A)}(\Delta_2)\, E_{B^2}(\Delta_3) H_0^{\perp}$$

($\Delta_1, \Delta_2, \Delta_3$ are compact in $\mathbb{R}^1$), we get that $B\,\Phi_0 = \{0\}$ and the set

$$B\, E_A(\Delta_1)\, E_{F(A)}(\Delta_2)\, E_{B^2}(\Delta_3) H_0^{\perp} = B\, E_{B^2}(\Delta_3)\, E_A(\Delta_1)\, E_{F(A)}(\Delta_2) H_0^{\perp}$$

is contained in $B_l\, E_A(\Delta_1)\, E_{F(A)}(\Delta_2)\, E_{B^2}(\Delta_3) H_0^{\perp}$ for some $l > 0$. Since on $H_0^{\perp}$ the operator B_l^{-1} exists, by Proposition 2, $F(\cdot)$ is one-to-one on $\sigma(A \restriction H_0^{\perp})$, and so it follows from Proposition 3 that

$$B_l\, E_A(\Delta_1) H_0^{\perp} = E_A(F(\Delta_1)) B_l H_0^{\perp},$$

$$B_l\, E_{F(A)}(\Delta_2) H_0^{\perp} = E_{F(A)}(F(\Delta_2)) B_l H_0^{\perp}.$$

Because $F(\Delta_1)$ and $F(\Delta_2)$ are compact subsets of $\mathbb{R}^1$

$$B_l\, E_A(\Delta_1)\, E_{F(A)}(\Delta_2)\, E_{B^2}(\Delta_3) H \subset E_A(F(\Delta_1)) \times$$

$$\times E_{F(A)}(F(\Delta_2))\, E_{B^2}([-l, l] \cap \Delta_3) B H_0^{\perp} \subset$$

$$\subset \bigcup_{\delta_1,\delta_2,\delta_3 \text{ compact}} E_A(\delta_1)\, E_{F(A)}(\delta_2)\, E_{B^2}(\delta_3) H = \Phi$$

i.e. $B : \Phi \to \Phi$ for all compact $\delta_1, \delta_2, \delta_3$. But since $H^b(B) = H^b(B^2)$, Φ consists of entire (even bounded) vectors for B.

In the proof of Theorem 4, we constructed for the operators A, $F(A)$, B the dense invariant subspace $\Phi = H^b(A) \cap H^b(F(A)) \cap H^b(B)$. We can introduce a nuclear topology on a dense invariant set $\mathrm{T} \subset \Phi$ such that A, $F(A)$, $B : \mathrm{T} \to \mathrm{T}$ are continuous, i.e. we have the following theorem.

THEOREM 5. There exists a nuclear rigging for the self-adjoint operators A, $F(A)$ and B which satisfy relation (1).

Proof. First of all note that there exists a nuclear rigging T_0 for the operators A, $F(A)$ and B in the invariant subspace $E_B(\{0\})H = H_0$. Now, for the bounded operators $A \restriction H_{nm}$, $B \restriction H_{nm}$, $F(A) \restriction H_{nm}$ defined on the invariant subspace

$$H_{nm} = E_{A \restriction H_0^{\perp}}(\delta_n \cup F(\delta_n)) \times$$

$$\times E_{B^2 \restriction H_0^{\perp}}([m, m-1])H \quad (n \in \mathbb{Z}\,,\ \bigcup_{n \in \mathbb{Z}} (\delta_n \cup F(\delta_n)) = \sigma(A \restriction H_0^{\perp})\,;\ m = 1,2,\ldots)$$

there exists a nuclear rigging T_{nm}. The subspace $\mathrm{T} = \mathrm{T}_0 \cup (\bigcup_{n \in \mathbb{Z}} \bigcup_{m=1}^{\infty} \mathrm{T}_{nm})$ endowed with the projective limit topology gives a nuclear rigging for the operators A, $F(A)$ and B in space $H = H_0 \oplus \sum_{n \in \mathbb{Z}} \sum_{m=1}^{\infty} H_{nm}$. ∎

Let us describe the structure of a pair of unbounded operators A and B such that $AB = BF(A)$.

The theorem given below is a generalization of two theorems, one being the classic spectral theorem for two commuting unbounded self-adjoint operators A and B in terms of their joint resolution of the identity, the other one being the structure theorem 3.8 for a pair of anti-commuting self-adjoint operators (for $F(\lambda) = -\lambda$).

THEOREM 6. Let A and B be self-adjoint operators satisfying relation (1). Then there exists a uniquely defined decomposition

$$H = H_0 \oplus H_1 \oplus H_2 = H_0 \oplus H_1 \oplus \mathbb{C}^2 \otimes H_+$$

and orthogonal resolutions of the identity:

a) $dE_0(\lambda)$ is defined on $\mathbb{R}^1$ with as its values projections into H_0;

b) $dE_1(\lambda, b)$ is defined on the set

$$M_1 = \{(\lambda, b) \in \mathbb{R}^2 \mid F(\lambda) = \lambda,\ b \neq 0\}$$

with as its values projections into H_1;

c) $dE_2(\lambda,b)$ is defined on the set

$$M_2 = \{(\lambda,b) \in \mathbb{R}^2 \mid F(F(\lambda)) = \lambda, F(\lambda) > \lambda, b > 0\}$$

with as its values projections into H_+ such that

$$A = \int_{\mathbb{R}^1} \lambda\, dE_0(\lambda) + \int_{M_1} \lambda\, dE_1(\lambda,b) + \int_{M_2} \begin{bmatrix} \lambda & 0 \\ 0 & F(\lambda) \end{bmatrix} \otimes dE_2(\lambda,b) \tag{11}$$

$$B = \int_{M_1} b\, dE_1(\lambda,b) + \begin{bmatrix} 0 & 1 \\ 1 & 0 \end{bmatrix} \otimes \int_{M_2} b\, dE_2(\lambda,b).$$

Proof. If the operators A and B are bounded, then Theorem 6 follows from Theorem 2.

For unbounded A and B, the theorem follows from the equalities:

$$(A_0)_m = (A_m)_0\,,\ (A_1)_m = (A_m)_1\,,\ (A_2)_m = (A_m)_2$$

and correspondingly

$$(B_1)_l = (B_l)_1\,,\ (B_2)_l = (B_l)_2.$$

□

Remark 3. For $F(\lambda) = \lambda$, the set $M_2 = \varnothing$. Introduce an orthogonal projection valued measure $E(\cdot)$ on $\mathbb{R}^2 \supset \Delta$:

$$E(\Delta) = E_0(\Delta \cap \{\mathbb{R}^1 \times \{0\}\}) + E_1(\Delta \cap \{\mu \neq 0\}).$$

Then $A = \int_{\mathbb{R}^2} \lambda\, dE(\lambda,\mu)$ and $B = \int_{\mathbb{R}^2} \mu\, dE(\lambda,\mu)$ establish the classic spectral representation for a pair of commuting self-adjoint operators A and B.

If $F(\lambda) = -\lambda$, we have

$$M_1 = \{(\lambda,\mu) \in \mathbb{R}^2 \mid \lambda = 0, \mu \neq 0\}$$

and

$$M_2 = \{(\lambda,\mu) \in \mathbb{R}^2 \mid \lambda > 0, \mu > 0\}.$$

Since

$$\int_{M_2} \begin{bmatrix} \lambda & 0 \\ 0 & -\lambda \end{bmatrix} \otimes dE_2(\lambda,\mu) = \begin{bmatrix} 1 & 0 \\ 0 & -1 \end{bmatrix} \otimes \int_{M_2} \lambda\, dE_2(\lambda,\mu)$$

the structure theorem 3.8 for a pair of anti-commuting self-adjoint operators follows from Theorem 6.

Remark 4. To get the structure theorem 6 for any $F(\cdot)$ the self-adjointness of the operators A and B was essential. Thus it is difficult at this time to get a structure theorem for a pair of unitary operators satisfying $UV = e^{i2\pi\alpha} VU$ ($\alpha \in (0,1)$ is irrational) since it would be a structure theorem for representations for a C^*-algebra generated by these operators that is not of type I (see, for example, M. Rieffel [2]).

10.3. Collections $\mathbf{A} = (A_k)$ and $\mathbf{B} = (B_j)$ satisfying general commutation relations.

Consider a family of commuting self-adjoint operators $\mathbf{A} = (A_k)_{k=1}^{\infty}$ which together with a single self-adjoint operator B satisfies the relations

$$A_k B = B F_k(A_1, \ldots, A_n, \ldots) \quad (k = 1,2,\ldots) \tag{12}$$

($F_k : \mathbb{R}^{\infty} \to \mathbb{R}^1$ are $\mathbf{B}(\mathbb{R}^{\infty})$-measurable functions).

We define relations (12) also for unbounded operators $(A_k)_{k=1}^{\infty}$ and B. Let $\mathbf{B}(\mathbb{R}^{\infty}) \ni \Delta \mapsto E_{\mathbf{A}}(\Delta)$ be a joint resolution of the identity for the commutative family $\mathbf{A}$,

$$F_k(\mathbf{A}) = F_k(A_1, \ldots, A_n, \ldots) = \int_{\mathbb{R}^{\infty}} F_k(\lambda)\, dE_{\mathbf{A}}(\lambda).$$

Set

$$\mathbb{R}^{\infty} \ni \lambda \mapsto F(\lambda) = (F_1(\lambda), \ldots, F_n(n), \ldots) \in \mathbb{R}^{\infty}.$$

DEFINITION 2. We will say that unbounded self-adjoint operators $(A_k)_{k=1}^{\infty}$ and B satisfy relations (12) if $\forall \Delta \in \mathbf{B}(\mathbb{R}^{\infty})$ and $\forall l \geq 0$

$$E_{\mathbf{A}}(\Delta) B_l = B_l E_{\mathbf{A}}(F^{-1}(\Delta))$$

where $B_l = \int_{-l}^{l} \mu\, dE_B(\mu)$, $E_B(\cdot)$ is a resolution of the identity for the operator B, $F^{-1}(\Delta)$ is the full pre-image of Δ.

For bounded $(A_k)_{k=1}^{\infty}$ and B the given definition is equivalent to relations (12).

PROPOSITION 8. For bounded $(A_k)_{k=1}^{\infty}$ and B, the following conditions are equivalent:

(i) $A_k B = B F_k(\mathbf{A}) \quad (k = 1,2,\ldots)$;

(ii) $\forall l \geq 0$, $\forall \Delta \in \mathbf{B}(\mathbb{R}^{\infty})\, E_{\mathbf{A}}(\Delta) B_l = B_l E_{\mathbf{A}}(F^{-1}(\Delta))$;

(iii) $\forall$ bounded measurable $f(\cdot) : \prod_{k=1}^{\infty} [-\|A_k\|, \|A_k\|] \to \mathbb{C}^1$ and $g(\cdot) : [-\|B\|, \|B\|] \to \mathbb{C}^1$

$$f(\mathbf{A})\, g(B) = g_e(B) f(\mathbf{A}) + g_o(B) f(F(\mathbf{A})).$$

The proof of Proposition 8 is similar to the proof of Proposition 7. Assuming that $F(\cdot) : \mathbb{R}^{\infty} \to \mathbb{R}^{\infty}$ maps compact sets with respect to the product topology into compact sets, following the proofs of Theorems 3,4,5 we get

THEOREM 7. For unbounded self-adjoint operators $(A_k)_{k=1}^{\infty}$ and B the following relations are equivalent.

1) $\forall l \geq 0,\ \forall \Delta \in \mathbf{B}(\mathbb{R}^\infty)\ E_{\mathbf{A}}(\Delta) B_l = B_l\, E_{\mathbf{A}}(F^{-1}(\Delta))$,

2) $\forall t \in \mathbb{R}^1,\ \forall \Delta \in \mathbf{B}(\mathbb{R}^\infty)\ E_{\mathbf{A}}(\Delta) \sin tB = \sin tB\ E_{\mathbf{A}}(F^{-1}(\Delta))$,

3) $\forall$ bounded measurable $f(\cdot) : \mathbb{R}^\infty \to \mathbb{C}^1$ and $g(\cdot) : \mathbb{R}^1 \to \mathbb{C}^1$
$f(\mathbf{A})\, g(B) = g_e(B)\, f(\mathbf{A}) + g_o(B)\, f(F(\mathbf{A}))$,

4) $AB\, u = BF(A)\, u,\ \forall u \in \mathbf{T}$

on a nuclear linear topological space $\mathbf{T}$ which is topologically dense in H invariant relatively to $(A_k, F_k(\mathbf{A}))_{k=1}^\infty$, B and consisting of entire vectors for these operators.

Similarly, we have also the corresponding structure theorem.

THEOREM 8. Let the self-adjoint operators $(A_k)_{k=1}^\infty$ satisfy (12). Then there is uniquely defined an orthogonal decomposition, $H = H_0 \oplus H_1 \oplus H_2$, $H_2 = \mathbb{C}^2 \otimes H_+$ and resolutions of the identity

1) $dE_0(\lambda)$ defined on $\mathbb{R}^\infty \ni \lambda$ with as its values projections into H_0;

2) $dE_1(\lambda,\mu)$ defined on

$$M_1 = \{(\lambda,\mu) \in \mathbb{R}^\infty \times \mathbb{R}^1 \mid F(\lambda) = \lambda,\ \mu \neq 0\}$$

with as its values projections into H_1;

3) $dE_2(\lambda,\mu)$ defined on

$$M_2 = \{(\lambda,\mu) \in \mathbb{R}^\infty \times \mathbb{R}^1 \mid F(F(\lambda)),\ F(\lambda) > \lambda,\ \mu > 0\}$$

($\mathbb{R}^\infty$ is ordered lexicographically) with as its values projections into H_+ such that

$$A_k = \int_{\mathbb{R}^\infty} \lambda_k\, dE_0(\lambda) + \int_{M_1} \lambda_k\, E_1(\lambda,\mu) + \int_{M_2} \begin{bmatrix} \lambda_k & 0 \\ 0 & F_k(\lambda) \end{bmatrix} \otimes dE_2(\lambda,\mu)$$

$$B = \int_{M_1} b\, dE_1(\lambda,\mu) + \begin{bmatrix} 0 & 1 \\ 1 & 0 \end{bmatrix} \otimes \int_{M_2} b\, E_2(\lambda,\mu).$$

For such collections of operators $(A_k)_{k=1}^\infty$ and B, the irreducible representations are still either one-dimensional (which corresponds to the case that B is degenerate or to the fixed points of the mapping $F(\cdot) : \mathbb{R}^\infty \to \mathbb{R}^\infty$) or two-dimensional (which correspond to pairs of the points $\lambda^{(1)} \neq \lambda^{(2)} \in \mathbb{R}^\infty$ such that $F(\lambda^{(1)}) = \lambda^{(2)}$ and $F(\lambda^{(2)}) = \lambda^{(1)}$). □

Now consider a finite collection of CSO $(B_j)_{j=1}^n$ which together with a self-adjoint operator A satisfies the relations

$$A\, B_j = B_j\, F_j(A)$$
$$(F_j(\cdot) : \mathbb{R}^1 \to \mathbb{R}^1;\ j = 1, \ldots, n). \tag{13}$$

Since the CSO $(B_j^2)_{j=1}^n$ commute with A, the subspaces

$$H_{0,i_1\cdots i_k}=E_{B[i_1]}(\{0\})\cdots E_{B[i_k]}(\{0\})E_{B[j_1]}(\mathbb{R}^1\setminus\{0\})\cdots E_{B[j_{n-k}]}(\mathbb{R}^1\setminus\{0\})H$$

$$(i_1,\ldots,i_k=1,\ldots,n\,;\ i_1<\cdots<i_k\,;\ k=1,\ldots,n)$$

and

$$\mathbf{H}=\prod_{k=1}^{n}E_{B_k}(\mathbb{R}^1\setminus\{0\})H$$

on which a part of the operators $B_{i_1},\ldots,B_{i_k}$ are degenerate and the rest are not, are invariant relatively to $(B_j)_{j=1}^n$ and A, and

$$(B_j\restriction H_{0,i_1,\ldots,i_k})_{j=1}^n\,,\ A\restriction H_{0,i_1,\ldots,i_k}$$

satisfy relations (13) with

$$B_{i_1}\restriction H_{0,i_1,\ldots,i_k}=\cdots=B_{i_k}\restriction H_{0,i_1\cdots i_k}=0.$$

The decomposition $H=\oplus\sum_{\substack{k=1\\ i_1<\cdots<i_k}}H_{0,i_1,\ldots,i_k}\oplus\mathbf{H}$ reduces the study of the structure of the collection $(B_j)_{j=1}^n$, A to the study of the structure of the collection $B_{j_1},\ldots,B_{j_{n-k}}$, A on $H_{0,i_1\cdots i_k}$, where the operators also satisfy (13). So in what follows we will assume that the operators of the CSO $(B_j)_{j=1}^n$ are invertible. We have the following lemma.

LEMMA 2. For any collection of indices $j_1,\ldots,j_k$ $(j_i=1,\ldots,n)$ and any permutation of these indices $\sigma(j_1,\ldots,j_k)=(\sigma(j_1),\ldots,\sigma(j_k))$

$$F_{j_1}(F_{j_2}(\cdots(F_{j_k}(A))\cdots))=F_{\sigma(j_1)}(F_{\sigma(j_2)}(\cdots(F_{\sigma(j_k)}(A))\cdots))\,,$$
$$F_j(F_j(A))=A. \tag{14}$$

Proof. The condition $F_j(F_j(A))=A$ has been proved in Proposition 2b). Since

$$AB_{j_1}B_{j_2}\cdots B_{j_k}=B_{j_1}F_{j_1}(A)B_{j_2}\cdots B_{j_k}=B_{j_1}B_{j_2}F_{j_1}(F_{j_2}(A))\cdots B_{j_k}=$$
$$=B_{j_1}\cdots B_{j_k}F_{j_1}(F_{j_2}(\cdots(F_{j_k}(A))\cdots))\,,$$
$$AB_{\sigma(j_1)}B_{\sigma(j_2)}\cdots B_{\sigma(j_k)}=B_{\sigma(j_1)}F_{\sigma(j_1)}(A)B_{\sigma(j_2)}\cdots B_{\sigma(j_k)}=$$
$$=B_{\sigma(j_1)}\cdots B_{\sigma(j_k)}F_{\sigma(j_1)}(F_{\sigma(j_2)}(\cdots(F_{\sigma(j_k)}(A))\cdots))\,,$$
$$AB_{j_1}\cdots B_{j_k}=AB_{\sigma(j_1)}\cdots B_{\sigma(j_k)}$$

and the operators $B_{j_1},\ldots,B_{j_k}$ are invertible, equality (14) holds. $\square$

We describe up to unitary equivalence the irreducible collections of operators A and CSO $(B_j)_{j=1}^n$ satisfying (13).

First of all, choose $\lambda\in\mathbb{R}^1$ such that

$$F_j(F_j(\lambda)) = \lambda \quad (j = 1, \ldots, n)$$

$$F_{j_1}(F_{j_2}(\cdots (F_{j_k}(\lambda)) \cdots)) = F_{\sigma(j_1)}(F_{\sigma(j_2)}(\cdots (F_{\sigma(j_k)}(\lambda)) \cdots)) \tag{15}$$

for any collection of the indices $j_1, \ldots, j_k$ $(j_i = 1, \ldots, n)$ and any permutation of these indices σ.

Then the notation

$$F_{i_1 \cdots i_n}(\lambda) = F_1^{i_1}(F_2^{i_2}(\cdots (F_n^{i_n}(\lambda)) \cdots)) \quad (i_k \in \{0,1\},\ k = 1, \ldots, n)$$

is correct. In the series of numbers $(F_k(\lambda))_{k=1}^n$ choose the first $F_{j_1}(\lambda) \neq \lambda$, then the second $F_{j_2}(\lambda) \neq \lambda$ and $F_{j_2}(\lambda) \neq F_{j_1}(\lambda)$ (note that $F_{j_1}(F_{j_2}(\lambda))$ will then be different from any of the numbers $\lambda, F_{j_1}(\lambda), F_{j_2}(\lambda)$) and so on. Let now the indices $j_1, \ldots, j_k$ be taken such that all the numbers

$$\lambda, F_{j_1}(\lambda), \ldots, F_{j_k}(\lambda), F_{j_1}(F_{j_2}(\lambda)), \ldots, F_{j_{k-1}}(F_{j_k}(\lambda)), \ldots, F_{j_1}(F_{j_2}(\cdots (F_{j_k}(\lambda)) \cdots))$$

are different.

If there is $F_{j_{k+1}}(\lambda)$ different from all these numbers, then in the series

$$\lambda, F_{j_1}(\lambda), \ldots, F_{j_k}(\lambda), F_{j_{k+1}}(\lambda), F_{j_1}(F_{j_2}(\lambda)), \ldots, F_{j_1}(F_{j_{k+1}}(\lambda)),$$

$$F_{j_2}(F_{j_3}(\lambda)), \ldots, F_{j_2}(F_{j_{k+1}}(\lambda)), \ldots, F_{j_1}(F_{j_2}(\cdots (F_{j_{k+1}}(\lambda)) \cdots))$$

all the numbers are different (if

$$F_{j_1}^{i_1}(F_{j_2}^{i_2}(\cdots (F_{j_k}^{i_k}(\lambda)) \cdots)) = F_{j_1}^{i_1'}(F_{j_2}^{i_2'}(\cdots (F_{j_k}^{i_k'}(F_{j_{k+1}}(\lambda))) \cdots))$$

$$(i_1, \ldots, i_k, i_1', \ldots, i_k' \in \{0,1\})$$

we get a contradiction, $F_{j_1}^{i_1+i_1'}(\cdots (F_{j_k}^{i_k+i_k'}(\lambda)) \cdots) = F_{j_{k+1}}(\lambda)$), so we will be completing this series of numbers until we arrive at $F_{j_m}(\lambda)$ such that the collection of different numbers is equal to the whole collection $(F_{i_1 \cdots i_n}(\lambda))_{i_1, \ldots, i_n = 0}^1$.

Now construct an irreducible representation of the relations (13). Set

$$H_\lambda = \mathrm{LS}\,\{(e[F_{i_1 \cdots i_n}(\lambda)] \mid (i_1, \ldots, i_n \in \{0,1\})\}$$

and define the representation operators $A : H_\lambda \to H_\lambda$ and $B_j : H_\lambda \to H_\lambda$ $(j = 1, \ldots, n)$ accordingly:

$$A\, e[F_{i_1 \cdots i_n}(\lambda)] = F_{i_1 \cdots i_n}(\lambda)\, e[F_{i_1 \cdots i_n}(\lambda)],$$

$$B_j\, e[F_{i_1 \cdots i_n}(\lambda)] = \mu_j\, e[F_j(F_{i_1 \cdots i_n}(\lambda))] \quad (\mu_j > 0, \text{ if } j = j_k \tag{16}$$

$$(k = 1, \ldots, m),\ \mu_j \in \mathbb{R}^1 \setminus \{0\} \text{ if } j \neq j_k \ (k = 1, \ldots, m)).$$

The representation is irreducible. Indeed, if an operator C on H_λ commutes with A and $(B_j)_{j=1}^n$,

$$C\, e[F_{i_1 \cdots i_n}(\lambda)] = c[F_{i_1 \cdots i_n}(\lambda)]\, e[F_{i_1 \cdots i_n}(\lambda)]$$

and since

$$c[F_{i_1 \cdots i_n}(\lambda)]\, e[F_{i_1 \cdots i_n}(\lambda)] = \frac{1}{\mu_{j_1} \cdots \mu_{j_k}} C B_{j_1} \cdots B_{j_k} e[\lambda] = \frac{1}{\mu_{j_1} \cdots \mu_{j_k}} B_{j_1} \cdots B_{j_k}$$

we have that $C = c[\lambda] I$.

The dimension of this representation is $\dim H_\lambda = 2^m$ and a basis could be formed with the vectors

$$e[F_{j_1}^{i_{j_1}}(\cdots (F_{j_m}^{i_{j_m}}(\lambda)) \cdots)] \quad (i_{j_1}, \ldots, i_{j_m} \in \{0,1\}).$$

The representations corresponding to the collection of numbers $(\mu_1, \ldots, \mu_n)$ on H_λ and corresponding to the collection $(\mu_1', \ldots, \mu_m')$ on H_λ are unitarily equivalent if and only if $\exists\, i_1, \ldots, i_n$ such that $F_{i_1 \cdots i_n}(\lambda) = \lambda'$ and $\mu_j = \mu_j'$ $(j = 1, \ldots, n)$.

THEOREM 9. Any irreducible representation of the relations (13) is unitarily equivalent to a representation of the form (16) for some

$$b_1, \ldots, b_n (b_j \neq 0\,; b_k > 0\,, k = 1, \ldots, m)$$

and some $\lambda \in \mathbb{R}^1$ which satisfies system (15).

To prove the theorem we will need the following lemma.

LEMMA 3. If a representation of the relations (13) is irreducible, then the spectrum of the operator A is discrete and finite.

Proof. Denote $F_{i_1 \cdots i_n}(A) = F_1^{i_1}(\cdots (F_n^{i_n}(A)) \cdots)$. Since the representation is irreducible

$$\sum_{i_1, \ldots, i_n = 0}^{1} F_{i_1 \cdots i_n}(A) = d_0 I\,,$$

$$(\sum_{i_1, \ldots, i_n = 0}^{1} (-1)^{i_k} F_{i_1 \cdots i_n}(A))^2 = d_k^2 I\,, \quad (k = 1, \ldots, n)$$

$$(\sum_{i_1, \ldots, i_n = 0}^{1} (-1)^{i_k + i_l} F_{i_1 \cdots i_n}(A))^2 = d_{kl}^2 I\,, \quad (k < l)$$

....

$$(\sum_{i_1, \ldots, i_n = 0}^{1} (-1)^{i_1 + \cdots + i_n} F_{i_1 \cdots i_n}(A))^2 = d_{12 \cdots n}^2 I.$$

But then, for $\lambda \in \sigma(A)$, we have

$$\sum_{i_1,\ldots,i_n=0}^{1} F_{i_1\cdots i_n}(\lambda) = d_0\,,$$

$$\sum_{i_1,\ldots,i_n=0}^{1} (-1)^{i_k} F_{i_1\cdots i_n}(\lambda) = \pm d_k\,,\quad (k=1,\ldots,n)$$

....

$$\sum_{i_1,\ldots,i_n=0}^{1} (-1)^{i_1+\cdots+i_n} F_{i_1\cdots i_n}(\lambda) = \pm d_{i_1\cdots i_n}\,.$$

But this system has only a finite number of solutions $\lambda = F_{0\cdots 0}(\lambda)$. ▯

Denote by $e = e[\lambda]$ a unit eigenvector of the operator A corresponding to the eigenvalue λ. Since the representation is irreducible, it follows that $B_k^2 = \mu_k^2 I$ $(k=1,\ldots,n)$. Introduce unitary (and self-adjoint) commuting operators

$$U_k = B_k/\mu_k \;\; (\mu_k > 0\,,\; k=1,\ldots,n).$$

The space $H = \mathrm{LS}\,\{U_1^{i_1}\,\cdots\,U_n^{i_n}\,e\}$ is invariant relatively to A and $(B_j)_{j=1}^n$.

LEMMA 4. If $F_{i_1\cdots i_n}(\lambda) = F_{i_1'\cdots i_n'}(\lambda)$ then

$$U_1^{i_1}\,\cdots\,U_n^{i_n} = c\left[\!\left[\begin{pmatrix} i_1,\ldots,i_n \\ i_1',\ldots,i_n'\end{pmatrix}\right]\!\right] U_1^{i_1'}\,\cdots\,U_n^{i_n'}\,,\; c\left[\!\left[\begin{pmatrix} i_1,\ldots,i_n \\ i_1',\ldots,i_n'\end{pmatrix}\right]\!\right] = \pm 1.$$

Proof. It follows from (15) that

$$F_{i_1+i_1'\cdots i_n+i_n'}(F_{j_1}(\cdots(F_{j_k}(\lambda))\cdots)) = F_{j_1}(\cdots(F_{j_k}(\lambda))\cdots)$$

$$(j_1,\ldots,j_k \in \{0,1\}\;;\; k \le m).$$

Then the operator $U_1^{i_1+i_1'}\,\cdots\,U_n^{i_n+i_n'}$ defined on H commutes with the operator A. Indeed

$$A\,U_1^{i_1+i_1'}\,\cdots\,U_n^{i_n+i_n'} = U_1^{i_1+i_1'}\,\cdots\,U_n^{i_n+i_n'}\,F_{i_1+i_1'\cdots i_n+i_n'}(A) =$$

$$= U_1^{i_1+i_1'}\,\cdots\,U_n^{i_n+i_n'}\,A.$$

Consequently,

$$U_1^{i_1+i_1'}\,\cdots\,U_n^{i_n+i_n'} = c\left[\!\left[\begin{pmatrix} i_1,\ldots,i_n \\ i_1',\ldots,i_n'\end{pmatrix}\right]\!\right] I\,,\;\text{ i.e.}$$

$$U_1^{i_1}\,\cdots\,U_n^{i_n} = c\left[\!\left[\begin{pmatrix} i_1,\ldots,i_n \\ i_1',\ldots,i_n'\end{pmatrix}\right]\!\right] U_1^{i_1'}\,\cdots\,U_n^{i_n'}.$$

▯

According to Lemma 4, the vectors

$$(e[F_{j_1}^{i_1}(\cdots(F_{j_m}^{i_m}(\lambda))\cdots)])_{i_1,\ldots,i_m=0}^{1} = (U_{j_1}^{i_1} \cdots U_{j_m}^{i_m} e)_{i_1,\ldots,i_m=0}^{1}$$

form an orthonormal basis in H, and for this basis

$$A\, e[(F_{j_1}^{i_1}(\cdots(F_{j_m}^{i_m}(\lambda))\cdots)] = F_{j_1}^{i_1}(\cdots(F_{j_m}^{i_m}(\lambda))\cdots)\, e[F_{j_1}^{i_1}(\cdots(F_{j_m}^{i_m}(\lambda))\cdots)]\,,$$

$$B_{j_k}\, e[(F_{j_1}^{i_1}(\cdots(F_{j_m}^{i_m}(\lambda)))] = \mu_{j_k}\, e[F_{j_k}^{i_1}(\cdots(F_{j_k}^{i_k+1}(\cdots(F_{j_m}^{i_m}(\lambda))\cdots)] \quad (\mu_{j_k} > 0,\ k = 1, \ldots,$$

The operators B_k $(k \neq j_1, \ldots, j_m)$ for which

$$F_k(\lambda) = F_{j_1}^{i_{1,k}}(F_{j_2}^{i_{2,k}}(\cdots(F_{j_m}^{i_{m,k}}(\lambda))\cdots)$$

are equal to

$$B_k = \mu_k U_{j_1}^{i_{1,k}} \cdots U_{j_m}^{i_{m,k}} \quad (\mu_k \neq 0)$$

i.e. the representation has the form (16). ▯

For a collection of self-adjoint operators A, $(B_j)_{j=1}^n$ satisfying relations (13), there is also a structure theorem similar to Theorems 6 and 8. We formulate it under the assumption that the numbers $(F_{i_1} \cdots {}_{i_n}(\lambda))_{i_1,\ldots,i_n=0}^{1}$ are different for all the points $\lambda \in \sigma(A)$.

THEOREM 10. If the operators $(B_j)_{j=1}^n$ are non-degenerate and the spectrum of the operator A contains only such λ that $(F_{i_1} \cdots {}_{i_n}(\lambda))_{i_1,\ldots,i_n=0}^{1}$ are all distinct, then for any representation of the relations (13) there correspond a unique decomposition

$$H = \mathbb{C}^{2^n} \otimes H_+ = \underbrace{\mathbb{C}^2 \otimes \ldots \otimes \mathbb{C}^2}_{n} \otimes H_+$$

and an orthogonal operator measure $dE(\lambda, \mu_1, \ldots, \mu_n)$ on

$$M = S \times \mathbb{R}^n_{+\cdots+} = \{\lambda \in \mathbb{R}^1 \mid \lambda \leq F_{i_1} \cdots {}_{i_n}(\lambda)\,;$$

$F_{i_1} \cdots {}_{i_n}(\lambda)$ satisfy (15) $(i_1, \ldots, i_n = 0, 1)\} \times \{\mu_1 > 0, \ldots, \mu_n > 0\}$ with as its values projections into H, such that

$$A = \int_M \begin{bmatrix} \lambda & & & 0 \\ & F_{10\cdots0}(\lambda) & & \\ & & \ddots & \\ 0 & & & F_{1\cdots1}(\lambda) \end{bmatrix} \otimes dE(\lambda, b_1, \ldots, b_n)\,,$$

$$B_k = 1 \otimes \cdots \otimes \begin{bmatrix} 0 & 1 \\ 1 & 0 \end{bmatrix} \otimes \cdots \otimes 1 \otimes \int_M b_k\, dE(\lambda, b_1, \ldots, b_n).$$

Consider now a countable collection of CSO $(A_k)_{k=1}^{\infty}$ and a finite collection of CSO $(B_j)_{j=1}^n$ which satisfy the relations

$$A_k B_j = B_j F_{kj}(\mathbf{A}) \tag{17}$$

$$(k = 1,2,\ldots;\ j = 1, \ldots, n;\ F_{kj}(\cdot) : \mathbb{R}^\infty \to \mathbb{R}^1;\ F_j(\cdot) : \mathbb{R}^\infty \to \mathbb{R}^\infty).$$

In this case, irreducible representations of (17) have the dimension $\leq 2^n$ and can also be described by a theorem similar to Theorem 9 using solutions of the system

$$F_j(F_j(\lambda)) = \lambda, \quad (\lambda \in \mathbb{R}^\infty)$$

$$F_{j_1}(\cdots(F_{j_k}(\lambda))\cdots) = F_{\sigma(j_1)}(\cdots(F_{\sigma(j_k)}(\lambda))\cdots).$$

There is also a corresponding structure theorem.

10.4. Commutative models.

If one only assumes that the mappings $F_{kj} : \mathbb{R}^\infty \to \mathbb{R}^1$ are measurable, then to have the structure theorems for the CSO $\mathbf{A} = (A_k)_{k=1}^\infty$ and the CSO $\mathbf{B} = (B_j)_{j=1}^n$ satisfying

$$A_k B_j = B_j F_{kj}(\mathbf{A}) \quad (k = 1,2,\ldots;\ j = 1, \ldots, n)$$

it is essential that the collection $\mathbf{B}$ is finite. For countable collections $(B_j)_{j=1}^\infty$, for example, if $F_{kj}(\lambda_1, \ldots, \lambda_n, \ldots) = (-1)^{\delta_{kj}} \lambda_k$ we are lead to the difficult problem of describing the structure of representations of a spin system with infinitely many degrees of freedom (see Chapter 9, Section 9.3).
In what follows, for a countable collection of CSO $\mathbf{A} = (A_k)_{k=1}^\infty$ and a countable collection of CSO $\mathbf{B} = (B_j)_{j=1}^\infty$ such that

$$A_k B_j = B_j F_{kj}(\mathbf{A}) \tag{18}$$

$$(k,j = 1,2,\ldots;\ F_{kj}(\cdot) : \mathbb{R}^\infty \to \mathbb{R}^1)$$

we construct commutative models similar to the ones constructed for representations of some "large" groups and of the algebra of local observables in Chapter 4, 5, 6 and 9.

The methods used to construct commutative models for infinite collections of operators goes back to L. Gårding, A. Wightman [1,2], I.M. Gel'fand, N.Ya. Vilenkin [1], V.Ya. Golodets [3] etc. The term "commutative model" was introduced in A.M. Vershik, I.M. Gel'fand, M.I. Graev [6].

The first step is to give an exact meaning to the relations (18) for unbounded $(A_k)_{k=1}^\infty$ and such $(B_j)_{j=1}^\infty$ that B_j maps the joint eigenspaces H_λ of the commutative family $(A_k)_{k=1}^\infty$ corresponding to the eigenvalue $\lambda \in \mathbb{R}^\infty$ into an eigenspace

$$H_{F_j(\lambda)} \quad (F_j(\lambda) = (F_{1j}(\lambda), F_{2j}(\lambda), \ldots) \in \mathbb{R}^\infty).$$

This needs to be done in such a way that it would allow at the second step to construct a commutative model for the operators $(A_k)_{k=1}^\infty$, B_j i.e. to define an operator B_j on the space of Fourier images of the family **A**. The third step is to construct a commutative model for all the operators $(A_k)_{k=1}^{\infty}$, $(B_j)_{j=1}^{\infty}$ taking into account the relations satisfied by $(B_j)_{j=1}^{\infty}$.

We make the following definition.

DEFINITION 3. The collections of CSO $\mathbf{A}=(A_k)_{k=1}^{\infty}$ and $\mathbf{B}=(B_j)_{j=1}^{\infty}$ are said to satisfy (18) if $\forall j=1,2,\ldots;\ l>0$ and $\Delta\in \mathbf{B}(\mathbb{R}^{\infty})$

$$B_{j,l}\, E_{\mathbf{A}}(\Delta)=E_{\mathbf{A}}(F_j(\Delta))\, B_{j,l}$$

($E_{\mathbf{A}}(\cdot)$ is the joint resolution of the identity of the collection $\mathbf{A}=(A_k)_{k=1}^{\infty}$, and we have $B_{j,l}=\int_{-l}^{l}\lambda_j\, dE_{B_j}(\lambda j))$.

Everywhere we assume that the mappings $F_j(\cdot):\mathbb{R}^{\infty}\to\mathbb{R}^{\infty}$ map the compact sets with respect to product topology into compact sets, and for any compact $K\subset\mathbb{R}^{\infty}$, the image $F_j(K)$ is contained in a compact set which does not depend on $j=1,2,\ldots$.

THEOREM 11. The following conditions are equivalent

I) the CSO $(A_k)_{k=1}^{\infty}$ and the CSO $(B_j)_{j=1}^{\infty}$ satisfy relations (18),

II) there exists a rigging $\Phi\subset H$ standardly related to the collections $(A_k, F_{kj}(\mathbf{A}), B_j)_{k,j=1}^{\infty}$. It consists of entire vectors for every operator of these collections such that $\forall u\in\Phi$, $j=1,2,\ldots$

$$A_k\, B_j\, u=B_j\, F_{kj}(\mathbf{A})\, u\ ,$$

III) there exists a rigging $\Phi\subset H$ standardly related to the operators $(A_k, F_{kj}(\mathbf{A}), B_j)_{k,j=1}^{\infty}$. It consists of entire vectors for the operators of these collections such that $\forall u,v\in\Phi$, $j=1,2,\ldots$, $\Delta\in\mathbf{B}(\mathbb{R}^{\infty})$

$$(B_j\, u,\, E_{\mathbf{A}}(\Delta)\, v)=(E_{\mathbf{A}}(F_j^{-1}(\Delta))\, u,\, B_j\, v).$$

Proof. (I) $\Rightarrow$ (III). Since the representation space can be decomposed into a direct sum of invariant subspaces, on which the operators B_j are non-degenerate, to prove that (I) $\Rightarrow$ (III) we can consider the families of the CSO $(A_k)_{k=1}^{\infty}$ and the CSO $(B_j)_{j=1}^{\infty}$ under the assumption that the operators B_j $(j=1,2,\ldots)$ are non-degenerate. Decompose the space H into a direct sum of invariant subspaces H_n on which the operators $(A_k, F_{kj}(\mathbf{A}), B_j)_{k,j=1}^{\infty}$ are bounded. To do this, choose the sets

$$K_n=\prod_{k=1}^{\infty}[-a_k^{(n)}, a_k^{(n)}]\subset\mathbb{R}^{\infty}\quad(0\cdots<a_k^{(n)}<a_k^{(n+1)}<\cdots\to\infty)$$

such that $E_{\mathbf{A}}(\bigcup_{n=1}^{\infty}K_n)=I$,

$$M_m = \prod_{k=1}^{\infty} [0, b_k^{(m)}] \subset \mathbb{R}^\infty \quad (\cdots < b_k^{(m)} < b_k^{(m+1)} < \cdots \to \infty)$$

such that $E_{(B_j)^2}(\bigcup_{m=1}^{\infty} M_m) = I$ and set

$$\delta_n = (K_n \setminus K_{n-1}) \bigcup_{j=1}^{\infty} F_j(K_n \setminus K_{n-1}) \subset \mathbb{R}^\infty, \quad H_{nm} = E_{\mathbf{A}}(\delta_n) \cdot E_{B_j^2}(M_n) H .$$

Rigging every space H_{nm} which is standardly related to the operators $(A_k, F_{kj}(\mathbf{A}), B_j)_{k,j=1}^{\infty}$ with the spaces Φ_{nm}, construct $\Phi = \oplus \Sigma \Phi_{nm}$ which consists of entire (even bounded) vectors. Since

$$E_{\mathbf{A}}(\Delta) B_j u = B_j E_{\mathbf{A}}(F_j^{-1}(\Delta)) u$$

on the bounded vectors $u \in \Phi \; \forall j = 1, 2, \ldots, \Delta \in \mathbf{B}(\mathbb{R}^\infty)$, (III) is proved.

Condition (II) is a particular case of (III). The proof of (III) $\Rightarrow$ (I) is similar to the proof of Theorem 4. ▯

So, speaking about algebraic relations in the form (18), for unbounded operators $(A_k)_{k=1}^{\infty}$ and $(B_j)_{j=1}^{\infty}$, we can suppose that the relations

$$(B_j u, E_{\mathbf{A}}(\Delta) v) = (E_{\mathbf{A}}(F_j(\Delta)) u, B_j v), \quad u, v \in \Phi, \; \Delta \in \mathbf{B}(\mathbb{R}^\infty)$$

hold on a rigging Φ which is dense in H, consists of entire vectors and which is invariant with respect to $(A_k, F_{kj}(\mathbf{A}), B_j)_{k,j=1}^{\infty}$.

First, consider a single operator B such that

$$A_k B = B \, F_k(\mathbf{A}) \quad (k = 1, 2, \ldots).$$

If $e[\lambda]$ is a joint eigenvector for the family $\mathbf{A}$, $A_k e[\lambda] = \lambda_k e[\lambda]$ and $e[\lambda] \in \Phi$, $B e[\lambda] \in \Phi$ is a joint eigenvector of $\mathbf{A}$ with eigenvalue $F(\lambda) = (F_1(\lambda), F_2(\lambda), \ldots)$. To make a similar statement in the case of an arbitrary joint spectrum of $\mathbf{A}$, it is necessary to use the notion of generalized joint eigenvector.

Since the space Φ is nuclear, the joint resolution of the identity of the family $\mathbf{A}$, $E_{\mathbf{A}}(\cdot)$ can be differentiated with respect to the spectral measure $\rho(\cdot) : dE_{\mathbf{A}}(\lambda) = P(\lambda) \, d\rho(\lambda)$ where $P(\lambda) : \Phi \to \Phi'$ are generalized projections. Under the made assumptions, the image of the projection $\mathbf{R}(P(\lambda)) \subset \Phi'$ consists for $\rho(\cdot)$-almost all $\lambda \in \mathbb{R}^\infty$ of generalized joint eigenvectors of the family $\mathbf{A}$ with eigenvalue $\lambda \in \mathbb{R}^\infty$ (recall that $\alpha \in \Phi'$ is a generalized joint eigenvector of the family $\mathbf{A}$ if

$$\forall u \in \Phi, \; \forall k = 1, 2, \ldots \quad (\alpha, A_k u) = \lambda_k (\alpha, u)).$$

Define the operator $\tilde{B} : \Phi' \to \Phi'$ by setting

$$(\tilde{B} \alpha, u) = (\alpha, B u) \quad \text{for} \quad \alpha \in \Phi', \; u \in \Phi.$$

THEOREM 12. Let there exist a nuclear rigging Φ of the space H with the properties stated above such that for all

$$u, v \in \Phi,\ \Delta \in \mathbf{B}(\mathbb{R}^\infty)\ (B\,u,\ E_{\mathbf{A}}(\Delta)\,v) = (E_{\mathbf{A}}(F(\Delta))\,u,\ B\,v).$$

Then there exists exists a set $\tau \in \mathbf{B}(\mathbb{R}^\infty)$ of full measure $\rho(\cdot)$ such that for all $\lambda \in \tau$, $\mathbf{R}(P(\lambda))$ consists of generalized joint eigenvectors of the family $\mathbf{A}$ with eigenvalue λ and $\tilde{B}\,\mathbf{R}(P(\lambda))$ consists of generalized joint eigenvectors of the family $\mathbf{A}$ with value $F(\lambda)$.

Proof. First of all, we prove the following lemma.

LEMMA 5. There exists a set of full measure $\tau \in (\mathbb{R}^\infty)$ such that for all $\lambda \in \tau$, $\mathbf{R}(P(\lambda))$ consists of generalized joint eigenvectors of the family $\mathbf{A}$ with eigenvalue λ. These vectors are also generalized joint eigenvectors of the family $F(\mathbf{A})$ with eigenvalue $F(\lambda)$.

Proof. Fix $u, v \in \Phi$, $k \in \mathbb{N}$. For any $\Delta \in \mathbf{B}(\mathbb{R}^\infty)$,

$$\int_\Delta (P(\lambda)\,u,\ F_k(\mathbf{A})\,v)\,d\rho(\lambda) = (E_{\mathbf{A}}(\Delta)\,u,\ F_k(\mathbf{A})\,v) =$$

$$= (F_k(\mathbf{A})\,E_{\mathbf{A}}(\Delta)\,u,\ v) = \int_\Delta F_k(\lambda)\,(P(\lambda)\,u,\ v)\,d\rho(\lambda)$$

so there exists a set $\Delta_{k;u,v}$ of full measure such that for $\lambda \in \Delta_{k;u,v}$

$$(P(\lambda)\,u,\ F_k(\mathbf{A})\,v) = F_k(\lambda)\,(P(\lambda)\,u,\ v).$$

Choose in Φ a countable everywhere dense set L and set $\Delta_k = \bigcap_{u,v \in L} \Delta_{k;u,v}$. Then the set $\tau_0 = \bigcap_{k=1}^{\infty} \Delta_k$ has full measure. Let τ_1 be a similarly constructed set for the collection $\mathbf{A}$. The intersection $\tau = \tau \cap \tau_1$ has full measure and is the wanted set. $\square$

For any $u, v \in \Phi$, $k \in \mathbb{N}$, by Lemma 5 for $\lambda \in \tau$ it follows that

$$(\tilde{B}\,P(\lambda)\,u,\ A_k\,v) = (P(\lambda)\,u,\ B^*\,A_k\,v) = (P(\lambda)\,u,\ F_k(\mathbf{A})\,B^*\,v) =$$

$$= F_k(\lambda)\,(P(\lambda)\,u,\ B^*\,v) = F_k(\lambda)\,(\tilde{B}\,P(\lambda)\,u,\ v).$$

$\square$

First, we construct a commutative model for a single operator B i.e. we realize the commutative family $\mathbf{A}$ in the space of Fourier images.

Let $H_- \supset H \supset H_+$ be a Hilbert rigging of the space H, let the imbedding $O : H_+ \to H$ be quasi-nuclear, $\rho(\Delta) = Tr\ O^+\,E(\Delta)\,O$ be a spectral measure, and let $P(\lambda) : H_+ \to H_-$ be generalized projections $(d\,O^+\,E(\lambda)\,O = P(\lambda)\,d\rho(\lambda))$. For $\rho(\cdot)$-almost all λ the operator $J^+\,P(\lambda)\,J : H \to H$ $(J : H \to H_+,\ J^+ : H_- \to H$ are isometries constructed using the rigging) is a Hilbert-Schmidt operator on H with the Hilbert-Schmidt norm $|J^+\,P(\lambda)\,J| = 1$. Let $\psi_\gamma(\lambda)$, $\gamma = 1, \ldots, N_\lambda \le \infty$ be a collection of normed eigenvectors of the operator $J^+\,P(\lambda)\,J$

corresponding to non-zero eigenvalues $\nu_\gamma(\lambda)$. For every vector $u \in H_+$, we consider its Fourier transform

$$\tilde{u}(\lambda) = (\tilde{u}_1(\lambda), \tilde{u}_2(\lambda), \ldots) \in l_2(N_\lambda)$$

$$\tilde{u}_\gamma(\lambda) = \nu_\gamma^{-\frac{1}{2}}(\lambda)\,(u, P(\lambda) J^{-1} \psi_\gamma(\lambda)).$$

The space H is isomorphic to a direct integral

$$H = \oplus \int_{I\!R^\infty} l_2(N_\lambda)\, d\rho(\lambda)$$

and Parseval's equality

$$(u, v) = \int_{I\!R^\infty} (u(\lambda), v(\lambda))_{l_2(N_\lambda)}\, d\rho(\lambda)$$

holds.

THEOREM 13. Let the relations $A_k B = B F_k(\mathbf{A})$ hold for the operators $(A_k)_{k=1}^\infty$, B. Then, in the Fourier image space for the family $\mathbf{A}$, for $u \in \Phi$,

$$(\tilde{B}\, u)(\lambda) = B(\lambda)\, \chi_{\Delta_0}(\lambda) \left[\frac{d\rho(F^{-1}(\lambda))}{d\rho(\lambda)} \right]^{\frac{1}{2}} \tilde{u}(F^{-1}(\lambda)). \tag{19}$$

Here $\Delta_0 = \{\lambda \in I\!R^\infty \mid P(\lambda) B \neq 0\}$, the measure $\chi_{\Delta_0}(\lambda)\, d\rho(F^{-1}(\lambda))$ is absolutely continuous with respect to $d\rho(\lambda)$, $B(\cdot)$ is a weakly measurable operator-valued function, $B(\lambda) : l_2(N_{F^{-1}\lambda}) \to l_2(N_\lambda)$.

If B is an invertible operator, $B : \Phi \to \Phi$, then for $\rho(\cdot)$-almost all $\lambda \in I\!R^\infty$, $N_{F^{-1}(\lambda)} = N_\lambda$. If the operator B is bounded, then (19) holds for all $f \in H$, the operator function $B(\lambda)$ is essentially bounded, and if B is unitary, then the values $B(\lambda)$ are unitary for ρ-almost all λ.

We prove this in several stages.

I. Let $\Delta_0 = \{\lambda \in I\!R^\infty \mid P(\lambda) B \neq 0\}$.

LEMMA 6. The set Δ_0 is measurable. The measure $\chi_{\Delta_0}(\lambda)\, d\rho(F^{-1}(\lambda))$ is absolutely continuous with respect to $d\rho(\lambda)$.

Proof. It follows that Δ_0 is measurable because $P(\lambda)$ is weakly measurable and Φ is separable.

Consider the measure $\chi_{F^{-1}(\Delta_0)}(\lambda)\, d\rho(F(\lambda))$ and show that it is absolutely continuous with respect to $d\rho(\lambda)$. For any $\Delta \in \mathbf{B}(I\!R^\infty)$, $u, v \in \Phi$

$$\int_\Delta (P(\lambda) u, B v)\, d\rho(\lambda) = (E_{\mathbf{A}}(\Delta) u, B v) = (E_{\mathbf{A}}(F(\Delta)) B u, v) =$$

$$= \int_{F(\Delta)} (P(\lambda) B\, u,\, v)\, d\rho(\lambda) = \int_{\Delta} (P(F(\lambda)) B\, u,\, v)\, d\rho(F(\lambda)).$$

So, it follows that

$$(P(\lambda)\, u,\, B\, v)\, d\rho(\lambda) = (P(F(\lambda)) B\, u,\, v)\, d\rho(F(\lambda))$$

and for $\lambda \in F^{-1}(\Delta_0)$

$$d\rho(F(\lambda)) = \frac{(P(\lambda)\, u,\, B\, v)}{(P(F(\lambda)) B\, u,\, v)}\, d\rho(\lambda). \tag{20}$$

Let for some $\Delta \in \mathrm{B}(I\!R^\infty)$, $\rho(\Delta) = 0$. Subdivide Δ into a union of non-intersecting sets $\Delta = \Delta_1 \cup \Delta_2$, where $\Delta_1 = \Delta \cap F^{-1}(\Delta_0)$. Then

$$\int_{\Delta} \chi_{F^{-1}(\Delta_0)}(\lambda)\, d\rho(F(\lambda)) = \rho(\Delta_1) + \int_{\Delta_2} \chi_{F^{-1}(\Delta_0)}(\lambda)\, d\rho(F(\lambda)) = 0.$$

Here, the first term equals zero because of (20) and the second one equals zero because Δ_2 and $F^{-1}(\Delta_0)$ are non-intersecting. Thus,

$$\chi_{F^{-1}(\Delta_0)}(\lambda)\, d\rho(\lambda)) \ll d\rho(\lambda).$$

$\square$

2. For any $u, v \in \Phi$ and $\rho(\cdot)$-almost all $\lambda \in I\!R^\infty$

$$(P(\lambda)\, u,\, B\, v) = \chi_{\Delta_0}(\lambda) = \frac{d\rho(F^{-1}(\lambda))}{d\rho(\lambda)}\, (P(F^{-1}(\lambda)) B\, u,\, v).$$

To show this, use the formulas

$$(E_{\mathbf{A}}(\Delta)\, u,\, B\, v) = (E_{\mathbf{A}}(F^{-1}(\Delta)) B\, u,\, v)\,,$$

$$(E_{\mathbf{A}}(\Delta)\, u,\, B\, v) = (E_{\mathbf{A}}(F(\Delta)) B\, u,\, v)$$

and the results on absolute continuity of measures.

3. For future constructions we will need a chain of densely imbedded Hilbert spaces $H_{--} \supset H_- \supset H \supset H_+ \supset H_{++} \supset \Phi$ where $H_{--} = H_{++}'$, $H_- = H_+'$, the imbedding $O_1 : H_+ \hookrightarrow H$ is quasi-nuclear, the imbeddings $O_2 : H_{++} \hookrightarrow H_+$, $O_3 : \Phi \hookrightarrow H_{++}$ are continuous, and the operator B operates continuously from H_{++} into H_+. Such chains exist because Φ is nuclear and the operator $B \restriction \Phi$ is continuous.

Each of the spaces H_+, H_{++} is imbedded in H in a quasi-nuclear way. This allows to construct a Fourier transform for every space. Let

$$\rho_+(\Delta) = Tr\, O_1^\dagger\, E_{\mathbf{A}}(\Delta) O_1\,, \quad \rho_{++}(\Delta) = Tr\, O_2^\dagger\, O_1^\dagger\, E_{\mathbf{A}}(\Delta) O_1\, O_2$$

be spectral measures,

$$d\,O_1^{\dagger}\,E_{\mathbf{A}}(\lambda)\,O_1 = P_+(\lambda)\,d\rho_+(\lambda),\ d\,O_2^{\dagger}\,O_1^{\dagger}\,E_{\mathbf{A}}(\Delta)\,O_1\,O_2 = P_{++}(\lambda)\,d\rho_{++}(\lambda).$$

There, for $\rho(\cdot)$-almost all λ, $P_{++}(\lambda) = \dfrac{d\rho_+(\lambda)}{d\rho_{++}(\lambda)}\,O_2^{\dagger}\,P(\lambda)\,O_2$. The Fourier transform of a vector $u \in \Phi$ will be denoted by $\tilde{u}_+(\lambda)$ if it corresponds to the chain $H_+ \supset H \supset H_+$ and by $\tilde{u}_{++}(\lambda)$ if it corresponds to the chain $H_{--} \supset H \supset H_{++}$.

4. Let us find out in which way the operator B acts in the Fourier image space. For $u \in \Phi$, $\rho(\cdot)$-almost all $\lambda \in \mathbb{R}^{\infty}$

$$\widetilde{(Bu)}_{++,j}(\lambda) = (\nu_{++,j}(\lambda))^{-½}\,(Bu,\ P_{++}(\lambda)\,J_{++}^{-1}\,\psi_{++,j}(\lambda)) =$$

$$= (\nu_{++,j}(\lambda))^{-½}\,(u,\ \bar{B}\,P_{++}(\lambda)\,J_{++}^{-1}\,\psi_{++,j}(\lambda)) =$$

$$= (\nu_{++,j}(\lambda))^{-½}\,\frac{d\rho_+(\lambda)}{d\rho_{++}(\lambda)}\,\chi_{\Delta_0}(\lambda)\,\frac{d\rho_+(F^{-1}(\lambda))}{d\rho_+(\lambda)}\,(u,\ O_2\,P_+(F^{-1}(\lambda))\,B\,J_{++}\,\psi_{++,j}(\lambda)) =$$

$$= \sum_{k=1}^{N_{F^{-1}(\lambda)}} \beta_{jk}(\lambda)\,\chi_{\Delta_0}(\lambda)\,\sqrt{\frac{d\rho_+(F^{-1}(\lambda))}{d\rho_+(\lambda)}}\ \tilde{u}_{+,k}(F^{-1}(\lambda))$$

where the set

$$\beta_{jk}(\lambda) = \left[\frac{\nu_{+,k}(F^{-1}(\lambda))}{\nu_{++,j}(\lambda)}\right]^{½}\ \frac{d\rho_+(\lambda)}{d\rho_{++}(\lambda)}\,\chi_{\Delta_0}\left[\frac{d\rho_+(F^{-1}(\lambda))}{d\rho_+(\lambda)}\right]^{½} \times$$

$$\times\,(J_+^{-1}\,\psi_{+,k}(F^{-1}(\lambda)),\ B\,J_{++}\,\psi_{++,j}(\lambda))_{H_+}.$$

Thus, if we set $\beta(\lambda) = (\beta_{jk}(\lambda)) : l_2(N_{F^{-1}(\lambda)}) \to l_2(N_\lambda)$ we get

$$\widetilde{(Bu)}_{++}(\lambda) = \chi_{\Delta_0}(\lambda)\left[\frac{d\rho_+(F^{-1}(\lambda))}{d\rho_+(\lambda)}\right]^{½}\ \beta(\lambda)\,\tilde{u}_+(F^{-1}(\lambda)).$$

For $\rho(\cdot)$-almost all λ consider an operator $C(\lambda) : l_2(N_\lambda) \to l_2(N_\lambda)$ such that $C(\lambda)\,\tilde{u}_{++}(\lambda) = \tilde{u}_+(\lambda)$. Setting $B(\lambda) = C(\lambda)\,\beta(\lambda)$ we get

$$\widetilde{(Bu)}_+(\lambda) = B(\lambda)\,\chi_{\Delta_0}(\lambda)\left[\frac{d\rho_+(F^{-1}(\lambda))}{d\rho_+(\lambda)}\right]^{½}\ \tilde{u}_+(F^{-1}(\lambda)).$$

Similarly, we can get a formula for $\widetilde{(B^* u)}(\lambda)$.

5. If the operator $B : \Phi \to \Phi$ is invertible, it follows from (4) that $l_2(N_\lambda)$ and $l_2(N_{F^{-1}(\lambda)})$ are isomorphic and $N_\lambda = N_{F^{-1}(\lambda)}$.

The remaining statements of the theorem could be obtained from the representation (19). □

Constructing a commutative model, we take the Fourier transform using the space H_+ which can be chosen independently of B (H_{++} is chosen using B). This allows to get a commutative model not only for a single operator B but as well for a family of operators

$(B_j)_{j=1}^{\infty}$ which together with the operators $(A_k)_{k=1}^{\infty}$ satisfy the relation $A_k B_j = B_j F_{kj}(\mathbf{A})$. In the latter case it should be required that the operators $(B_j)_{j=1}^{\infty}$ be standardly related to the rigging Φ.

Since for the collections of CSO $(A_k)_{k=1}^{\infty}$ and CSO $(B_j)_{j=1}^{\infty}$ which satisfy relations (18), there exists a rigging, Φ related in a standard way to these collections, there also exists a commutative model. We state the following theorem under the assumption that B_j $(j=1,2,\ldots)$ are invertible.

THEOREM 14. If the collections of CSO $\mathbf{A} = (A_k)_{k=1}^{\infty}$ and CSO $\mathbf{B} = (B_j)_{j=1}^{\infty}$ satisfy relations (18), then in the Fourier image space of the family $\mathbf{A}$

$$(\tilde{B_j u})(\lambda) = B_j(\lambda) \left[\frac{d\rho(F^{-1}(\lambda))}{d\rho(\lambda)} \right] \tilde{u}(F_j^{-1}(\lambda)) \quad (j = 1,2,\ldots)$$

for $u \in \Phi$. Here $\rho(\cdot)$ is a spectral F_j^{-1}-quasi-invariant measure of the family $\mathbf{A}$ $(j=1,2,\ldots)$, $N_\lambda = N_{F^{-1}(\lambda)}$ for $\rho(\cdot)$-almost all $\lambda \in \mathbb{R}^\infty$, $B_j(\lambda)$ are weakly measurable functions, $B_j(\lambda) : l_2(N_{F^{-1}(\lambda)}) \to l_2(N_\lambda)$ such that

$$B_j(\lambda)\, B_k(\lambda) = B_k(\lambda)\, B_j(\lambda) \quad (k,j = 1,2,\ldots).$$

Comments to Chapter 10.

1. This is an exposition of the article Yu.S. Samoĭlenko, A.M. Kharitonskiĭ [1].

2. We give a detailed exposition of the note V.L. Ostrovskiĭ, Yu.S. Samoĭlenko [2].

3. The results given in Section 10.3 are due to the author.

4. The relations (1), (12), (18) give necessary and sufficient conditions to reduce the study of CSO $\mathbf{A} = (A_k)_{k=1}^{\infty}$ and, generally speaking, not self-adjoint $\mathbf{B} = (B_j)_{j=1}^{\infty}$ to the study of quasi-invariant measures and cocycles of a dynamic system $(\sigma(\mathbf{A}), \mathbf{F} = (F_1, F_2, \ldots))$ (following A.M. Vershik, I.M. Gel'fand, M.I. Graev [6], such a system $(\sigma(\mathbf{A}), \mathbf{F})$ is called a commutative model). In some works (see the beginning of Section 10.4), such models were constructed for families of operators connected by concrete relations. To construct commutative models in Section 10.4 for families of unbounded operators connected by relations (1), (12), or (18) we use a decomposition into joint generalized eigenvectors of the families of CSO (see Yu.M. Berezanskiĭ [9,12]). It follows the papers Yu.M. Berezanskiĭ, V.L. Ostrovskiĭ, Yu.S. Samoĭlenko [1], V.L. Ostrovskiĭ, Yu.S. Samoĭlenko [1], Yu.M. Berezanskiĭ, Yu.G. Kondrat'ev [1]. If the commuting operators $\mathbf{B} = (\mathbf{B}_j)_{j=1}^{n}$ are self-adjoint, the dynamic system for the finite collection of CSO $(B_j)_{j=1}^{n}$ is simple, and this allows to obtain in Section 10.3 a structure theorem for $(\mathbf{A}, \mathbf{B})$.

PART IV
REPRESENTATIONS OF OPERATOR ALGEBRAS AND NON-COMMUTATIVE RANDOM SEQUENCES

In the previous Parts we did not essentially use the technique of C^*-algebras and their representations. However, considerations of inductive limits of locally compact groups in a general setting of algebras of local observables lead us to C^*-algebras, their inductive limits, their representations, and the relations between the representations theory of such algebras and noncommutative probability theory.

In this Part, we use the C^*-algebra technique. Firstly, we use them to study the construction of $*$-representations of certain inductive limits, $J-\underset{\rightarrow}{\lim}\,\mathbf{U}_n$ of C^*-algebras (they also include representations of the algebra of local observables of a spin system, and representations of a group C^*-algebra for the inductive limits of groups). One of them is the construction of a representation of $J-\underset{\rightarrow}{\lim}\,\mathbf{U}_n$ in the form of the inductive limit of representations of "pre-limiting" algebras $\mathbf{U}_n$ (Chapter 11, Section 11.2), and another is the construction of Gårding-Wightman type by giving a measure and a cocycle (Chapter 11, Section 11.3). Secondly, using the C^*-algebra techniques we can consider the introduction of measures and cocycles as a constructive way to define states on C^*-algebras (they can be regarded as non-commutative probability measures). Chapter 12, Section 12.1 is devoted to non-commutative probability measures and to the ways to define them. We finish this part by considering one more method to define non-commutative measures using moments. We state the non-commutative moment problem and solve it for the states on certain group C^*-algebras and, also, in a more general situation.

Chapter 11.
C^*-ALGEBRAS $\mathbf{U}_0^\infty$ AND THEIR REPRESENTATIONS

In this chapter, we give a brief introduction to the theory of C^*-algebras $\mathbf{U}$ of type I and their inductive limits. Then we study the simplest classes of representations of these algebras.

11.1. C^*-algebras of type I.

Recall that a C^*-algebra is an involutary algebra which is a Banach space over $\mathbb{C}^1$ and for which

1) $\|xy\| \le \|x\| \cdot \|y\| \quad \forall x, y \in \mathbf{U}$;

2) $\|x^*\| = \|x\| \quad \forall x \in \mathbf{U}$;

3) $\|x^* x\| = \|x^*\| \, \|x\| \quad \forall x \in \mathbf{U}$.

We will always suppose that $\mathbf{U}$ is a separable C^*-algebra with identity 1, $1^* = 1$ and $\|1\| = 1$.

DEFINITION 1. Two C^*-algebras $\mathbf{U}$ and $\tilde{\mathbf{U}}$ are $*$-isomorphic if there exists an algebra isomorphism $U : \mathbf{U} \to \mathbf{U}$ which preserves the operation of taking the adjoint, i.e.

$$(U x)^* = U x^* \quad \forall x \in \mathbf{U}.$$

A $*$-isomorphism U of a C^*-algebra $\mathbf{U}$ into a C^*-algebra $\mathbf{U}$ is automatically an isometry, i.e.

$$\|U x\| = \|x\| \quad \forall x \in \mathbf{U}.$$

Consider some examples of C^*-algebras and their classes.

Example 1. The algebra $C(K) \ni f(\cdot)$ of all continuous complex-valued functions on a compact set K with the norm $\|f\| = \max_{k \in K} |f(k)|$ and the involution $(f(\cdot))^* = \overline{f(\cdot)}$ is an example of a commutative C^*-algebra (it is a completely regular commutative normed ring). Moreover, any commutative C^*-algebra is $*$-isomorphic to the algebra of all continuous functions on a compact topological space which can be taken to be the space of its maximal ideals (see, for example, N. Bourbaki [1], I.M. Gel'fand, D.A. Raikov, G.E. Shilov [1], L. Loomis [1]). The algebras $C(K_1)$ and $C(K_2)$ are $*$-isomorphic if and only if the compact sets K_1 and K_2 are homeomorphic. □

Example 2. The algebra $L(H) \ni A$ of all bounded operators on H, $\|A\| = \sup_{\|f\|=1} \|A f\|$, A^* is the adjoint. Every C^*-algebra is $*$-isomorphic to a uniformly closed $*$-subalgebra of

$L(H)$ for some H (see I.M. Gel'fand, M.A. Naĭmark [1]). The C^*-algebras $L(H)$ and $L(H_1)$ are $*$-isomorphic if and only if $\dim H = \dim H_1$. □

Example 3. Every finite-dimensional C^*-algebra **U** can be decomposed into a direct sum $\mathbf{U} = \oplus \sum_{k=1}^{m} \mathbf{U}_k$ where each $\mathbf{U}_k$ is $*$-isomorphic with a full matrix algebra $L(\mathbb{C}^{n_k}) = M_{n_k}$ $(k=1,\ldots,n;\ n_1 \le n_2 \le \cdots \le n_m)$. A finite-dimensional C^*-algebra $\tilde{\mathbf{U}} = \oplus \sum_{j=1}^{\tilde{m}} \tilde{\mathbf{U}}_j$ ($\tilde{\mathbf{U}}_j$ are isometricly $*$-isomorphic to $M_{\tilde{n}_j}$; $j=1,\ldots,\tilde{m};\ \tilde{n}_1 \le \tilde{n}_2 \le \cdots \le \tilde{n}_m$) is $*$-isomorphic to **U** if and only if $m = \tilde{m}$ and $n_k = \tilde{n}_k$ $(k=1,\ldots,m)$. □

Example 4. W^*-algebras (von Neumann's algebras) **M** are $*$-algebras of bounded operators on H closed with respect to the weak operator topology. The norm and the adjoint of an operator define the structure of a C^*-algebra on **M**. The W^*-algebras are distinguished among all of the C^*-subalgebras of $L(H)$ by the property that $\mathbf{M} = \mathbf{M}''$ (here $\mathbf{M}'$ is a commutant of the algebra **M** i.e. the set of all bounded operators on H that commute with all of the operators $A \in \mathbf{M}$).

The $*$-isomorphism problem for W^*-algebras is not solved. It can be reduced to the classification of factors (a factor is a W^*-algebra such that $\mathbf{M} \cap \mathbf{M}' = cI$). A decomposition of the set of factors into the classes of factors of type I, II and III was given in E. Murray, J. von Neumann [1]. The structure of factors of type I is simple: they are all $*$-isomorphic to the algebra $L(H)$ where H is a separable Hilbert space fixed for a given factor **M** of type I.

Note, that if there is a $*$-isomorphism $U : \mathbf{M} \to \tilde{\mathbf{M}}$ of W^*-algebras of operators on H and $\tilde{H}$ correspondingly, it does not necessarily mean that these algebras are isomorphic spaces, i.e. that there exsits a unitary operator $V : H \to \tilde{H}$ such that $A = V^* \, U(A) \, V \quad \forall A \in \mathbf{M}$.

For the $*$-isomorphism problem of the factors of type II and III see R. Powers [1], the monographs Sh. Sakai [1], M. Takecaki [3]. S. Strătilă, L. Zsido [1] and their bibliography. □

The study of C^*-algebras can roughly be splitted into two parts: the study of the inner structure of C^*-algebras, and the study of their representations.

DEFINITION 2. A representation of a C^*-algebra **U** is a $*$-homomorphism $\pi(\cdot)$ into the C^*-algebra $L(H)$. The Hilbert space H is called the representation space. A representation $\pi(\cdot)$ is called a factor-representation if $\{\pi(x) \mid x \in \mathbf{U}\}''$ is a factor.

DEFINITION 3. Two representations of **U**, $\pi(\cdot)$ on H and $\tilde{\pi}(\cdot)$ on $\tilde{H}$ are called unitarily equivalent if the families of operators $(\pi(x))_{x \in \mathbf{U}}$ and $(\tilde{\pi}(x))_{x \in \mathbf{U}}$ are unitarily equivalent.

The mentioned two parts of the theory of C^*-algebras are related, and the algebraic structure of a C^*-algebra can be studied using its different representations. In particular, the C^*-algebras can be divided with respect to the their representation structure into two classes: GCR-algebras (algebras of type I) and NGCR-algebras (algebras not of type I).

DEFINITION 4. A C^*-algebra is type I if any of its factor-representations is type I.

There is a fairly detailed information on the structure of C^*-algebras of type I and their representations (see J. Dixmier [3], Sh. Sakai [1]). The C^*-algebras of type I are important for the theory of locally compact groups G. $C^*(G)$ is the group C^*-algebra which is defined using a normed group algebra $L_1(G, dg)$ (dg is a left invariant Haar measure on G with the multiplication in $L_1(G, dg)$ being: $(f_1 \circ f_2)(g) = \int_G f_1(h) f_2(g^{-1} h)\, dh$ and $f^*(g) = \Delta(g)\, \overline{f(g^{-1})}$, where $\Delta(g)$ is a modular function on G). $C^*(G)$ is the completition of $L_1(G, dg)$ with respect to the norm

$$\|a\|_{C^*(G)} = \sup_{\pi} \|\pi(a)\|$$

where the supremum is taken over all $*$-representations $\pi(\cdot)$ of the algebra $L_1(G, dg)$, adding an identity if the obtained C^*-algebra does not contain it.

Any unitary representation of a group $G \ni g \mapsto T_g$ on H generates a $*$-representation of the algebra $L_1(G, dg)$ (and $C^*(G)$):

$$L_1(G, dg) \ni f(\cdot) \mapsto \int_G f(g)\, T_g\, dg.$$

Conversely, for any non-degenerate $*$-representation of $L_1(G, dg)$ (and $C^*(G)$) there is a unique unitary representation of the group G (since the representation $L_1(G, dg) \ni a \mapsto \pi(a)$ is non-degenerate on H, CLS $\{\pi(a)\phi \mid \phi \in H,\ a \in L_1(G, dg)\} = H$).

If $C^*(G)$ is a C^*-algebra of type I or which is the same thing, if any unitary representation of the group G generates a W^*-algebra $((T_g)_{g \in G})''$ of type I, then the group G is also called a group of type I. In particular, the group $SU(2)$ (being simultaneously a compact group and a semi-simple Lie group) and the group of upper triangular matrices

$$\begin{pmatrix} 1 & t_1 & \alpha_1 \\ 0 & 1 & s_1 \\ 0 & 0 & 1 \end{pmatrix}$$

(as any finite-dimensional nilpotent Lie group) are groups of type I.

11.2. Inductive limits of C^*-algebras and their representations.

In this section we consider examples of inductive limits of C^*-algebras and their representations. These are C^*-algebras that are closures of an increasing sequence of its subalgebras $(\mathrm{U}_\alpha)_{\alpha \in \Lambda}$ with respect to a norm (Λ is a directed partially ordered set of indices, and for all the subalgebras $\mathrm{U}_\alpha \ni 1_\alpha = 1 \in \mathrm{U}$).

As in Chapter 2, Section 2.1, we will assume that the indices take only integer values.

Example 5. A C^*-algebra U is called a uniformly hyperfinite (*UHF*) algebra of class $\{n_k\}$ if there exist an increasing sequence of full matrix algebras

$$M_{n_1} \subset \cdots \subset M_{n_k} \subset \cdots \subset \mathrm{U}$$

such that U is the closure of $\bigcup_{k=1}^{\infty} M_{n_k}$ with respect to a norm.

The *UHF* C^*-algebras were introduced and studied by J. Glimm [2]. In particular, Glimm gave a classification up to a $*$-isomorphism of *UHF*-algebras: if U and $\tilde{\mathrm{U}}$ are *UHF*-algebras of the classes $\{n_k\}$ and $\{\tilde{n}_k\}$ correspondingly, then they are $*$-isomorphic if and only if for every $k \in \mathbb{N}$ there is $j_k \in \mathbb{N}$ $(j_k \geq k)$ such that n_k is a divisor of $\tilde{n}_{j_k}$ and $\tilde{n}_k$ is a divisor of n_{j_k}.

The algebra of quasi-local observables of a one-dimensional spin system considered in Chapter 9 is a *UHF*-algebra of class $\{2^k\}$. □

Example 6. The *UHF*-algebras establish a particular example of the approximately finite dimensional C^*-algebras (*AF*-algebras) considered by O. Bratteli [1]. An *AF*-algebra U is a uniform closure of an increasing sequence of finite-dimensional C^*-subalgebras U_k.

A group C^*-algebra, $C^*(G)$, was constructed in D. Voĭculescu, S. Strătilă [1] for the group $G = \bigcup_{n=1}^{\infty} G_n$. Here $G_1 \subset \cdots \subset G_n \subset \cdots$ is an increasing sequence of compact subgroups, G_n is closed in G_{n+1} and has zero Haar measure. G has the inductive limit topology. They also proved that $C^*(G)$ is approximately finite dimensional.

For finite dimensional C^*-algebras (see Example 3), $\mathrm{U}_k = \oplus \sum_{l=1}^{m_k} \mathrm{U}_{l,k}$ where $\mathrm{U}_{l,k}$ are $*$-isomorphic to full matrix algebras $M_{n_{l,k}}$. As in O. Bratteli [1], the *AF*-algebras could be given by means of a diagram

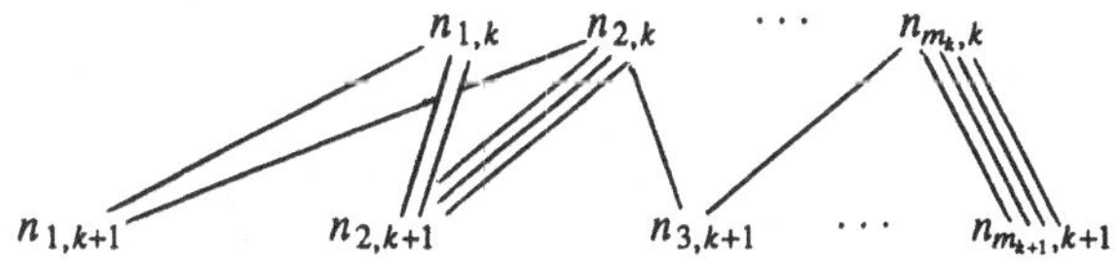

where the numbers $n_{l,k}$ and $n_{p,k+1}$ are joined with $s_{n_{l,k},n_{p,k+1}}$ edges if $\mathrm{U}_{l,k}$ is contained in $\mathrm{U}_{p,k+1}$ with multiplicity $s_{n_{l,k},n_{p,k+1}}$. In particular, the diagram of the *UHF*-algebra of class $\{n_k\}$ (Example 5) will have the form

$$\begin{array}{c} \cdots \\ n_k \\ \| \cdots \| \\ n_{k+1} \\ \cdots \end{array}$$

Here, the number of lines between n_k and n_{k+1} equals n_{k+1} / n_k.

If two AF-algebras U and $\tilde{\mathrm{U}}$ have the same diagram, then they are $*$-isomorphic. However, in an AF-algebra U one can choose systems of finite-dimensional subalgebras U_n and $\tilde{\mathrm{U}}_n$ in such a way that the diagrams generally speaking will not be the same. The question of the classification of AF-algebras up to a $*$-isomorphism was studied by O. Bratteli [1], G. Elliot [1]. ▯

Example 7. Let U be a fixed C^*-algebra of operators on H containing the identity I. Define a C^*-algebra

$$\otimes \mathrm{U}^n = \underbrace{\mathrm{U} \otimes \cdots \otimes \mathrm{U}}_{n \text{ times}}$$

of operators on

$$\otimes H^n = \underbrace{H \otimes \cdots \otimes H}_{n \text{ times}}$$

by completing the algebraic tensor product with respect to the operator norm on $\otimes H^n$ and setting $\mathrm{U}_0^\infty = \overline{J - \lim\limits_{\rightarrow} \otimes \mathrm{U}^n}$ where $J_{n+1}^n (\otimes \mathrm{U}^n) = \otimes \mathrm{U}^n \otimes I$ and $\bar{\ }$ is the closure. ▯

Let us now look at $*$-representations of the algebras

$$\mathrm{U} = \overline{\bigcup_{r=1}^{\infty} \mathrm{U}_r} \quad (\mathrm{U}_1 \subset \cdots \subset \mathrm{U}_n \subset \cdots)$$

and let us give a simple construction of irreducible $*$-representations of the algebra U in the form of an inductive limit of irreducible $*$-representations of U_n.

DEFINITION 5. A $*$-representation $\pi(\cdot)$ on H of an algebra U is called the inductive limit of the representations $\pi_n(\cdot)$ on M_n of C^*-algebras U_n if there exists a sequence of subspaces of $H, H_1 \subset \cdots \subset H_n \subset \cdots$ such that $H = \overline{\bigcup_{n=1}^{\infty} H_n}$, H_n is invariant with respect to the restriction $\pi(\cdot) \restriction \mathrm{U}_n$ of the representation $\pi(\cdot)$ to the subalgebra U_n and the representation $\pi(\cdot) \restriction \mathrm{U}_n$ on H_n is unitarily equivalent to the representation $\pi_n(\cdot)$ on M_n of the C^*-algebra U_n.

PROPOSITION 1. The inductive limit of irreducible $*$-representations of the algebras U_n is an irreducible $*$-representation of the inductive limit $\mathrm{U} = \overline{\bigcup_{n=1}^{\infty} \mathrm{U}_n}$.

Proof. Let $P_{H_k} = P_k$ be the orthogonal projection onto H_k. Then, for any bounded operator A on H which commutes with all the representation operators $\pi(a)$ $(a \in \mathbf{U})$ the operators $P_k A P_k$ on H_k commute with the irreducible $*$-representation $\pi(a) \restriction H_k$ $(a \in \mathbf{U}_n)$ and so $P_k A P_k = C_k I_{H_k}$. But since $\overline{\bigcup_{k=1}^{\infty} H_k} = H$, $c_k = c$ and $A = cI$. □

Using this proposition, we can construct irreducible $*$-representations (unitary strongly continuous representations) for the inductive limits of C^*-algebras (inductive limits of groups, $G = j-\lim\limits_{\rightarrow} G_n$). To do this, we define irreducible $\mathbf{U}_k$ and isometric imbeddings $i_n^k : M_k \hookrightarrow M_n$ for all $n > k$ such that the diagram

$$\begin{array}{ccc} M_k & \xrightarrow{\pi_k(a)} & M_k \\ i_n^k \downarrow & & \downarrow i_n^k \\ M_n & \xrightarrow{\pi_n(a)} & M_n \end{array} \tag{1}$$

be commutative for all $a \in \mathbf{U}_k$ $(k = 1,2,...)$ and $n > k$. The continuous extension of the operators $\lim\limits_{n \to \infty} \pi_n(a)$ from the dense subset $\bigcup_{k=1}^{\infty} M_k$ to the whole $M = \overline{i-\lim\limits_{\rightarrow} M_n}$ defines an irreducible $*$-representation of the C^*-algebra $\mathbf{U}$ on the Hilbert space M.

Going from $*$-representations of the group algebra $C^*(G_0^\infty)$ to the unitary representations of the group G_0^∞ we can construct a rather large collection of irreducible representations of the group G_0^∞ using diagrams similar to (1).

For example, these representations separate the points of the group G_0^∞ i.e. for $g_1 \neq g_2$ using diagram (1), we can construct an irreducible representation $G_0^\infty \ni g \mapsto T_g$ such that $T_{g_1} \neq T_{g_2}$. Nevertheless, such representations make only a small part of the set of all irreducible representations.

We can get a wider class of representations of $\mathbf{U}$ if we replace the commutative diagram (1) by the diagram

$$\begin{array}{ccc} M_{k+l_k} & \xrightarrow{\pi_{k+l_k}(a)} & M_{k+l_k} \\ i_{k+l_k+n}^{k+l_k} \downarrow & & \downarrow i_{k+l_k+n}^{k+l_k} \\ M_{k+l_k+n} & \xrightarrow{\pi_{k+l_k+n}(a)} & M_{k+l_k+n} \end{array} \tag{2}$$

which is commutative for all $a \in \mathbf{U}_k$ and $n = 1,2 \cdots$. Then the $*$-representation

$$\mathbf{U}_n \ni a \mapsto \pi(a) = \lim_{n \to \infty} \pi_n(a)$$

is defined on $M = \overline{i-\lim\limits_{\rightarrow} M_k}$.

However, irreducibility of the "pre-limiting" representations

$$\mathrm{U}_n \ni a \mapsto \pi_n(a) : M_n \to M_n$$

does not, generally speaking, imply that the constructed representation of U is irreducible.

Example 8. Let $\mathrm{U} = \mathrm{A}_0^\infty$ and $\pi(\cdot)$ be a fixed irreducible representation on H of the C^*-algebra A. Set $M_n = \otimes H^n$,

$$\pi_n(a_1 \otimes \cdots \otimes a_n) = \pi(a_1) \otimes \cdots \otimes \pi(a_n)$$

is an irreducible representation on $\otimes H^n$ of $\otimes \mathrm{A}^n$. Now, choose a fixed unit vector $e \in H$ and define

$$i_{n+1}^n(\otimes H^n) = \otimes H^{n-1} \otimes e \otimes H \quad (n = 1,2,...).$$

The diagrams (2) are commutative for $l_k = 1$ $(k=1,2,...)$ so that there is a $*$-representation of A_0^∞ on $M = \overline{i - \lim\limits_{\to} M_n}$. But it will not be irreducible. Indeed, if P_e is the projection onto the subspace spanned by the vector e then the sequence of projections $P^{(k)} = 1 \otimes \cdots \otimes 1 \otimes P_e$ on $M_k = \otimes H^k$ commutes with the imbeddings $i = (i_{k+1}^k)_{k=1}^\infty$, and so it defines a non-trivial projection which commutes with the representation. □

11.3. Models for representations of C^*-algebras.

In Chapter 9, following L. Gårding, A. Wightman [1] the representations of the C^*-algebra of local observables, $\mathrm{U} = \lim\limits_{\to} \mathrm{U}_n$, with the property that the "pre-limiting" algebras U_n $(n=1,2,...)$ had only one irreducible representation, were put into the form that expressed the representations of U using measures and cocycles, i.e. the study of representations of U was reduced to the study of quasi-invariant measures and cocycles of a certain dynamical system.

We note that a description of C^*-algebras with a finite number of irreducible representations was given by A.N. Vasil'ev [1], H. Behncke, H. Leptin [1,2,3].

Irreducible representations of the Heisenberg-Weyl group G (the group C^*-algebra $C^*(G)$) have a more complicated structure. However, if one uses CCR-representations, then for the CCR-systems with finite degrees of freedom, we have the uniqueness theorem, and for the CCR-systems with countably many degrees of freedom, we can also use measures and cocycles to describe them (see Chapter 4). We had the same thing for the unitary representations of the group $SU(2)_0^\infty$. Considering only factor-representations, we could always reduce them into the Gårding-Wightman form (see Chapter 5).

In this section we give the Gårding-Wightman's form for the factor-representations of some C^*-algebras. Let $\mathrm{U}_1, \ldots, \mathrm{U}_n, \ldots$ be separable type I C^*-algebras of operators on $H_1, \ldots, H_n, \ldots$ with identity. Consider the C^*-algebra $\mathrm{U} = j - \lim\limits_{\to} \mathrm{U}_1 \otimes \cdots \otimes \mathrm{U}_n$ where

$U_1 \otimes \cdots \otimes U$ is a C^*-algebra of operators on

$$H_1 \otimes \cdots \otimes H_n$$

and

$$j_{n+1}^n : \bigotimes_{k=1}^{n} U_k \to \bigotimes_{k=1}^{n+1} U_k, \; j_{n+1}^n \, x = x \otimes 1_{n+1}.$$

We describe the factor-representation of the algebra U using measures and cocycles.

Let $\pi^{(k)}$ be irreducible representations of the algebras U_k on the spaces H_k. Use the elements of the cyclic group $\mathbb{Z}_{N_k}$, where N_k is finite or infinite, to number a basis of the space H_k. Let us also have a probability measure $\mu(\cdot)$ on $\prod_{k=1}^{\infty} \mathbb{Z}_{N_k}$ quasi-invariant and ergodic with respect to the shifts by the elements of $\bigoplus_{k=1}^{\infty} \mathbb{Z}_{N_k}$. Let $\pi_{ml}^{(k)}$ be the matrix elements of the representations $\pi^{(k)}$ $(m, l \in \mathbb{Z}_{N_k})$.

We define a certain representation of the algebra U. Let $H = \oplus \int_{\prod_k \mathbb{Z}_{N_k}} H_\lambda \, d\mu(\lambda)$ with

$$H_\lambda = H_{\lambda + t}, \quad \lambda \in \prod_k \mathbb{Z}_{N_k}, \quad t \in \bigoplus_k \mathbb{Z}_{N_k}$$

(since $\mu(\cdot)$ is ergodic, the dimension of H_λ is constant on a set of full measure).

Let $U_t(\lambda)$ be a function of the variables $t \in \bigoplus_k \mathbb{Z}_{n_k}$ and $\lambda \in \mathbb{Z}_{N_k}$ such that $U_t(\lambda)$ is a unitary operator form $H_{\lambda - t}$ into H_λ satisfying the condition

$$U_{t_1 + t_2}(\lambda) = U_{t_1}(\lambda) \, U_{t_2}(\lambda - t_1)$$

(i.e. it is a cocycle of the group $\oplus \mathbb{Z}_{N_k}$ in the group of the measurable mappings that for a $\lambda \in \mathbb{Z}_{N_k}$ give a unitary operator on H_λ).

Let $A_n \in j_\infty^n (1_1 \otimes \cdots \otimes 1_{n-1} \otimes U_n)$. Construct the operators

$$(T(A_n) f)(\lambda) = \sum_{i \in \mathbb{Z}_{N_n}} \pi_{\lambda_n, \lambda_n - i}^{(n)} \sqrt{\frac{d\mu(\lambda - i_n)}{d\mu(\lambda)}} \; U_{i_n}(\lambda) \, f(\lambda - i_n),$$

where $i_n = (0, \ldots, 0, i, 0, \ldots) \in \mathbb{Z}_{N_k}$. If $\mathbb{Z}_{N_n}$ is infinite, the sum is understood as the limit in the weak operator topology. One can show that the family of the operators $T(A_n)$ generates a representation of the C^*-algebra U. Denote it by $T(\mu, U, \pi^{(1)}, \pi^{(2)}, \ldots)$.

PROPOSITION 2. Any cyclic factor-representation of the C^*-algebra U is unitarily equivalent to a representation $T(\mu, U, \pi^{(1)}, \pi^{(2)}, \ldots)$ for some $\mu(\cdot)$, U and $\pi^{(1)}, \pi^{(2)}, \ldots$.

The reason that one can construct a model for factor-representations of the algebra U is that one can choose generators satisfying a relation of the form (18), Chapter 10 for any factor-representation of the algebra U.

Generalized Gårding-Wightman's constructions for factor-representations of *AF* C^*-algebras are given in S. Strătilă, D. Voĭculescu [1] and in a more general setting - in G.N. Zholtkevich [1]. For any $*$-representations of an *AF* C^*-algebra U there is a model because in any *AF* C^*-algebra there are generators which satisfy relations similar to (18), Chapter 10.

Comments to Chapter 11.

1. We give the most well known examples of C^*-algebras. The following monographs are devoted to the theory of C^*-algebras: I.M. Gel'fand, D.A. Raikov, G.E. Shilov [1], J. Dixmier [1,3], M.A. Naimark [4], Sh. Sakai [1], M. Takesaki [3], S. Strătilă, L. Zsido [1], etc. Applications to physics are described in N.N. Bogolubov, A.A. Logunov and I.T. Todorov [1], G. Emch [1], O. Bratteli and D. Robinson [1] and others.

2 The simplest inductive limits of finite-dimensional C^*-algebras, *UHF*-algebras, were introduced and studied by J. Glimm [2]. Some important results on the structure of representations were obtained by R. Powers [1]. $UHF-C^*$-algebras are a particular case of *AF*-algebras which were introduced by O. Bratteli [1]. He also generalized for this case some results of J. Glimm and R. Powers. The proof of Proposition 1 follows the paper V.I. Kolomytsev and Yu.S. Samoĭlenko [1]. Example 8 was constructed by the author.

3. The Gårding-Wightman's construction for *AF* C^*-algebras is given in D. Voĭculescu and S. Strătilă [1] and, more generally, in G.N. Zholtkevich [1]. The exposition for the algebras $\lim\limits_{\to} \mathrm{U}_1 \otimes \cdots \otimes \mathrm{U}_n$ follows B.L. Tsigan.

Chapter 12.
NON-COMMUTATIVE RANDOM SEQUENCES AND METHODS FOR THEIR CONSTRUCTION

We can use the terminology of the theory of operator algebras and of states on these algebras to generalize the notion of random value and random sequence to the non-commutative case. In Section 12.1, we give the Gel'fand-Naimark-Segal theorem that describes the relation between representation theory and the theory of non-commutative random values and non-commutative random sequences, which are states on C^*-algebras and on inductive limits of C^*-algebras. In Chapter 11 we considered methods to define representations of the inductive limits of C^*-algebras and choosed a cyclic vector in the representation space. These are methods to define non-commutative random sequences. Non-commutative random values and sequences can be fixed through their moments. In Section 12.2, we give necessary conditions on a sequence of numbers, so that one can reconstruct a non-commutative measure with these numbers as its moments.

12.1. Non-commutative random sequences.

Before we get into the non-commutative probability theory using the terminology of states on C^*-algebras, using this terminology we define a random value (a probability measure on $(\mathbb{R}^1, \mathbf{B}(\mathbb{R}^1)))$ and a random sequence (a probability measure on $(\mathbb{R}^\infty, \mathbf{B}(\mathbb{R}^\infty)))$.

A probability measure $\mu(\cdot)$ on $(\mathbb{R}^1, \mathbf{B}(\mathbb{R}^1))$ can be given through a cyclic strongly continuous unitary representation $\mathbb{R}^1 \ni t_1 \mapsto e^{it_1\lambda_1}$ on $L_2(\mathbb{R}^1, d\mu(\lambda))$ with a fixed cyclic vector Ω (Ω is a $\mathbf{B}(\mathbb{R}^1)$-measurable function, whose absolute value equals 1 almost everywhere) such that $\mu(\Delta) = (E(\Delta)\,\Omega, \Omega)$ for all $\Delta \in \mathbf{B}(\mathbb{R}^1)$. For this measure, there is a uniquely defined characteristic function

$$k(t_1) = \int_{\mathbb{R}^1} e^{it_1\lambda_1}\, d\mu(\lambda_1)$$

which is a continuous, positive definite, normed function on the group $\mathbb{R}^1$, or there is a non-negative linear functional on a Banach group algebra $L_1(\mathbb{R}^1, dt_1) \ni f(\cdot)$:

$$\mathbf{K}(f) = \int_{\mathbb{R}^1} k(t_1)\, f(t_1)\, dt_1\,,$$

which generates a non-negative linear normed, functional on the group C^*-algebra $C^*(\mathbb{R}^1)$ (the states on $C^*(\mathbb{R}^1)$).

Let now G be a locally compact group, dg be a left invariant measure on G. Given a cyclic strongly continuous unitary representation $G \ni g \mapsto U_g$ on H with a cyclic vector Ω, we can determine:

1) a continuous positive definite function, $k(\cdot)$ on the group G

$$k(g) = (U_g \Omega, \Omega) ;$$

or

2) a cyclic $*$-representation of the C^*-algebra $C^*(G) \ni a \mapsto \pi(a)$ with a cyclic vector, Ω

$$\pi(a) = \int_G a(g)\, U_g\, dg ;$$

or

3) a positive normed functional $\mathbf{K}(\cdot)$ (a state) on $C^*(G) \ni a$

$$\mathbf{K}(a) = (\pi(a)\, \Omega, \Omega).$$

We also have a converse statement.

THEOREM 1. For every continuous normed positive definite function $k(\cdot)$ on the group G (or, for a state $\mathbf{K}(\cdot)$ on $C^*(G)$) there corresponds a cyclic strongly continuous unitary representation $G \ni g \mapsto U_g$ (or a $*$-representation $C^*(G) \ni a \mapsto \pi(a)$) on H with a cyclic vector Ω such that $k(g) = (U_g \Omega, \Omega)$ (or $\mathbf{K}(a) = (\pi(a)\, \Omega, \Omega)$).

The proof of this theorem is based on a construction of a cyclic unitary strongly continuous representation of the group $G \ni g \mapsto U_g$ on H using a positive definite normed function on the group (or of the corresponding $*$-representation of $C^*(G)$ with a given cyclic vector, Ω using the states on $C^*(G)$). This construction is a generalization of Bochner's construction of a cyclic unitary strongly continuous representation $\mathbb{R}^1 \ni t_1 \mapsto U_{t_1}$ on H with a marked cyclic vector Ω using a positive definite normed function on $\mathbb{R}^1$. Note, that for a C^*-algebra $\mathbf{U}$ which is not, generally speaking, a group algebra, this construction also allows to determine the relationship between its cyclic $*$-representation with a cyclic vector Ω and positive definite normed functionals on $\mathbf{U}$. The construction is called the Gel'fand-Naimark-Segal (GNS) construction.

The normed positive definite functions on the group (the states on $\mathbf{U}$) are usually considered as a possible non-commutative analogue of the probability measures on $(\mathbb{R}^1, \mathbf{B}(\mathbb{R}^1))$ (more exactly, as their characteristic functions). So defining a representation $G \ni g \mapsto U_g$ on H with a cyclic vector Ω (or $\mathbf{U} \ni a \mapsto \pi(a)$ on H with a cyclic vector Ω) is a possible way to define a non-commutative analogue of the probability measures on $(\mathbb{R}^1, \mathbf{B}(\mathbb{R}^1))$.

Consider now the random sequences (the measures on $(\mathbb{R}^\infty, \mathbf{B}(\mathbb{R}^\infty))$). The probability measures $\rho(\cdot)$ on $(\mathbb{R}^\infty, \mathbf{B}(\mathbb{R}^\infty))$ can be defined through cyclic strongly continuous unitary representations $\mathbb{R}_0^\infty \ni t \mapsto e^{i(t,\lambda)}$ with cyclic vector $\Omega = 1$. The measures $\rho(\cdot)$ are uniquely determined by their characteristic functions

$$k(t) = \int_{\mathbb{R}^\infty} e^{i(t,\lambda)}\, d\rho(\lambda)\,,$$

which are positive definite normed functions on the group $\mathbb{R}_0^\infty$.

The simplest non-commutative analogue of random sequences (more exactly, of their characteristic functions) is brought about the positive definite functions on G_0^∞ (or the states on $\mathbf{U}_0^\infty$).

Defining a cyclic representation $G_0^\infty \ni g \mapsto U_g$ on H with a marked cyclic vector Ω is a way to define a non-commutative random sequence.

Example 1. The Weyl's form of a CCR representation of a system with countable degrees of freedom, under the assumption that the CSO $(Q_k)_{k=1}^\infty$ have a joint simple spectrum, is realized using a probability measure $\rho(\cdot)$ on $(\mathbb{R}^\infty, \mathbf{B}(\mathbb{R}^\infty))$ and a cocycle $\alpha_s(\cdot)$ (see Chapter 4, Section 4.3). Set $\Omega = 1 \in L_2(\mathbb{R}^\infty, d\rho(\lambda))$. Then the pair $(\rho(\cdot), \alpha_s(\cdot))$ defines a non-commutative (CCR) random sequence. □

12.2. Non-commutative moment problem.

The starting point to pose and solve the non-commutative moment problem is the classic n-dimensional power moment problem of Hamburger: find a probability measure $\mu(\cdot)$ on $(\mathbb{R}^n, \mathbf{B}(\mathbb{R}^n))$ such that

$$s_{k_1 \cdots k_n} = \int_{\mathbb{R}^n} \lambda_1^{k_1} \cdots \lambda_n^{k_n}\, d\mu(\lambda_1, \ldots, \lambda_n) \quad (k_1, \ldots, k_n = 0, 1, \ldots)\,,$$

for a given sequence of real numbers $(s_{k_1,\ldots,k_n})_{k_1,\ldots,k_n=0}^\infty$ $(s_{0,\ldots,0} = 1)$.

A unitary cyclic strongly continuous representation

$$\mathbb{R}^n \ni (t_1, \ldots, t_n) \mapsto U_{(t_1,\ldots,t_n)} = \exp(i \sum_{k=1}^n t_k A_k)$$

with a cyclic vector Ω defines a measure on $(\mathbb{R}^n, \mathbf{B}(\mathbb{R}^n))$

$$\rho(\cdot) = (E(\cdot)\Omega, \Omega).$$

The moments of this measure, if they exist, have the form

$$m_{k_1 \ldots k_n} = \int_{\mathbb{R}^n} \lambda_1^{k_1} \cdots \lambda_n^{k_n}\, d\rho(\lambda_1, \ldots, \lambda_n) = (A_1^{k_1} \cdots A_n^{k_n}\Omega, \Omega) \quad (k_1, \ldots, k_n = 0, 1, \ldots).$$

They define a linear functional ω on $\mathbf{U}(\mathbb{R}^n)$ which is the complex enveloping Lie algebra of the group $\mathbb{R}^n$ (the algebra of all polynomials of n real variables with complex coefficients):

$$\omega\Big(\sum_{k_1=0}^{l_1} \cdots \sum_{k_n=0}^{l_n} c_{k_1 \cdots k_n} \lambda_1^{k_1} \cdots \lambda_n^{k_n}\Big) =$$

$$= \int_{\mathbb{R}^n} \Big(\sum_{k_1=0}^{l_1} \cdots \sum_{k_n=0}^{l_n} c_{k_1 \cdots k_n} \lambda_1^{k_1} \cdots \lambda_n^{k_n}\Big)\, d\rho(\lambda_1, \ldots, \lambda_n) =$$

$$= \sum_{k_1=0}^{l_1} \cdots \sum_{k_n=0}^{l_n} c_{k_1 \cdots k_n} (A_1^{k_1} \cdots A_n^{k_n} \Omega, \Omega) =$$

$$= \sum_{k_1=0}^{l_1} \cdots \sum_{k_n=0}^{l_n} c_{k_1 \cdots k_n} m_{k_1 \cdots k_n}.$$

The functional ω is non-negative on non-negative polynomials.

In these terms, the n-dimensional classic moment problem is the problem of reconstruction of a unitary strongly continuous representation $\mathbb{R}^n \ni (t_1, \ldots, t_n) \mapsto U_{(t_1, \ldots, t_n)}$ on H with a cyclic vector Ω from the functional ω on a commutative $*$-algebra $\mathrm{U}(\mathbb{R}^n)$ $(P^*(\cdot) = \overline{P(\cdot)})$ such that

$$\Omega \in \bigcap_{k_1=0}^{\infty} D(A_1^{k_1}) \cap \cdots \cap \bigcap_{k_n=0}^{\infty} D(A_n^{k_n}) = H^\infty(A_1, \ldots, A_n)$$

and

$$s_{k_1 \cdots k_n} = (A_1^{k_1} \cdots A_n^{k_n} \Omega, \Omega).$$

A necessary and sufficient condition for the classic n-dimensional moment problem to be solvable is that the functional

$$\phi\Big(\sum_{k_1=0}^{l_1} \cdots \sum_{k_n=0}^{l_n} c_{k_1 \cdots k_n} \lambda_1^{k_1} \cdots \lambda_n^{k_n}\Big) = \sum_{k_1=0}^{l_1} \cdots \sum_{k_n=0}^{l_n} c_{k_1 \cdots k_n} s_{k_1 \cdots k_n}$$

be non-negative on non-negative polynomials, i.e. on the polynomials

$$P(\lambda_1, \ldots, \lambda_n) = \sum_{k_1=0}^{l_1} \cdots \sum_{k_n=0}^{l_n} c_{k_1 \cdots k_n} \lambda_1^{k_1} \cdots \lambda_n^{k_n} \geq 0 \;\; (\lambda_1, \ldots, \lambda_n \in \mathbb{R}^n)$$

(Riesz theorem). In particular, for the moment problem to be solvable, it is necessary that the sequence of numbers $(s_{k_1 \cdots k_n})_{k_1 \cdots k_n=0}^{\infty}$ be positive definite (p.d.):

$$\sum_{j_1,k_1=0}^{m_1} \sum_{j_n,k_n=0}^{m_n} s_{j_1+k_1 \cdots j_n+k_n} c_{j_1 \cdots j_n} \bar{c}_{k_1 \cdots k_n} \geq 0$$

for all finite sequences $(c_{k_1 \cdots k_n})_{k_1, \ldots, k_n=0}^{\infty}$, or, which is the same that the functional $\phi(\cdot)$ be positive on all $|P(\lambda_1, \ldots, \lambda_n)|^2$.

LEMMA 1. Sufficient conditions for the moment problem to be uniquely solvable are:

a) the sequence $(s_{k_1 \cdots k_n})_{k_1,\ldots,k_n=0}^{\infty}$ is p.d.;
b) $|s_{k_1 \cdots k_n}| \le CD^{k_1+\cdots+k_n}(k_1+\cdots+k_n)!$ $(k_1,\ldots,k_n=0,1,\ldots)$.

Proof. Defining a quasi-scalar product on the set of polynomials,

$$(P_1(\cdot), P_2(\cdot)) = \phi(P_2^*(\cdot)P_1(\cdot)),$$

factoring and completing it, we construct a Hilbert space, H. The polynomials $P(\cdot)$ (or the equivalence classes of polynomials) form a dense invariant set in H of analytic (using the estimate b)) vectors for the symmetric commuting operators of multiplication $\lambda_1,\ldots,\lambda_n$. By Nelson's theorem the closures of these operators form a collection of CSO $(A_k)_{k=1}^n$. As the vector $\Omega \equiv 1$ is cyclic for this collection of CSO, the spectral measure for this collection is $\mu(\cdot) = (E(\cdot)\Omega,\Omega)$. This gives a unique solution of the moment problem

$$\int_{\mathbb{R}^n} \lambda_1^{k_1} \cdots \lambda_n^{k_n}\, d\mu(\lambda_1,\ldots,\lambda_n) = (A_1^{k_1} \cdots A_n^{k_n}\Omega,\Omega) =$$

$$= \phi(\lambda_1^{k_1} \cdots \lambda_n^{k_n}) = s_{k_1 \cdots k_n} \quad (k_1,\ldots,k_n=0,1,\ldots)$$

with given $(s_{k_1 \cdots k_n})_{k_1,\ldots,k_n=0}^{\infty}$. □

Let now G be a connected simply connected real Lie group, $\mathbf{g}$ be its complex Lie algebra, and $X_1,\ldots,X_n$ be a basis in $\mathbf{g}$ with

$$[X_l, X_m] = -i \sum_{k=1}^{n} c_{lm}^k X_k \quad (c_{lm}^k \in \mathbb{R}^1)$$

being the commutation relations. The non-commutative moment problem (NMP) is the following: Find conditions on a sequence of numbers $(s_{i_1 \cdots i_n})_{i_1,\ldots,i_n=0}^{\infty}$ such that there exists a unitary representation U of the group G on a Hilbert space H and a vector $\Omega \in H$ for which

$$s_{i_1 \cdots i_n} = (dU(X_1)^{i_1} \cdots dU(X_n)^{i_n}\Omega,\Omega)$$

where dU is the corresponding representation of the Lie algebra $\mathbf{g}$.

Let L be the complex enveloping $*$-algebra of the Lie algebra $\mathbf{g}$ ($*$ is introduced in such a way that the elements of $\mathbf{g}$ are considered to be skew-self-adjoint in L). A sequence $(s_{k_1 \cdots k_n})_{k_1,\ldots,k_n=0}^{\infty}$ defines a linear functional ϕ on L:

$$\phi(X_1^{i_1} \cdots X_n^{i_n}) = s_{i_1 \cdots i_n}.$$

If (U,Ω) is a solution of NMP,

$$\phi(x) = (dU(x)\Omega,\Omega)$$

(from this point on we denote a representation of the Lie algebra and the corresponding representation of the enveloping algebra by one symbol. So the functional ϕ is positive definite ($\phi(x^* x) \geq 0$, $x \in \mathbf{L}$) i.e. it is a state on $\mathbf{L}$.

Conversely, given a state ϕ following the GNS construction, we can construct a Hilbert space H, a cyclic representation of the $*$-algebra $\mathbf{L}$ on H and a cyclic vector $\Omega \in H$ such that

$$\phi(x) = (T(x)\,\Omega, \Omega).$$

So solvability of the NMP is reduced to the problem of the integrability of the representation T to a unitary representation of the corresponding ϕ.

For the first time, the NMP was considered on the algebra of canonical commutation relations in S. Woronovich [2,3] and R. Powers [2] (the quantum moment problem). They generalized the Riesz' method. For CCR-systems, the enveloping algebra $\mathbf{L}$ is isomorphic to a $*$-algebra of differential operators with polynomial coefficients, and a criteria for the solvability of NMP is that the functional ϕ be positive definite on the cone of positive operators.

A further generalization of the Riesz' method and its application to the NMP on an arbitrary finite-dimensional Lie algebra is given in K. Schmüdgen [6], R. Powers [2], P. Richter [2]. In Schmüdgen's paper, the cone S of positive elements is constructed as follows: $s = \{x \in \mathbf{L} \mid \pi(x) \geq 0$ for any irreducible integrable representation $\pi\}$.

A criteria for the NMP to be solvable is that the functional ϕ be non-negative on the cone s. Similar results were obtained in R. Powers [2] and P. Richter [2].

Note, that it is rather difficult to check these conditions even in the case of a CCR-moment problem. The condition that ϕ is positive definite is weaker and simpler to check, so the sufficient conditions for NMP to be solvable that would generalize Lemma 1 are of some interest.

First, for a CCR-moment problem, we carry over the sufficient conditions for its unique solvability. They involve estimates on the growth of the sequence $(s_k)_{k=1}^{\infty}$ and positive definiteness. For the sake of simplicity, we take the CCR-systems with one degree of freedom.

Let $(s_{n,m})_{n,m=0}^{\infty}$ be a sequence of complex numbers. We formulate sufficient conditions on $(s_{n,m})_{n,m=0}^{\infty}$ so that there exists a cyclic representation of the CCR such that

$$\Omega \in H^{\infty}(Q) \cap H^{\infty}(P)$$

and

$$(Q^n P^m \Omega, \Omega) = s_{n,m}$$

(H is the representation space, $U(t_1)$, $V(s_1)$ are the representation operators, Ω is a cyclic vector of the representation).

Similar to the commutative case, we define a linear functional

$$\phi\left[(\lambda_1)^n\left[-i\frac{d}{d\lambda_1}\right]^m\right]$$

on the $*$-algebra of differential operators on $S(\mathbb{R}^1)$ with polynomial coefficients.

Introduce a sequence of numbers:

$$b_{(n,m)(n_1,m_1)}=\sum_{k=1}^{r} i^k\frac{(n+n_1)!\,m_1!}{(n+n_1-k)!\,k!(m_1-k)!}s_{n+n_1-k,m+m_1-k}$$

$$(r=\min(n+n_1,m_1)).$$

LEMMA 2. The functional $\phi(\cdot)$ is positive definite if and only if the kernel $b_{(n,m)(n_1,m_1)}$ is positive definite, i.e. if for any sequence of complex numbers

$$z_{(0,0)},\ldots,z_{(p,l)}\quad(p,l\in\mathbb{N})$$

$$\sum_{n,n_1=0}^{p}\sum_{m,m_1=0}^{l} z_{(n,m)}\bar{z}_{(n_1,m_1)}b_{(n,m)(n_1,m_1)}\geq 0.$$

The proof follows from the equality

$$b_{(n,m)(n_1,m_1)}=\phi\left[\left[-i\frac{d}{d\lambda_1}\right]^{m_1}(\lambda_1)^{n+n_1}\left[-i\frac{d}{d\lambda_1}\right]^m\right].$$

Indeed, since

$$\left[-i\frac{d}{d\lambda_1}\right]^m(\lambda_1)^n=$$

$$=\sum_{k=0}^{\min(n,m)}(-i)^k\frac{n!\,m!}{(n-k)!\,k!(m-k)!}(\lambda_1)^{n-k}\left[-i\frac{d}{d\lambda_1}\right]^{m-k},\tag{1}$$

then

$$\phi\left[\left[-i\frac{d}{d\lambda_1}\right]^{m_1}\lambda^{n+n_1}\left[-i\frac{d}{d\lambda_1}\right]^m\right]=$$

$$=\phi\left[\sum_{k=0}^{r}(-i)^k\frac{(n+n_1)!\,m_1!}{(n+n_1-k)!\,k!(m_1-k)!}\lambda^{n+n_1-k}\left[-i\frac{d}{d\lambda_1}\right]^{m+m_1-k}\right]=$$

$$=\sum_{k=0}^{r}(-i)^k\frac{(n+n_1)!\,m_1!}{(n+n_1-k)!\,k!(m_1-k)!}s_{n+n_1-k,m+m_1-k}=b_{(n,m)(n_1,m_1)}.$$

□

A necessary condition for the CCR-moment problem to be solvable is that $\phi(\cdot)$ be positive definite.

THEOREM 2. If

1) the functional $\phi(\cdot)$ (or, which is the same thing, the sequence $b_{(n,m)(n_1,m_1)}$) is positive definite,

2) the estimates

$$|s_{n,0}| \leq CD^n n!, \quad |s_{0,m}| \leq CD^m m!$$

$$(C,D>0, \quad n,m=0,1,2,\dots) \tag{2}$$

hold, then the CCR-moment problem has a unique solution.

Proof. On the $*$-algebra $\mathbf{L}$ of differential operators with polynomial coefficients, introduce the scalar product $(x,y)=\phi(y^* x)$.

According to the GNS-construction, the operators $\pi(x)y=xy$ form a representation of the $*$-algebra $\mathbf{L}$ on a space H obtained by completing and factorizing $\mathbf{L}$ with respect to the scalar product. $\Omega=1$ is a cyclic vector of this representation, $(\pi(x)\,\Omega,\Omega)=\phi(x)$, and so

$$(\pi((\lambda_1)^n)\,\pi\left(\left[-i\frac{d}{d\lambda_1}\right]^m\right)\Omega,\Omega)=s_{n,m}.$$

Now we show that by the estimates (2), the vectors $\mathbf{L}\subset H$ are analytic vectors for the operators $\pi(-i\dfrac{d}{d\lambda_1})$ and $\pi(\lambda_1)$.

LEMMA 3. If $|s_{n,0}|\leq CD^n n!$ and $|s_{0,m}|\leq CD^m m!$, then for a fixed m

$$|b_{(n,m)(n,m)}|\leq C_m D_m^n\,(2n)! \tag{2}$$

and for a fixed n

$$|b_{(n,m)(n,m)}|\leq C_n D_n^m\,(2m)!. \tag{4}$$

Proof. We show that for a fixed m

$$b_{(n,m)(n,m)}|\leq C_m D_m^n\,(2n)!.$$

For $n>m$, we have:

$$|b_{(n,m)(n,m)}|\leq C\sum_{k=0}^{m}\frac{(2n)!\,m!}{(2n-k)!\,k!(m-k)!}\,|s_{2n-k,\,2m-k}|\leq$$

$$\leq C_1\,2n\,\cdots\,(2n-m+1)\sum_{k=0}^{m}|s_{2n-k,\,2m-k}|.$$

Since

$$s_{n,m} = \phi\left(\lambda_1^n\left(-i\frac{d}{d\lambda_1}\right)^m\right) = \left(\pi\left(-i\frac{d}{d\lambda_1}\right)^m\Omega,\, \pi(\lambda_1)^n\Omega\right) \leq$$

$$\leq \|\pi\left(-i\frac{d}{d\lambda_1}\right)^m\Omega\|\ \|\pi(\lambda_1)^n\Omega\| \leq K_m\sqrt{s_{2n,0}} \leq K_m D_1^n\, n!\,,$$

we have that

$$|b_{(n,m)(n,m)}| \leq C_m D_m^n\,(2n)!.$$

One similarly proves inequality (4) for a fixed n. □

To show that the vectors of $\mathbf{L} \subset H$ are analytic for the operator $\pi(\lambda_1)$ it is sufficient to show that the series

$$\sum_{n=0}^{\infty}\frac{\|\pi(\lambda_1)^n\,\pi\left(-i\frac{d}{d\lambda_1}\right)^k\pi(\lambda_1)^l\,\Omega\|}{n!}\,s^n$$

converges for small $s > 0$.

But

$$\|\pi(\lambda_1)^{n+k}\,\pi\left(-i\frac{d}{d\lambda_1}\right)^l\Omega\| =$$

$$= \sqrt{\phi\left(\left(-i\frac{d}{d\lambda_1}\right)^l\lambda_1^{2(n+k)}\left(-i\frac{d}{d\lambda_1}\right)^l\right)} = \sqrt{b_{(k+n,l)(k+n,l)}}$$

and by Lemma 3

$$\|\pi(\lambda_1)^{n+k}\,\pi\left(-i\frac{d}{d\lambda_1}\right)^l\Omega\| \leq K_{(k,l)}\,L_{(k,l)}^n\,n!.$$

One can similarly prove that the vectors of $\mathbf{L}$ are analytic for the operator $\pi(-i\frac{d}{d\lambda_1})$.

Since it was proved in M. Flato, J. Simon, H. Snellman and D. Sternheimer [1] (see also A. Barut, R. Rączka [1]) that the existence of an invariant dense set of analytic vectors in H only for the representation operators of a basis of a Lie algebra guarantees that it is integrable to a unitary representation of the corresponding group, the constructed representation of the algebra $\mathbf{L}$ can be extended to a cyclic CCR-representation with a cyclic vector Ω such that

$$\Omega \in H^{\infty}(\pi(\lambda_1)) \cap H^{\infty}(\pi\left(-i\frac{d}{d\lambda_1}\right))$$

and

$$s_{n,m} = (\pi(\lambda_1)^n \, \pi\left[-i \frac{d}{d\lambda_1}\right]^m \Omega, \Omega).$$

Consider now an algebra g, $\{X_1, \ldots, X_n\}$ a basis of g (it is enough to consider a collection of generators of g). Let T be a representation of **L** constructed using a state ϕ.

For T to be integrable, it is sufficient that there existed a dense set D of analytic vectors for the operators $T(X_k)$, $k = 1, \ldots, n$ in H invariant relatively to T. In this case, the construction of the corresponding representation of the group is unique.

Set $D = \{T(X)\Omega \mid X \in \mathbf{L}\}$. It is obvious that D is invariant, and since Ω is a cyclic vector, it is dense in H. Besides, let

$$|\phi(X_k^m)| \leq C^m m!, \quad k = 1, \ldots, n; m = 1, 2, \ldots$$

where C is a constant. From these estimates it follows that Ω is an analytic vector for the operators $T(X_k)$. It is known (see M. Flato and J. Simon [1]) that the set of analytic vectors for one representation operator of a Lie algebra is invariant with respect to all the representation operators. Consequently, all the elements of D are analytic vectors for the operators $T(X_k)$. So, the representation T can be uniquely integrated to a unitary representation of the group G and thus we have proved the following result.

THEOREM 3. For a G-moment problem to be solvable it is sufficient that the sequences $(s_{i_1, \ldots, i_n})$ define a positive definite functional on **L** and that the estimates

$$|s_{0,\ldots,0,i_k,0,\ldots,0}| \leq C^{i_k} i_k!; \quad k = 1, \ldots, n; i_k = 0, 1, \ldots$$

hold.

Note that if, posing the NMP, require that Ω be cyclic, the conditions given above guarantee that the solution is unique.

However, the non-commutative moment problem may be related not to a group but to a more general object.

Here, we will pose and solve this problem for a pair of anticommuting self-adjoint operators (an anticommutative moment problem).

Consider a complex $*$-algebra generated by two generators, $a_1 = a_1^*, a_2 = a_2^*$ and the relation $a_1 a_2 + a_2 a_1 = 0$. As a vector space, it is the same as the space of complex polynomials of two real variables but with the following multiplication of terms

$$a_1^{k_1} a_2^{k_2} \cdot a_1^{l_1} a_2^{l_2} = (-1)^{k_2 l_1} a_1^{k_1+l_1} a_2^{k_2+l_2}$$

and involution

$$(a_1^{k_1} a_2^{k_2})^* = (-1)^{k_1 k_2} a_1^{k_1} a_2^{k_2}.$$

Let now $(s_{k_1,k_2})_{k_1,k_2=0}^{\infty}$ be a sequence of complex numbers. We need to formulate sufficient conditions for $(s_{k_1,k_2})_{k_1,k_2=0}^{\infty}$ so that there exists a unique pair (A_1, A_2) of, generally speaking, not bounded anticommuting (see Chapter 8) self-adjoint operators and a vector $\Omega \in H$ cyclic for this pair, such that

$$s_{k_1,k_2} = (A_1^{k_1} A_2^{k_2} \Omega, \Omega) \quad (k_1, k_2 = 0,1,...).$$

THEOREM 4. For an anticommutative moment problem to have a unique solution it is sufficient that

a) for any sequence of complex numbers $(z_{kl})_{k,l=0}^{p}$ $(p = 0,1,...)$

$$\sum_{k,l=0}^{p} \sum_{m,n=0}^{p} (-1)^{m(n+k)} s_{k+n,l+m} z_{kl} \bar{z}_{mn} \geq 0;$$

b) $|s_{n,0}| \leq CD^n n!$, $|s_{0,m}| \leq CD^m m!$ for $n,m = 0,1,...$ and some $C,D > 0$.

Proof. Using a), we construct a representation of the $*$-algebra generated by the pair of self-adjoint anticommuting generators. Using b), we check that they anticommute on a dense set of analytic vectors. Applying Theorem 1 from Chapter 8, we finish the proof. ▯

If a $*$-algebra **A** is the inductive limit of finitely generated $*$-subalgebras $\mathbf{A}_n$

$$(\mathbf{A}_1 \hookrightarrow A_2 \hookrightarrow ..., \mathbf{A} = \bigcup_{n=1}^{\infty} \mathbf{A}_n)$$

for a countable dimensional moment problem to be uniquely solvable, as in the classic situation, it is sufficient that it is uniquely solvable for all $n = 1,2,....$

Comments to Chapter 12.

1. The theorem on the correspondence between cyclic unitary representations of a locally compact group G (of a group algebra $C^*(G)$) and the states (p.d. functions on the group) is due to I.M. Gel'fand, M.A. Naimark [2]. The theorem on the correspondence between the states on a C^*-algebra and its cyclic representations can be found in I. Segal [1]. The construction of a representation of a $*$-algebra using a state is called the GNS (Gel'fand-Naimark-Segal)-construction.

Describing the methods of defining non-commutative random sequences, the author tried to follow the commutative situation (see A.V. Scorokhod [1]).

2. Posing the quantum problem and giving criteria for its solvability is due to S. Woronovich [2,3], R. Powers [2]. The further application of Riesz' method to a NMP is due to R. Powers [2], K. Schmüdgen [6], P. Richter [2].

The given form of the sufficient condition for a NMP to be solvable is due to be author, its proof for a quantum moment problem is given by A.Yu. Daletskiĭ [3], and for the NMP - by A.Yu. Daletskiĭ and the author [1]. Some generalizations of the NMP for non-group cases are given by M. Dubois-Violette [1].

Besides the monographs and papers used in the book, the given bibliography also contains some references to the works close to the contents of the book. We give some monographs and papers on theoretical physics, theory of $*$-algebras and their representations, noncommutative probability theory closely related to the spectral theory of the operators and their countable collections.

BIBLIOGRAPHY

ACCARDI, L.

1 On a noncommutative Markov property, Funktsional. Anal. i Prilozhen. **9** (1975), No. 1, pp. 1-8; English transl. in Functional Anal. Appl. **9** (1975).

2 Nonrelativistic quantum mechanics as a noncommutative Markov process, Adv. Math. **20** (1976), No. 3, pp. 329-366.

3 Recent developments in quantum probability, The first world congress of the society of mathematical statistics and probability theory, "Nauka", Moscow, 1986, pp. 683-686.

AKHIEZER, N.I.

1 Classical moment problem and some related analysis problems, "Fizmatgiz", Moscow, 1961 (Russian)

AKHIEZER, N.I., and I.M. GLAZMAN

1 The theory of linear operators in Hilbert space, 2nd rev. ed., "Nauka", Moscow, 1966; English transl. of 1 st ed., Vol. I, II, Ungar, New York, 1961.

ALBEVERIO, S., G. GALLAVOTTI and R. HÖEGH-KROHN

1 Some results for the exponential interaction in two or more dimensions, Communs. Math. Phys. **70** (1970), No. 2, pp. 187-192.

ALBEVERIO, S., and R. HÖEGH-KROHN

1 The energy representation of Sobolev-Lie groups, Compos. Math. **36** (1978), No. 1, pp. 37-52.

2 Some Markov processes and Markov fields in quantum theory, group theory, hydrodynamics and C^*-algebras, Lec. Notes Math. **851** (1981), pp. 497-540.

ALBEVERIO, S., R. HÖEGH-KROHN and D. TESTARD

1 Irreducibility and reducibility for the energy representations of the group of mappings of a Riemannian manifold into a compact semisimple Lie group, J. Funct. Anal. **41** (1981), No. 3, pp. 378-396.

ALBEVERIO, S., R. HÖEGH-KROHN, D. TESTARD and A.M. VERSHIK

1 Factorial representations of path groups, J. Funct. Anal. **51** (1983), No. 1, pp. 115-131.

ANDERSON, R.

1 The Weyl functional calculus, J. Funct. Anal. **4** (1969), No. 2, pp. 240-267.

ANSHELEVICH, V.V.

1 A central limit theory for "noncommutative" stationary processes, Uspekhi Mat. Nauk **28** (1973), No. 5, pp. 227-228.

2 A central limit theorem in noncommutative probability theory, Dokl. Akad. Nauk SSSR **208** (1973), No. 6, pp. 1265-1267.

ARAKI, H.

1 Hamiltonian formalism and canonical commutation relations in quantum field theory, J. Math. Phys. **1** (1960), pp. 492-504.

2 Factorizable representation of current algebra. Non commutative extension of the Levi-Khinchin formula and cohomology of a solvable group with values in a Hilbert space, Publ. Res. Inst. Math. Sci. Kyoto Univ. **5** (1970), pp. 361-422.

3 One-dimensional quantum lattice systems, Matematika **15** (1971), pp. 103-113.

4 On quasifree states of the canonical commutation relations. II, Publ. Res. Inst. Math. Sci. Kyoto Univ. **7** (1971), pp. 121-152.

ARAKI, H., and Y. NAKAGAMI

1 A remark on an infinite tensor product of von Neumann algebras, Publ. Res. Inst. Math. Sci. Kyoto Univ. **8** (1972), pp. 363-374.

ARAKI, H., and M. SHIRAISHI

1 On quasifree states of the canonical commutation relations. I., Publ. Res. Inst. Math. Sci. Kyoto Univ. **7** (1971), pp. 105-120.

ARAKI, H., and E.J. WOODS

1 Representations of the canonical commutation relations describing a nonrelativistic infinite free Bose gas, J. Math. Phys. **4** (1963), pp. 637-662.

ARENS, R.

1 Representations of $*$-algebras, Duke Math. J. **14** (1947), pp. 269-282.

ARNAL, D., and J.P. JURZAK

1 Topological aspects of algebras of unbounded operators, J. Funct. Anal. **24** (1977), pp. 397-405.

ARVESON, W.B.

1 Unitary invariants for compact operators, Bull. Amer. Math. Soc. **76** (1970), pp. 88-91.

2 Operator algebras and invariant subspaces, Ann. Math. **100** (1974), PP. 433-532.

AVERBUKH, V.I., O.G. SMOLYANOV and S.V. FOMIN

1 Generalized functions and differential equations in linear spaces. I: Differentiable measures, Trudy Moskov. Mat. Obshch. **24** (1971), pp. 133-174; English transl. in Trans. Moscow Math. Soc. **24** (1971).

2 Generalized functions and differential equations in linear spaces. II: Differential operators and their Fourier transforms, Trudy Moskov. Mat. Obshch. **27** (1972), pp. 247-262; English transl. in Trans. Moscow Math. Soc. **27** (1972).

AYUPOV, Sh.A.

1 Statistical ergodic theorems in Jordan algebras, Uspekhi Mat. Nauk. **36** (1981), No. 6, pp. 201-202; English transl. in Russian Math. Surveys **36** (1981).

2 On the construction of Jordan algebras of self-adjoint operators, Dokl. Acad. Nauk SSSR **267** (1982), No. 3, pp. 521-524; English transl. in Soviet Math. Dokl. **267** (1982).

BAKER, B.M., and R.T. POWERS

1 Product states and C^*-dynamical system of product type, J. Funct. Anal. **50** (1983), pp. 229-266.

BALSLEV, E., J. MANUCEAU and A. VERBEURE

1 Representations of anticommutation relations and Bogolyubov transformations, Communs Math. Phys. **8** (1968), pp. 315-326.

BALSLEV, E., and A. VERBEURE

1 States on Clifford algebras, Communs Math. Phys. **7** (1968), pp. 55-76.

BARUT, A.O., and R. RĄCZKA

1 Theory of group representations and applications, PWN, Warszawa, 1977.

BEHNCKE, H., and H. LEPTIN

1 C^*-algebras with two point duals, J. Funct. Anal. **10** (1972), pp. 330-335.

2 C^*-algebras with finite duals, J. Funct. Anal. **14** (1973), pp. 253-268.

3 Classification of C^*-algebras with a finite dual, J. Funct. Anal. **16** (1974), pp. 241-257.

BEREZANSKIĬ, Yu.M.

1 Some classes of continual algebras, Dokl. Akad. Nauk SSSR **72** (1950), No. 2, pp. 237-240 (Russian).

2 Hypercomplex systems with discrete basis, Dokl. Akad. Nauk SSSR **81** (1951), No. 3, pp. 329-332 (Russian).

3 On some normed rings constructed from orthogonal polynomials, Ukrain, Mat. Zh. **3** (1951), No. 4, pp. 412-432 (Russian).

4 On hypercomplex systems constructed from Sturm-Liouville equations on the half-axis, Dokl. Akad. Nauk SSSR **91** (1953), No. 6, pp. 1245-1248 (Russian).

5 Expansion in eigenfunctions of self-adjoint operators, "Naukova Dumka", Kiev, 1965; English transl., Amer. Math. Soc., Providence, R.I., 1968.

6 Expansion in generalized eigenvectors and integral representation of positive definite kernels in the form of a functional integral, Sibirsk. Mat. Zh. **9** (1968), pp. 998-1013; English transl. in Sibirian Math. J. **9** (1968).

7 On expansions in joint generalized eigenvectors of an arbitrary family of commuting normal operators. Dokl. Akad. Nauk SSSR **229** (1976), pp. 531-533; English transl. in Soviet Math. Dokl. **17** (1976).

8 The expansion in simultaneous generalized eigenvectors of an arbitrary family of commuting normal operators, Proc. Intern. conf. operator algebras, ideals and their appl. in theor. physics, Leipzig, 1978, pp. 265-276.

9 Self-adjoint operators in spaces of functions of infinitely many variables, "Naukova Dumka", Kiev, 1978; English transl., Amer. Math. Soc., Providence, R.I., 1986.

10 Spectral representations of solutions of certain classes of functional and differential equations, Dokl. Akad. Nauk Ukrain, SSR Ser. A 1978, pp. 579-583 (Russian).

11 Self-adjoint operators in spaces of functions of infinitely many variables, Partial differential equations, "Nauka", Novosibirsk, 1978, pp. 579-583 (Russian).

12 The spectral projection theorem, Uspekhi Mat. Nauk **39** (1984), No. 4 (238), pp. 3-52 (Russian).

13 On the spectral projection theorem, Ukrain. Mat. Zh. **37** (1985), No. 2, pp. 146-154 (Russian).

BEREZANSKIĬ, Yu.M., I.M. GALI and Yu.G. KONDRAT'EV

1 Stone's theorem for the additive group of a Hilbert space. Funktsional. Anal. i Prilozhen. **11** (1977), No. 4, pp. 68-69 (Russian).

BEREZANSKIĬ, Yu.M., and A.A. KALYUZHNYĬ

1 Representations of hypercomplex systems with locally compact base, Ukrain. Mat. Zh. **36** (1984), No. 4, pp. 417-421 (Russian).

2 Spectral decompositions of representations of hypercomplex systems, Spectral operator theory and infinite-dimensional analysis, Inst. Mat. Akad. Nauk Ukrain. SSR, Kiev, 1988, pp. 4-19 (Russian).

3 Hypercomplex systems with locally compact bases, Selecta Math. Sovietica **4** (1985).

BEREZANSKIĬ, Yu.M., and Yu.G. KONDRAT'EV

1 Spectral methods in infinite-dimensional analysis, "Naukova Dumka", Kiev, 1988 (Russian).

BEREZANSKIĬ, Yu.M., Yu.G. KONDRAT'EV and Yu.S. SAMOĬLENKO

1 Generalized functions of infinitely many variables and their application in spectral theory, Generalized functions and their applications in Mathematical Physics (Proc. Internat. Conf., Moscow, 1980), Vychisl. Tsentr Akad. Nauk SSSR, Moscow, 1981., pp. 50-70 (Russian).

BEREZANSKIĬ, Yu.M., and S.G. KREĬN

1 Continual algebras, Dokl. Akad. Nauk SSSR **72** (1950), No. 1, pp. 5-8 (Russian).

2 Hypercomplex systems with a compact base, Ukrain. Mat. Zh. **3** (1951), No. 2, pp. 184-203 (Russian).

3 Hypercomplex systems with a continuous base, Uspekhi Mat. Nauk **12** (1957), No. 1 (73), pp. 147-152 (Russian).

BEREZANSKIĬ, Yu.M., G. LASSNER and V.S. YAKOVLEV

1 On decomposition of positive functionals on commutative nuclear $*$-algebras, Ukrain. Mat. Zh. **39** (1987), No. 5, pp. 638-641 (Russian).

BEREZANSKIĬ, Yu.M., V.L. OSTROVSKIĬ and Yu.S. SAMOĬLENKO

1 Decomposition in eigenfunctions of families of commuting operators and representations of commutation relations, Ukrain. Mat. Zh. **40** (1988), No. 1, pp. 106-109 (Russian).

BEREZANSKIĬ, Yu.M., and Yu.S. SAMOĬLENKO

1 Nuclear spaces of functions of infinitely many variables, Ukrain. Mat. Zh. **25** (1973), No. 6, pp. 723-737 (Russian).

BEREZANSKIĬ, Yu.M., Yu.S. SAMOĬLENKO and G.F. US

1 Self-adjoint operators in spaces of functions of infinitely many variables, Operator Theory in Function Spaces (Proc. School, Novosibirsk, 1975; G.P. Akilov, editor), "Nauka", Novosibirsk, 1977, pp. 20-41 (Russian).

BEREZANSKIĬ, Yu.M., and S.N. SHIFRIN

1 The generalized symmetric power moment problem, Ukrain. Mat. Zh. **23** (1971), pp. 291-306 (Russian).

BEREZANSKIĬ, Yu.M., and G.F. US

1 On expansions in eigenfunctions of self-adjoint operators admitting separation of infinitely many variables, Dokl. Akad. Nauk SSSR **213** (1973), pp. 1005-1008 (Russian).

2 Eigenfunction expansions of operators admitting separation of an infinite number of variables, Rep. Math. Phys. **7** (1975), pp. 103-126.

BEREZIN, F.A.

1 A secondary quantization method, "Nauka", Moscow, 1965 (Russian).

2 On a certain representation of operators using functionals, Trudy Moskov, Mat. Obshch. **17** (1967), pp. 117-196 (Russian).

3 Several remarks on representations of commutation relations, Uspekhi Mat. Nauk. **24** (1969), No. 4, pp. 65-88 (Russian).

4 Wick and anti-Wick operator symbols, Mat. Sb. **86** (1971), pp. 578-610 (Russian).

5 Non-Wiener continual integrals, Teor. Mat. Phys. **6** (1971), No. 4, pp. 578-610 (Russian).

6 Covariant and contravariant symbols of operators, Izvestia Acad. Nauk SSSR, Ser. Mat. **36** (1972), No. 5, pp. 1134-1167 (Russian).

7 Representation of the infinite direct product of universal coverings of isometry groups of the complex ball, Repts. Math. Phys. **9** (1976), pp. 15-30.

8 Lie superalgebras, ITEP-66, Moskow, 1977.

9 Introduction into algebra and analysis of anticommuting variables, Izdat. Moskov. Univ., Moscow, 1983 (Russian).

BEREZIN, F.A., and G.I. KATS

1 Lie groups with commuting and anticommuting parameters, Mat. Sb. **82** (1970), No. 3, pp. 343-359 (Russian).

BEREZIN, F.A., and M.A. SHUBIN

1 Lectures on quantum mechanics, Izdat. Moskov. Univ., Moscow, 1972 (Russian).

BIRMAN, M.Sh., and M.Z. SOLOMYAK

1 The spectral theory of self-adjoint operators in Hilbert spaces, Izdat. Leningrad. Univ., Leningrad, 1980 (Russian).

BIRMAN, M.Sh., A.M. VERSHIK and M.Z. SOLOMYAK

1 A product of commutative spectral measures can fail to be countably additive, Funktsional. Anal. i Prilozhen. **13** (1979), No. 1, pp. 61-62 (Russian).

BLACKADAR, B.E.

1 Infinite tensor products of C^*-algebras, Pacif. J. Math. **72** (1977), pp. 313-334.

BLYUMIN, S.L.

1 Convolution algebras related to compact nonabelian groups, Automat. i telemechan. **12** (1978), pp. 48-53 (Russian).

BOGOLYUBOV, N.N., A.A. LOGUNOV and I.T. TODOROV

1 Foundations of an axiomatic approach to quantum field theory, "Nauka", Moscow, 1969 (Russian).

BOGOLYUBOV, N.N. (Jr), and A.K. PRIKARPATSKIĬ

1 A quantum method of Bogolyubov generating functionals in statistical physics: Lie algebras of currents, its representations and functional equations, Element. chastits. i atom. yadra **17** (1986), No. 4, pp. 423-468 (Russian).

2 A quantum method of Bogolyubov generating functionals in statistical physics: Lie algebras of currents, its representations and functional equations, Ukrain. Mat. Zh. **38** (1986), No. 3, pp. 284-289 (Russian).

BORISOV, N.V.

1 The structure of canonical variables in the theory of quantum systems with a finite or infinite degree of freedom, Teoret. Mat. Phys. **19** (1984), pp. 27-36 (Russian).

BOSECK, H., G. CZICHOWSKI and K.-P. RUDOLPH

1 Analysis on topological groups - general Lie theory, Teubner, texte zur Math. **37**, 1981.

BOURBAKI, N.

1 Integration, Chaps. 6-8, Actualitées Sci. Indust. nos 1281, 1306, Hermann, Paris, 1959, 1963.

2 Integration, Chaps. 3-5, 9. Actualitées Sci. Indust. nos 1175 (2nd ed.), 1244 (2nd ed.), 1343, Hermann, Paris, 1965, 1967, 1969.

3 Theories spectrales, Fascicule XXXII, Hermann, Paris, 1967.

BOYER, R.P.

1 Representation theory of the Hilbert-Lie groups, Duke Math. J. **47** (1980), pp. 325-344.

BRATTELI, O.

1 Inductive limits of finite dimensional C^*-algebras, Trans. Amer. Math. Soc. **171** (1972), pp. 195-234.

2 Structure spaces of approximately finite dimensional C^*-algebras, J. Funct. Anal. **16** (1974), pp. 192-204.

3 The center of approximately finite-dimensional C^*-algebras, J. Funct. Anal. **21** (1976), pp. 295-302.

BRATTELI, O., and D.W. ROBINSON

1 Operator algebras and quantum statistical mechanics, Text and Monographs in Physics, Springer-Verlag, 1979.

BULDIGIN, V.V.

1 Convergence of random elements in topological spaces, "Naukova Dumka", Kiev, 1980 (Russian).

BULDIGIN, V.V., and V.S. DONCHENKO

1 On a certain class of probability measures in the space of sequences, Preprint 76.1, Inst. Mat. Akad. Nauk Ukrain. SSR, Kiev, 1976 (Russian).

2 Oscillation theorem for Gaussian sequences in a Banach space, Probability distributions in infinite-dimensional spaces, Kiev, 1978, pp. 19-25 (Russian).

BURES, D.

1 Certain factors constructed as infinite tensor product, Compos. Math. **15** (1963), pp. 169-191.

2 Representations of infinite weak product groups, Compos. Math. **22** (1970), pp. 7-18.

BUTSAN, G.P.

1 An example of a quasi-invariant measure on a non-locally-compact non-commutative group, Questions of statistics and control of random processes, Instit. Mat. Akad. Nauk Ukrain. SSSR, Kiev, 1973, pp. 32-36 (Russian).

CAMERON, R.H., and N.T. MARTIN

1 Fourier-Wiener transforms of analytic functionals, Duke Math. J. **12** (1945), pp. 489-507.

CAREY, A.L.

1 Some infinite dimensional groups and bundles, Publ. Res. Inst. Math. Sci. Kyoto Univ. **20** (1984), pp. 1103-1118.

2 Projective representations of the Hilbert Lie group $U(H)_2$ via quasifree states on the CAR algebra, J. Funct. Anal. **55** (1984), pp. 277-296.

CARTIER, P., and J. DIXMIER

1 Vecteurs analyticus dans les representations des groupes de Lie, Amer. J. Math. **80** (1958), pp. 131-145.

CHAIKEN, J.M.

1 Finite particle representations and states of the canonical commutation relations, Ann. Phys. **42** (1967), pp. 23-80.

2 Number operators for representations of the canonical commutation relations, Communs. Math. Phys. **8** (1968), pp. 164-184.

CHEREDNIK, I.V.

1 On some finite-dimensional representations of generalized Sklyanin algebras, Funktsional. Anal. i Prilozhen. **19** (1985), No. 1, pp. 89-90 (Russian).

2. Functional realizations of basis representations of factorizible Lie groups and Lie algebras, Funktsional. Anal. i Prilozhen. **19** (1985), No. 3, pp. 36-52 (Russian).

CORDESSE, A., and G. RIDEAU

1 On some representations of anticommutation relations. I, II, III. Nuovo Cimento **45** (1966), pp. 1-14, **46** (1966), pp. 624-636, **50** (1967), pp. 244-255.

COURBAGE, M., S. MIRACLE-SOLE and D.W. ROBINSON

1 Normal states and representations of the canonical commutation relations, Ann. Inst. H. Poincare **14** (1971), pp. 171-178.

CUCULESCU, I.

1 Spectral families and stochastic integrals, Rev. Roum. Math. Pures Appl. **15** (1970), pp. 201-221.

2 Martingales on von Neumann algebras, J. Multiv. Anal. **1** (1971), pp. 17-27.

CUNTZ, J.

1 On the continuity of seminorms on operator algebras, Math. Anal. **220** (1966), pp. 171-184.

2 Simple C^*-algebras generated by isometries, Communs. Math. Phys. **57** (1977), pp. 173-185.

DALETSKIĬ, A.Yu.

1 Representations of a group of finite shifts in the space of functions of a countable number of variables, Mathematical models of Statistical Physics, Tyumen', 1982, pp. 148-151 (Russian).

2 Representation of translation groups of a nuclear space, Dokl. Akad. Nauk Ukrain. SSR Ser. A. 1982, No. 12, pp. 9-11 (Russian).

3 On quantum moment problems, Probability theory and statistics, 1983, No. 28, pp. 17-25 (Russian).

4 The problem on moments of finite-dimensional Lie algebras, Spectral operator theory in problems of Mathematical Physics, Inst. Mat. Akad. Nauk Ukrain. SSR, 1983, pp. 93-97 (Russian).

5 Integration of representation of nuclear Lie algebras of smooth currents, Spectral operator theory and infinite-dimensional analysis, Inst. Mat. Akad. Nauk Ukrain. SSR, 1984, pp. 77-92 (Russian).

DALETSKIĬ, A.Yu., and Yu.S. SAMOĬLENKO

1 A noncommutative moment problem, Funktsion. Anal. i Prilozhen. **21** (1987), No. 2, pp. 72-73 (Russian).

DALETSKIĬ, Yu.L.

1 Integration in functional spaces, VINITI Mat. Anal., 1966, pp. 83-124 (Russian).

2 Infinite-dimensional elliptic operators and the related parabolic equations, Uspekhi Mat. Nauk **22** (1967), No. 4, pp. 3-54 (Russian).

DALETSKIĬ, Yu.L., and S.V. FOMIN

1 Measures and differential equations in infinite-dimensional linear spaces, "Nauka", Moscow, 1983 (Russian).

DALETSKIĬ, Yu.L., and S.G. KREĬN

1 Integration and differentiation of operators depending on a parameter, Uspekhi Mat. Nauk **12** (1957), No. 1, pp. 182-186 (Russian).

DALETSKIĬ, Yu.L., and Ya.I. SHNAĬDERMAN

1 Diffusion and quasi-invariant measures on infinite dimensional Lie groups, Funktsion. Anal. i Prilozhen. **3** (1969), No. 2, pp. 88-90 (Russian).

DAMASKINSKIĬ, E.V.

1 *O*-invariant *U*-cyclic Weyl systems, Teor. Mat. Phys. **15** (1973), No. 1, pp. 70-77 (Russian).

2 Invariant Weyl systems which are not *U*-cyclic (notes on the work of Hegerfeldt, Melsheimer), Teor. Mat. Phys. **15** (1973), No. 2, pp. 221-226 (Russian).

3 Euclidean convariant representations of the group of non-relativistic currents, Teor. Mat. Phys. **20** (1974), No. 2, pp. 170-176 (Russian).

DATE, E., M. JIMBO, M. KASHIWARA and T. MIWA

1 Tranformation groups for solution equations, Euclidean Lie algebras and reduction of the *KP*-hierarchy, Publ. Res. Inst. Math. Sci., Kyoto Univ. **18** (1982), pp. 1077-1110.

DAVIS, E.B.

1 Quantum stochastic processes I, II, III, Communs. Math. Phys. **15** (1969), pp. 277-304, **19** (1970), pp. 83-105, **20** (1971), pp. 51-70.

2 Hilbert space representations of Lie algebras, Communs. Math. Phys. **23** (1971), pp. 159-168.

DEBACKER-MATHOT, F.

1 Some operator algebras in nested Hilbert spaces, Communs. Math. Phys. **42** (1975), pp. 183-193.

2 Integral decomposition of unbounded operator families, Communs. Math. Phys. **71** (1980), pp. 47-58.

3 Distributionlike representations of $*$-algebras, J. Math. Phys. **22** (1981), pp. 1386-1389.

DE LA HARPE, P.

1 Classical Banach-Lie algebras and Banach-Lie groups of operators in Hilbert spaces, Lecture Notes in Math. **285**, 1972.

DELL'ANTONIO, G.F.

1 Structure of the algebras of some free system, Communs. Math. Phys. **9** (1968), pp. 81-117.

DELL'ANTONIO, G.F., S. DOPLICHER and D. RUELLE

1 A theorem on canonical commutation and anticommutation relations, Communs. Math. Phys. **2** (1966), pp. 223-230.

DELSARTE, J.

1 Sur une extension de formule de Taylor, J. Math. Pures et Appl. **17** (1938), pp. 213-230.

DELORME, P.

1 Sur la 1-cohomologie de groupes localement compacts et produits tensoriels continus de representation, C.R. Acad. Sci. Paris **280** (1975), pp. 1101-1104.

2 Sur la 1-cohomologie des representations unitaires des groupes de Lie semi-simples et resolubles, C.R. Acad. Sci. Paris **282** (1976), pp. 499-501.

DEVINATZ, A., and A.E. NUSSBAUM

1 On the permutability of normal operators, Ann. Math. **65** (1957), pp. 144-152.

DITKIN, V.A., and A.P. PRUDNIKOV

1 Integral transforms and operational calculus, "Nauka", Moscow, 1974 (Russian).

DIXMIER, J.

1 Les algèbres d'operateurs dans l'espace Hilbertien (Algèbres de von Neumann), Deuxieme edition, Gauthier-Villars Ed., Paris, 1969.

2 On some C^*-algebras considered by Glimm, J. Funct. Anal. **1** (1967), pp. 182-203.

3 Les C^*-algebras et leur representations, Gauthier-Villars, Paris, 1969.

DIXON, P.G.

1 Unbounded operator algebras, Proc. London Math. Soc. **23** (1971), pp. 53-69.

DOBRUSHIN, R.L., and R.A. MINLOS

1 Polynomials of a generalized random field and its moments, Teor. Veroyatn. i Primenen. **23** (1978), No. 4, pp. 715-729 (Russian).

DRINFEL'D, V.G.

1 Hamiltonian structures on Lie groups, Lie bialgebras and the geometric meaning of Yang-Baxter equations, Doklady Akad. Nauk SSSR **268** (1983), No. 5, pp. 285-287 (Russian).

2 Quantum groups, Differential geometry, Lie groups and Mechanics, VIII (Notes of the seminar LOMI, vol. **155**) "Nauka", Leningrad, 1986, pp. 18-49 (Russian).

DRINFEL'D, V.G., and V.V. SOKOLOV

1 Lie algebras and equations of Korteweg-De Vries type, Itogy Nauki i Tekniki **24**, 1984, pp. 81-180 (Russian).

DUBOIS-VIOLETTE, M.

1 A generalization of the classical moment problem on $*$-algebra with applications to relativistic quantum theory. I, Communs. Math. Phys. **43**, 1975, pp. 225-254; II, **54**, 1977, pp. 151-172.

DUNFORD, N., and J.T. SCHWARTZ

1 Linear operators, Wiely-Interscience, 1971.

DYE, H.A.

1 On groups of measure preserving transformations. I, Amer. J. Math. **81** (1959), pp. 119-159.

DYNKIN, E.B.

1 Normed Lie algebras and analytical groups, Uspekhi Mat. Nauk **5** (1950), No. 1, pp. 135-187 (Russian).

EDWARDS, C.M.

1 Spectral theory for $A(X)$, Math. Ann. **207** (1974), pp. 67-85.

2 The spectrum of a real Banach space, Proc. London Math. Soc. **28** (1974), pp. 654-670.

EFFROS, E.G.

1 A decomposition theory for representations of C^*-algebras, Trans. Amer. Soc. **107** (1963), pp. 83-106.

2 Transformation groups and C^*-algebras Ann. Math. **81** (1965), pp. 38-55.

EFFROS, E.G., and E. STÖRMER

1 Jordan algebras of self-adjoint operators, Trans. Amer. Math. Soc. **127** (1967), pp. 313-316.

ELLIOT, G.A.

1 On the classification of inductive limits of sequences of semisimple finite-dimensional algebras, J. Algebra **38** (1976), pp. 29-44.

EMBRY, M.R.

1 Self-adjoint strictly cyclic operator algebras, Pacif. J. Math. **52** (1974), pp. 53-57.

EMCH, G.G.

1 Algebraic methods in statistical mechanics and quantum field theory, Wiley-Interscience, 1972.

ENOCK, M., and J.-M. SCHWARTZ

1 Le dualite dans les algebras de von Neumann, Bull. Soc. Math. France, Mem. **44**, 1975, pp. 1-144.

ERNEST, J.

1 A new group algebra for locally compact groups I. II., Amer. J. Math. **86** (1964), pp. 467-492; Canad. J. Math. **17** (1965), pp. 604-605.

2 The enveloping algebra of a covariant system, Communs. Math. Phys. **17** (1970), pp. 149-153.

3 On the topology of the spectrum of a C^*-algebra, Math. Anall. **216** (1975), pp. 149-153.

4 Charting the operator terrain, Mem. Amer. Math. Soc. **127** (1967), pp. 313-316.

FANNES, M., and A. VERBEURE

1 Gauge transformations and normal states of the CCR algebra, J. Math. Phys. **18** (1975), pp. 2086-2088 (Erratum) ibid. **17** (1976), pp. 284.

FARIS, W.G.

1 Self-adjoint operators, Lecture Notes in Math. **433** (1975), Springer-Verlag, 1975.

FEIGIN, B.L., and D.B. FUKS

1 Skew-symmetric invariant differential operators on the line and Verma modules over Virasoro algebra, Funktsional. Anal. i Prilozhen. **16** (1982), pp. 47-63 (Russian).

2 Verma modulus over Virasoro algebra, Funktsional. Anal. i Prilozhen. **17** (1983), pp. 91-92 (Russian).

FEYNMAN, R.P.

1 An operator calculus having applications in quantum electrodynamics, Phys. Rev. **86** (1951), pp. 108-128.

FEYNMAN, R.P., and A.R. HIBBS

1 Quantum mechanics and path integrals, McGraw-Hill, N.Y., 1965.

FIL', I.A.

1 On the spectrum of infinite dimensional differential operators with constant coefficients, Mathematical models of Statistical Physics, Tumen', 1982 (Russian).

2 On a set of cocycles with $\mathbb{R}_0^\infty$-quasi-invariant measure, Application of methods of functional analysis to problems of mathematical physics, Inst. Mat. Akad. Nauk Ukrain, SSR, Kiev, 1987, pp. 77-85 (Russian).

FLATO, M., and J. SIMON

1 Separate and joint analyticity in Lie groups representations, J. Funct. Anal. **12** (1973), pp. 268-276.

FLATO, M., J. SIMON, H. SNELLMAN and D. STERNHEIMER

1 Simple facts about analytic vectors and integrability, Ann. Sci. Ec. Norm. Sup. 4 serie **5** (1972), pp. 424-234.

FLATO, M., J. SIMON and D. STERNHEIMER

1 Sur l'integrabilité des representations antisymetriques des algèbres de Lie compacts, C.R. Acad. Sci. Paris, ser. A. **277** (1973), pp. 939-942.

FLATO, M., and D. STERNHEIMER

1 On the infinite-dimensional group, Communs. Math. Phys. **14** (1969), pp. 5-12.

FORELLI, F.

1 Analytic and quasi-invariant measures, Acta Math. **118** (1967), pp. 33-59.

FRENKEL, I.B.

1 Two constructions of affine Lie algebra representations and Boson-Fermion correspondence in quantum field theory, J. Funct. Anal. **44** (1981), pp. 259-327.

FRENKEL, I.B., and V.G. KAC

1 Basic representation of affine Lie algebras and dual resonance models, Invent. Math. **62** (1980), pp. 23-66.

FRÖHLICH, J.

1 Application of commutator theorems to the integration of representations of Lie algebras and commutation relations, Communs. Math. Phys. **54** (1977), pp. 135-150.

FRIEDRICHS, K.O.

1 Spectral theory of operators in Hilbert space, Springer-Verlag, 1980.

FUTORNYĬ, V.M.

1 On two constructions of representations of affine Lie algebras, Applications of methods of functional analysis to problems of mathematical physics, Inst. Mat. Akad. Nauk Ukrain. SSR, Kiev, 1987 (Russian).

GALLAGHER, P.X., and R.J. PROULX

1 Orthogonal and unitary invariants of families of subspaces, A collection of Papers Dedicated to Ellis Kolchin, Academic Press, 1977, pp. 157-164.

GÅRDING, L.

1 Notes on continuous representations of Lie groups, Proc. Nat. Acad. Sci. USA **33** (1949), pp. 331-332.

2 Vecteurs analytiques dans les representations des groupes de Lie, Bull. Soc. Math. France **88** (1960), pp. 73-93.

3 Decomposition into eigenfunctions, Bers, L., Johs, F., Schechter, M.: Partial differential equations, Moscow, pp. 309-332.

GÅRDING, L., and A. WIGHTMAN

1 Representation of the anticommutation relations, Proc. Nat. Acad. Sci. USA **40** (1954), pp. 617-621.

2 Representation of the commutation relations, Proc. Nat. Acad. Sci. USA **40** (1954), pp. 622-626.

GEL'FAND, I.M.

1 On normed rings, Doklady Akad. Nauk SSSR **23** (1939), No. 5, pp. 430-432 (Russian).

2 Normierte Ringe, Mat. Sb. **9** (1941), No. 1, pp. 3-24.

3 Some problems of functional analysis, Uspekhi Mat. Nauk **11** (1956), No. 6, pp. 3-12 (Russian).

GEL'FAND, I.M., and M.I. GRAĬEV

1 Representations of the quaternion group over locally compact and functional fields, Funktsional Anal. i Prilozhen. **2** (1968), No. 1, pp. 20-35 (Russian).

2 On certain families of irreducible unitary representations of the group $U(\infty)$, Preprint 85.51 Izdat. Inst. Prikl. Mat. Akad. Nauk SSSR, Moscow, 1985 (Russian).

GEL'FAND, I.M., and NAĬMARK, M.A.

1 On the imbedding of normed rings into the rings of operators in Hilbert space, Mat. Sb. **12** (1943), No. 2, pp. 197-217.

2 Normed rings with involution and their representations, Izv. AN SSSR, Ser. Mat. **12** (1948), No. 5, pp. 445-480.

GEL'FAND, I.M., and V.A. PONOMAREV

1 Remarks on classification of a pair commuting linear transformation in a finite-dimensional space, Funktsional. Anal. i Prilozhen. **3** (1969), No. 4, pp. 81-82 (Russian).

GEL'FAND, I.M., D.A. RAIKOV and G.E. SHILOV

1 Commutative normed rings, "Fizmatgiz", Moscow, 1960 (Russian).

GEL'FAND, I.M., and G.E. SHILOV

1 Spaces of trial and generalized functions. Generalized functions, vol. 2., "Fizmatgiz", Moscow, 1958 (Russian).

GEL'FAND, I.M., and N.Ya. VILENKIN

1 Some applications of harmonic analysis, Rigged Hilbert spaces. Generalized functions. "Fizmatgiz", Moscow, 1961 (Russian).

GIKHMAN, I.I., and A.V. SKOROKHOD

1 Introduction to the theory of random processes, "Nauka", Moscow, 1969.

2 The theory of stochastic processes, I, "Nauka", Moscow, 1971.

GIKHMAN, I.I., A.V. SKOROKHOD and M.I. YADRENKO

1 Probability theory and Mathematical Statistics, "Vishcha shkola", Kiev, 1979.

GLEASON, A.M.

1 Measures on the closed subspaces of a Hilbert space, J. Rat. Mech. Anal. **6** (1957), pp. 885-894.

GLIMM, J.G.

1 On a certain class of operator algebras, Trans. Amer. Math. Soc. **95** (1960), pp. 318-340.

2 Type I C^*-algebras, Ann. Math. **73** (1961), pp. 572-612.

GOLDIN, A., R. MENICOFF and D.H. SHARP

1 Current algebras and their representations , J. Math. Phys. **21** (1980), pp. 650-664.

GOL'DSHTEIN, M.G.

1 On convergence of conditional mathematical expectations in von Neumann algebras, Funksional. Anal. i Prilozhen. **14** (1980), No. 3, pp. 75-76 (Russian).

GOLODETS, V.Ya.

1 On a question of irreducible representations of commutation and anticommutation relations, Uspekhi Mat. Nauk **20** (1965), No. 2, pp. 175-182 (Russian).

2 Quasi-invariant ergodic measures, Mat. Sb. **72** (1967), No. 4, pp. 558-572 (Russian).

3 Description of representations of anticommutations relations, Uspekhi Mat. Nauk **24** (1969), No. 4, pp. 3-64 (Russian).

4 Factor representations of anticommutation relations, Trudy Moskov. Mat. Obshchestva **22** (1970), pp. 3-62 (Russian).

GOLODETS, V.Ya., and G.H. ZHOLTKEVICH

1 Markov KMS - states, Teor. Mat. Fizika, **56** (1983), No. 1, pp. 29-36 (Russian).

GOODMAN, R.

1 Analytic and entire vectors for representation of Lie groups, Trans. Amer. Math. Soc. **143** (1969), pp. 55-76.

2 One-parameter groups generated by operators in an enveloping algebra, J. Funct. Anal. **6** (1970), pp. 218-236.

3 Complex Fourier analysis on a nilpotent Lie group, Trans. Amer. Math. Soc. **160** (1971), pp. 273-391.

GOODMAN, R., and N. WALLACH

1 Projective unitary positive-energy representations of Diff (S^1), J. Funct. Anal. **63** (1985), pp. 299-321.

GORBACHUK, V.I., Yu.S. SAMOĬLENKO and G.F. US

1 Spectral theory of self-adjoint operators and infinite-dimensional analysis, Uspekhi Mat. Nauk **31** (1976), No. 1, pp. 203-215 (Russian).

GORBACZEWSKI, P.

1 Representations of the CAR generated by the representations of the CCR. II., Bull. Acad. Sci. Pol. **24**, 1976, pp. 201-206.

GORBACZEWSKI, P., and Z. POPOVICH

1 Representations of the CAR generated by representations of the CCR. III., Repts. Math. Phys. II (1977), pp. 73-80.

GORIN, E.A.

1 Commutative Banach algebras generated by a group of unitary elements, Funktsional. Anal. i Prilozhen. **1** (1967), No. 3, pp. 86-87 (Russian).

GORODNIĬ, M.F., and G.B. PODKOLZIN

1 Irreducible representations of a graded Lie algebra, Spectral theory of operators and infinite-dimensional analysis, Instit. Mat. Acad. Nauk Ukrain. SSR, Kiev, 1984, pp. 66-77 (Russian).

GRABOVSKAYA, R.Ya., and S.G. KREĬN

1 On a formula for permutation of operator functions representing a Lie algebra, Funktsional. Anal. i Prilozhen. **7** (1973), No. 3, pp. 81 (Russian).

GROSS, L.

1 Measurable functions on a Hilbert space, Trans. Amer. Math. Soc. **105** (1962), pp. 372-390.

2 Classical analysis on a Hilbert space, Proc. conf. on the theory and appl. of Analysis in funct. space, Massachusetts, June 9-13, 1963, pp. 51-58.

3 Analytic vectors for representations of the canonical commutation relations and nondegeneracy of group states, J. Funct. Anal. **17** (1974), pp. 104-111.

GROSSMAN, A.

1 Elementary properties of Nested Hilbert spaces, Communs. Math. Phys. **2** (1975), pp. 1-30.

GUDDER, S.P., and R.L. HUDSON

1 A non-commutative probability theory, Trans. Amer. Math. Soc. **245** (1978), pp. 1-41.

GUDDER, S.P., and J.P. MARCHAND

1 Non-commutative probability on von Neumann algebras, J. Math. Phys. **13** (1972), pp. 799-806.

GUDDER, S.P., and W. SCRUGG

1 Unbounded representations of $*$-algebras, Pacif. J. Math. **70** (1977), pp. 369-382.

GUICHARDET, A.

1 Produits tensoriels infinis et representation de relation d'anticommutation, Ann. c. Norm. Sup. **83** (1966), pp. 1-52.

2 Tensor products of C^*-algebras, Lec. Notes Ser. Aarhus Univ. **12**, **13**, 1966.

3 Symmetric Hilbert spaces and related topics, Lect. Notes Math. **4**, 1972.

GUREVICH, D.I.

1 Generalized shift operators on Lie groups, Izvestia Acad. Nauk Armiansk. SSR Ser. Mat. **18** (1983), No. 4, pp. 305-317 (Russian).

GUSEĬNOV, R.V.

1 On the theory of Fourier-Wiener transformation, Vestnik Moskov. Univ. Ser. I. Mat. Mech. 1969, No. 4, pp. 17-25 (Russian).

2 Some properties of the Fourier-Wiener transformation, Vestnik Moskov. Univ. Ser. I. Mat. Mech. 1970, No. 4, pp. 39-49 (Russian).

HAAG, R., R.V. KADISON and D. KASTLER

1 Nets of C^*-algebras and classification of states, Communs. Math. Phys. **16** (1970), pp. 81-104.

HAAG, R., and D. KASTLER

1 An algebraic approach to quantum field theory, J. Math. Phys. **5** (1964), pp. 848-861.

HAGA, V.

1 On approximately finite algebras, Tohoku Math. J. **26** (1974), pp. 325-332.

HALMOS, P.R.

1 Measure theory, New York, 1950.

2 A Hilbert space problem book, Van Nostrand, 1967.

HALMOS, P.R., and J. McLAUGHLIN

1 Partial isometries, Pacif. J. Math. **13** (1962), No. 2, pp. 585-596.

HEGERFELDT, G.C.

1 On canonical commutation relations and infinite dimensional measures, J. Math. Phys. **13** (1972), pp. 45-50.

2 Gårding domain and analytic vectors for quantum fields, J. Math. Phys. **13** (1972), pp. 821-827.

3 Extremal decomposition of Wightman functions and of states on nuclear $*$-algebras by Choquet theory, Communs. Math. Phys. **45** (1975), pp. 133-135.

4 Prime field decompositions and infinitely divisible states on Borchers tensor algebra, Communs. Math. Phys. **45** (1975), pp. 137-151.

5 Noncommutative Analogs of Probabilistic Notions and Results, J. Funct. Anal. **64** (1985), pp. 436-456.

HEGERFELDT, G.C., and O. MELSHEIMER

1 The form of representations of CCR for Bose fields and connection with finitely many degrees of freedom, Communs. Math. Phys. **12** (1969), pp. 304-323.

HELSTROM, C.W.

1 Quantum detection and estimation theory, Academic Press, 1976.

HERMANN, R.

1 Infinite dimensional Lie algebras and current algebra, Lec. Notes Phys. **6** (1970), pp. 312-338.

HEWITT, E., and K.A. ROSS

1 Abstract Harmonic Analysis, Springer-Verlag, 1963.

HEWITT, E., and L.J. SAVAGE

1 Symmetric measures on Cartesian products, Trans. Amer. Math. Soc. **80** (1955), pp. 470-501.

HEYER, H.

1 Probability theory on hypergroups: A Survey, Confer. Proceedings of Prob. Measures on Groups, Springer-Verlag, 1984, pp. 481-550.

HIDA, T.

1 Note on the infinite dimensional Laplacian operator, Nagoya Math. J. **38** (1970), pp. 13-19.

2 Brownian Motion, Springer-Verlag, 1980.

HOLEVO, A.S.

1 On quantum characteristic functions, Problemy Peredachi Inform. **6** (1970), No. 4, pp. 42-48 (Russian).

2 Generalized free states of C^*-algebra of commutation relations. I, II, Teor. Mat. Phis. **6** (1971), pp. 3-20; **6** (1971), pp. 145-150 (Russian).

3 On quasi-equivalence of locally normal states, Teor. Mat. Phis. **13** (1972), pp. 184-199 (Russian).

4 On quasi-equivalence of generalized free states of C^*-algebra of anticommuting relations, Teor. Mat. Phis. **14** (1973), pp. 145-1561.

5 A research on general theory of statistical solutions, "Nauka", Moscow, 1976 (Russian).

6 Probabilistic and statistical aspects of quantum theory, "Nauka", Moscow, 1980 (Russian).

HORUŽIĬ, S.S.

1 Introduction into algebraic quantum field theory, "Nauka", Moscow, 1986 (Russian).

HOVANOVA, T.G.

1 Representation models and Clifford algebras, Funktsion. Anal. i Prilozhen. **16** (1982), pp. 90-91 (Russian).

HUDSON, R.L., and G.R. MODDY

1 Locally normal symmetric states and an analogue of de Finetti's theorem, Z. Warh. Verw. Gebiete, **33** (1975/76), pp. 343-351.

IBRAGIMOV, N.H.

1 Transformation groups in mathematical physics, "Nauka", Moscow, 1983 (Russian).

INOUE, A.

1 A class of unbounded operator algebras, I, II, III, Pacif. J. Math. **65** (1976), pp. 77-96, **66** (1976), pp. 411-431, **69** (1977), pp. 105-115.

IONESCU TULCEA, C.

1 Spectral representations of certain semigroups of operators, J. Math. Mech. **8** (1959), pp. 95-110.

ISMAGILOV, R.S.

1 On unitary representations of the group of diffeomorphisms of a circle, Funktsional. Anal. i Prilozhen. **5** (1971), No. 3, pp. 45-53 (Russian).

2 On unitary representations of the group of diffeomorphisms of a compact manifold, Funktsional. Anal. i Prilozhen. **6** (1972), No. 1, pp. 79-80 (Russian).

3 On unitary representations of the group of diffeomorphisms of a compact manifold, Izvestia Akad. Nauk SSSR Ser. Mat. **36** (1972), No. 1, pp. 180-208 (Russian).

4 On unitary representations of the group of diffeomorphisms of $\mathbb{R}^n$, $n \geq 2$ Funktsional. Anal. i Prilozhen. **9** (1975), No. 2, pp. 71-72 (Russian).

5 On unitary representations of the group $C_0^\infty(X,G)$, $G = SU_2$, Mat. Sb. **100** (1976), No. 1, pp. 117-131 (Russian).

6 Unitary representations of the group of measure preserving diffiomorphisms, Funktsional. Anal. i Prilozhen. **11** (1977), No. 3, pp. 80-81 (Russian).

7 On representation of the group of smooth mappings of an interval into a compact Lie group, Funktsion. Anal. i Prilozhen. **15** (1981), No. 2, pp. 73-74 (Russian).

8 Infinite-dimensional groups and their representations, Warsaw Intern. Symp., Warsaw, 1983, pp. 861-865.

JADCZYK, A.Z.

1 Note on canonical commutation relations, Bull. Acad. Pol. Sci. **22** (1974), pp. 963-965.

JIMBO, M.

1 A q-difference analogue of $U(g)$ and the Yang-Baxter equation, Lett. Math. Phys. **10** (1985), pp. 63-69.

JIMBO, M., and T. MIWA

1 Solitons and infinite-dimensional Lie algebras, Publ. Res. Inst. Math. Sci., Kyoto Univ. **19** (1983), pp. 943-1001.

JORDAN, P., J. VON NEUMANN and E. WIGNER

1 On an algebraic generalization of quantum mechanics formalism, Ann. Math. **35** (1934), pp. 29-64.

JORDAN, P., and E. WIGNER

1 Über das Paulische Äquivalenzverbot, Zs. f. Phys., **47**, 1928.

JURZAK, J.-P.

1 Simple facts about algebras of unbounded operators, J. Funct. Anal. **21** (1976), pp. 496-482.

KAC, V.G.

1 Infinite dimensional Lie algebras, Progress in Math. **44**, Birkhauser, 1983.

KAC, V.G., and D.A. KAZHDAN

1 Structure of representations with highest weight of infinite dimensional Lie algebra, Adv. in Math. **34** (1979), pp. 97-108.

KAC, V.G, D.A. KAZHDAN, J. LEPOWSKY and R.L. WILSON

1 Realization of the basic representations of the Euclidean Lie algebras, Adv. in Math. **42** (1981), pp. 83-112.

KAC, V.G., and I.T. TODOROV

1 Superconformal current algebras and their unitary representations, Communs. Math. Phys. **102** (1985), pp. 337-347.

KACZMARZ, S., and H. STEINHAUS

1 Theorie der Orthogonalreihen, Warszawa, 1935.

KADISON, R.V.

1 A representation theory for commutative topological algebras, Memoirs, Amer. Math. Soc., 1951.

2 Unitary invariants for representations of operator algebras, Ann. Math. **68** (1957), pp. 304-379.

3 Irreducible operator algebras, Proc. Nat. Acad. Sci. **47** (1957), pp. 273-276.

4 Theory of operators II. Operator algebras, Bull. Amer. Math. Soc. **64** (1958), pp. 61-85.

KAMEI, E.

1 Operators with skew commutative cartesian parts, Math. Jap. **25** (1980), pp. 431-432.

KAPLAN, S.

1 Extension of Pontryagin duality, I, II. Duke Math. J. **15** (1948), pp. 649-658, **17** (1950), pp. 419-435.

KAPLANSKY, I.

1 Algebras of type I, Ann. Math. **26** (1952), pp. 460-472.

KARASEV, M.V.

1 Some remarks on functions of ordered operators, Mat. Zametki **18** (1975), No. 2, pp. 267-277 (Russian).

2 On Weyl and ordered calculus for non-commuting operators, Mat. Zametki **26** (1979), No. 6, pp, 885-907 (Russian).

3 Operators of a regular representation for a certain class of non-Lie commutation relations, Funktsion. Anal. i Prilozhen. **13** (1979), No. 3, pp. 89-90 (Russian).

4 Analogs of the objects of Lie groups theory for non-linear Poisson brackets, Izvestia Akad. Nauk SSSR Ser. Mat. **50** (1986), No. 3, pp. 508-583 (Russian).

KARASEV, M.V., and V.P. MASLOV

1 Algebras with general commutation relations and their applications. II. Operator unitary - nonlinear equations, Itogy Nauk. i Techn., Sovrem. Probl. Mat. **13** (1979), pp. 145-267 (Russian).

KATS, G.I.

1 Generalized functions on locally compact group and decomposition of unitary representations, Trudy Moskov. Mat. Obshch. **10** (1960), pp. 3-40 (Russian).

2 Generalized functions on locally compact group and decomposition of unitary representations, Uspekhi Mat. Nauk **16** (1961), No. 1, pp. 190 (Russian).

3 Representations of compact ring groups, Dokl. Akad. Nauk SSSR **145** (1962), No. 5, pp. 989-992 (Russian).

4 Ring groups and a principle of duality. I, II, Trudy Moskov. Mat. Obshch. **12** (1963), pp. 289-301, **14** (1965), pp. 84-113 (Russian).

KATS, G.I., and V.G. PALYUTKIN

1 Finite ring groups, Trudy Moskov. Mat. Obshch. **15** (1966), pp. 224-261 (Russian).

KIRSCHBERG, E.

1 Darstellungen coinvolutiver Hopf W^*-Algebren und ihre Anwendung in der nicht-abelschen Dualitatstheorie lokalkompakter Gruppen, Ak. Wiss., Berlin, DDR, 1977.

KIRILLOV, A.A.

1 Unitary representations of nilpotent Lie groups, Uspekhi Mat. Nauk **17** (1962), No. 4, pp. 57-110 (Russian).

2 Dynamic systems, factors and group representations, Uspekhi Mat. Nauk **221** (1967), No. 4, pp. 67-80 (Russian).

3 Elements of representation theory, "Nauka", Moscow, 1972 (Russian).

4 Representations of an infinite-dimensional unitary group, Dokl. Akad. Nauk SSSR **212** (1973), No. 2, pp. 228-290 (Russian).

5 Unitary representations of a group of diffeomorphisms and of some of its subgroups, Preprint No. 82, Inst. Prikl. Mat. Akad. Nauk SSSR, Moscow, 1974 (Russian).

6 Representations of certain infinite-dimensional Lie groups, Vestnik Mosk. Univ. **1** (1974), pp. 75-83 (Russian).

7 Local Lie algebras, Uspekhi Mat Nauk **31** (1976), No. 4, pp. 57-76 (Russian).

8 Introduction into representation theory and non-commutative harmonic analysis, Itogi Nauki i Tekhn. **22** (1988), pp. 5-162 (Russian).

KLAUDER, J.R, J. McKENNA and E.J. WOODS

1 Direct product representations of the CCR, J. Math. Phys. **7** (1966), pp. 822-828.

KOLMOGOROV, A.N., and S.V. FOMIN

1 Elements of theory of functions and functional analysis, "Nauka", Moscow, 1976 (Russian).

KOLMOGOROV, A.N.

1 Fundamentals of Probability Theory, ONTI, Moscow, 1936 (Russian).

KOLOMYTSEV, V.I., and Yu.S. SAMOĬLENKO

1 On irreducible representations of inductive limits of groups, Ukrain. Mat. Zh. **29** (1977), No. 4, pp. 526-531.

2 On a countable set of commuting self-adjoint operators and the canonical commutation relations, Methods of functional analysis in problems of mathematical physics, Inst. Mat. Akad. Nauk Ukrain, SSR, Kiev, 1978, pp. 115-128 (Russian).

3 On a countable set of commuting self-adjoint operators and the algebra of local observables, Ukrain. Mat. Zh. **31** (1979), pp. 365-371 (Russian).

KONDRAT'EV, Yu.G., and Yu.S. SAMOǏLENKO

1 Integral representation of generalized positive-definite kernels in infinitely many variables, Dokl. Akad. Nauk SSSR **227** (1976), pp. 800-803 (Russian).

2 The spaces of trial and generalized function of infinite number of variables, Repts. Math. Phys. **14** (1978), pp. 325-350.

3 Generalized derivatives of probability measures on $\mathbb{R}^\infty$, Methods of Functional analysis in problems of mathematical physics, Inst. Mat. Akad. Nauk Ukrain. SSR, Kiev, 1978, pp. 159-176 (Russian).

KOSHMANENKO, V.D., and Yu.S. SAMOǏLENKO

1 On an isomorphism between Fock space and a space of functions of infinitely many variables, Ukrain. Mat. Zh. **27** (1975), pp. 669-674 (Russian).

KOSTYUCHENKO, A.G., and B.S. MITYAGIN

1 Positive definite functionals on nuclear spaces, Trudy Mosk. Mat. Obshch. **9** (1960), pp. 283-316 (Russian).

KOSYAK. A.V.

1 Analytic and entire vectors for families of operators, Spectral analysis of Differential operators, Inst. Mat. Akad. Nauk Ukrain. SSR, Kiev, 1980, pp. 3-18 (Russian).

2 Gårding domain for representations of canonical commutation relations, Ukrain. Mat. Zh. **36** (1984), No. 6, pp. 709-715 (Russian).

3 Extension of unitary representations of the group of finite upper triangular matrices of infinite order, Spectral operator theory and infinite-dimensional analysis, Inst. Mat. Akad. Nauk Ukrain. SSR, Kiev, 1984, pp. 102-111 (Russian).

4 Extension of unitary representations of inductive limits of finite dimensional Lie groups, Repts. Math. Phys. **26** (1987), pp. 129-146.

KOSYAK, A.V., and Yu.S. SAMOǏLENKO

1 Families of commuting self-adjoint operators with a joint simple spectrum, School on Operator theory in functional spaces, Minsk, 1978, pp. 70-71.

2 On families of commuting self-adjoint operators, Ukrain, Mat. Zh. **31** (1979), pp. 555-558 (Russian).

3 Gårding domain and entire vectors for inductive limits of commutative locally compact groups, Ukrain. Mat. Zh. **35** (1983), pp. 427-434 (Russian).

4 Quasi-invariant measures on "large" groups, Spectral theory of differential and integral equations, Inst. Mat. Akad. Nauk Ukrain. SSR, Kiev, 1986, pp. 64-49 (Russian).

KREĬN, M.G.

1 On a certain ring of functions defined on a topological group, Dokl. Akad. Nauk SSSR, **29** (1940), No. 4, pp. 275-280 (Russian).

2 On a certain special ring of functions, Dokl. Akad. Nauk SSSR, **29** (1940), No. 5-6, pp. 355-359 (Russian).

KREĬN, S.G.

1 Linear differential equations in Banach space, "Nauka", Moscow, 1967 (Russian).

2 (editor) Functional analysis, 2nd rev. ed., "Nauka", Moscow, 1972 (Russian).

KREĬN, S.G, and A.M. SHIKHVATOV

1 Linear differential equations on a Lie group, Funkts. Anal. i Prilozhen. **4** (1970), No. 1, pp. 52-61 (Russian).

KRUGLYAK, S.A.

1 Representations of free involutive quivers, Representations and quadratic forms, Kiev, 1979, pp. 149-151 (Russian).

2 Representations of involutive quivers, USSR algebraic conf., Leningrad, 1981, pp. 86-87 (Russian).

3 Representations of a quantum algebra related to the Yang-Baxter's equation, Spectral operator theory and infinite-dimensional analysis, Inst. Mat. Akad. Nauk Ukrain. SSR. Kiev, 1984, pp. 111-120 (Russian).

KRUGLYAK, S.A., and Yu.S. SAMOĬLENKO

1 On unitary equivalence of collections of self-adjoint operators, Funktsional. Anal. i Prilozhen. **14** (1980), No. 1, pp. 60-62 (Russian).

KRUSZYNSKI, P.

1 Probability measures on operator algebras, Repts. Math. Phys. **7** (1975), pp. 395-401.

LANCE, E.C.

1 Some properties of nest algebras, Proc. London. Math. Soc. **9** (1969), pp. 45-68.

2 On nuclear C^*-algebras, J. Funct. Anal. **12**, 1973, pp. 157-176.

LASSNER, G.

1 Topological algebra of operators, Repts. Math. Phys. **3** (1972), pp. 279-293.

LASSNER, G., and W. TIMMERMAN

1 Normal states on algebras of unbounded operators, Repts. Math. Phys. **3** (1972), pp. 295-305.

2 On essential self-adjointness of different algebras of field operators, Teor. Mat. Phisika **15** (19739, No. 3, pp. 311-314 (Russian).

3 Classification of domains of operator algebras, Repts. Math. Phys. **9** (1976), pp. 205-217.

LEBEDEV, D.P., and Yu.I. MANIN

1 Gel'fand-Dikiĭ Hamilton operator and the coadjoint representation of the Volterra group, Funktsional. Anal. i Prilozhen. **13** (1979), No. 4, pp. 40-46 (Russian).

LEĬTES, D.A.

1 Theory of supermanifolds, Karelian Dep. of Akad. Nauk of USSR, Petrozavodsk, 1983 (Russian).

LEPOWSKY, J., and R.L. WILSON

1 Construction of the affine Lie algebra $A_1^{(1)}$, Communs. Math. Phys. **62** (1978), pp. 43-53.

2 The structure of standard modules. I: Universal algebras and the Rogers-Ramanujan identities, Invent. Math. **77** (1984), pp. 199-290.

LEVITAN, B.M.

1 Normed rings generated by generalized shift, Dokl. Akad. Nauk SSSR **47** (1945), pp. 3-6 (Russian).

2 A generalization of the operation of translation and infinite hypercomplex system. I, II, III, Mat. Sb. **16** (1945), pp. 259-280, **171** (1945), No. 1, pp. 9-44, No. 2, pp. 163-192.

3 Operator rings and generalized shift operations, Dokl. Akad. Nauk SSSR **52** (1946), No. 2, pp. 99-102 (Russian).

4 Application of GSO to linear equations of second order, Uspekhi Mat. Nauk **4** (1949), No. 1, pp. 3-111 (Russian).

5 Theory of generalized shift operator, "Nauka", Moscow, 1973 (Russian).

LEZNOV, A.N., and M.V. SAVEL'EV

1 Group methods of integration of non-linear dynamic systems. "Nauka", Moscow, 1985 (Russian).

LINDBLAD, G.

1 Gaussian quantum stochastic processes on the CCR algebra, J. Math. Phys. **20** (1979), pp. 2081-2087.

LODKIN, A.A.

1 Approximation of dynamical systems and spectral theory in a factor of type II, Operators of Mathematical Physics and infinite-dimensional Analysis, Kiev, 1979, pp. 73-102 (Russian).

LOOMIS, L.H.

1 An introduction to abstract Harmonic Analysis, D. Van Nostrand Company, 1953.

LUBICH, Yu.I.

1 On spectrum of a representation of abelian groups, Dokl. Akad. Nauk SSSR **200** (1971), No. 4, pp. 777-779 (Russian).

MACKEY, G.W.

1 The mathematical foundation of quantum mechanics, W.A. Benjamin Inc., New York, 1963.

2 Ergodic theory and virtual groups, Math. Ann. **166** (1966), pp. 187-207.

MALTESE, G.

1 Spectral representations for solutions of certain abstract functional equations, Compos. Math. **15** (1962), pp. 1-22.

2 Spectral representations for some unbounded normal operators, Trans. Amer. Math. Soc. **110** (1964), pp. 79-97.

MANUCEAU, J., F. ROCCA and D. TESTARD

1 On the product form of quasi-free states, Communs. Math. Phys. **12** (1969), pp. 43-57.

MANUCEAU, J., M. SIRUGUE, D. TESTARD and A. VERBEURE

1 The smallest C^*-algebra for canonical commutation relations, Communs. Math. Phys. **32** (1973), pp. 231-244.

MANUCEAU, J., and A. VERBEURE

1 Quasi-free states of the CCR algebra and Bogolyubov transformations, Communs. Math. Phys. **9** (1968), pp. 293-302.

2 Non factor quasi-free states of the CAR algebra, Communs. Math. Phys. **18** (1970), pp. 319-329.

MARCHENKO, A.V.

1 On inductive limits of linear spaces and operators and their applications, Vestnik Moskov, Univ. Ser. I Mat. Mekh. 1974, No. 2, pp. 26-33 (Russian).

2 Self-adjoint differential operators with infinitely many independent variables, Mat. Sb. **96** (138) (1975), pp. 276-293 (Russian).

MASLOV, V.P.

1 Operator methods, "Nauka", Moscow, 1973 (Russian).

2 Complex Markov chains and Feynman's continual integral, "Nauka", Moscow, 1976 (Russian).

3 Application of the method of ordered operators for getting exact solutions, Teoret. Mat. Phys. **33** (1977), No. 2, pp. 185-209 (Russian).

MASLOV, V.P., and A.M. CHEBOTAREV

1 Generalized measures and Feynman's continual integral, Teoret. Math. Phys. **28** (1976), pp. 291-307 (Russian).

MASLOV, V.P., and V.E. NAZAĬKINSKIĬ

1 Algebras with general commutation relations, and their applications, I. Pseudo-differential operators with growing coefficients, Modern problems in Mathematics, **13**, VINITI, Moscow, 1979, pp. 5-144 (Russian).

MATHON, D., and R.F. STREATER

1 Infinitely divisible representations of Clifford algebras, Z. Wahrscheinlichkeitstheorie verw. Geb. **20** (1971), pp. 308-816.

MAURIN, K.

1 General eigenfucntion expansions and unitary representations of topological groups, PWN, Warszawa, 1968.

MENICOFF, R., and D.H. SHARP

1 Representations of a local current algebras: their dynamical determination, J. Math. Phys. **16** (1975), pp. 2341-2360.

MINLOS, R.A.

1 Generalized stochastic processes and their extension to a measure, Trudy Moskov. Mt. Obschch. **8** (1959), pp. 497-518 (Russian).

MITROPOL'SKIĬ, Yu.A, N.N. BOGOLYUBOV (Jr), A.K. PRIKARPATSKIĬ and V.G. SAMOĬLENKO

1 Integrable dynamical systems: spectral and differential geometry aspects, "Naukova dumka", Kiev, 1986 (Russian).

MOODY, R.V.

1 A new class of Lie algebras, J. Algebra **10** (1968), pp. 211-230.

MOROZOVA, E.A., and N.I. CHENTSOV

1 Unitary equivariants of a family of subspaces, Preprint 52, Inst. Priklad. Mat. Akad. Nauk SSSR, Moscow, 1975 (Russian).

2 Structure of a family of stationary states of a quantum Markov chain, Preprint 130, Inst. Priklad. Mat. Akad. Nauk SSSR, Moscow, 1976 (Russian).

3 Elements of stochactic quantum logic, Preprint Inst. Mat. Akad. Nauk SSSR, Novosibirsk, 1977 (Russian).

MORRIS, S.A.

1 Pontryagin duality and the structure of locally compact groups, Cambridge Univ. Press, 1977.

MOSOLOVA, M.V.

1 On functions of anti-commuting operators, generating a graded Lie algebra, Mat. Zametki **29** (1981), pp. 35-44 (Russian).

MURRAY. E.J., and J. VON NEUMANN

1 On rings of operators, I, II, IV, Annal. Math. **37** (1936), pp. 116-229, Trans. Amer. Math. Soc. **41** (1937), pp. 208-248, Annal. Math. **44** (1943), pp. 716-808.

NACHBIN, L.

1 Topology on spaces of holomorphic mappings, Springer-Verlag, Berlin, 1979.

NAǏMARK, M.A.

1 Rings with involution, Uspekhi Mat. Nauk **3** (1948), No. 5, pp. 52-146 (Russian).

2 Rings of operators on a Hilbert space, Uspekhi Mat. Nauk **4** (1948), No. 4, pp. 83-147 (Russian).

3 On a certain problem of theory of rings with involution, Uspekhi Mat. Nauk **6** (1951), No. 6, pp. 160-164 (Russian).

4 Normed rings, 2nd ed., "Nauka", Moscow, 1968 (Russian).

NAǏMARK, M.A., and S.V. FOMIN

1 Continuous direct sums of Hilbert spaces and some applications of them, Uspekhi Mat. Nauk **10** (1955), No. 2, pp. 11-142 (Russian).

NAKAGAMI, Y.

1 Infinite tensor product of von Neumann algebra, Kodai Math. Semin. Repts. **22** (1970), pp. 341-354.

2 Infinite tensor products of operators, Publ. Res. Inst. Math. Sci. Kyoto Univ. **10** (1974), pp. 111-145.

NAKAMURA, M., and T. TURUMARU

1 On the representation of positive definite functions and stationary functions on topological groups, Tohoku Math. J. **4** (1952), pp. 1-9.

NAKAMURA, M., and H. UMEGAKI

1 A remark on theorems of Stone and Bochner, Proc. Jap. Acad. **27** (1951), pp. 506-507.

2 Heisenberg commutation relations and the Plancherel theorem, Proc. Jap. Acad. **37** (1961), pp. 239-242.

NAKANO, H.

1 Reduction of Bochner's theorem to Stone's theorem, Ann. Math. **49** (1948), pp. 278-280.

NAPIORKOWSKI, K.

1 Continuous tensor products of Hilbert spaces and representations of CCR, Bull. Acad. Polon. Sci. Ser. Sci. Math. **18** (1970), pp. 267-271.

2 Continuous tensor products of Hilbert spaces and product operators, Studia Math. **39** (1971), pp. 307-327.

3 On a class of representations of CCR, Repts. Math. Phys. **3** (1972), pp. 235-243.

NAPIORKOWSKI, K., and W. PUSZ

1 Particle representations of CCR, Rept. Math. Phys. **3** (1972), pp. 221-225.

NELSON, E.

1 Analytic vectors, Ann. Math. **70** (1959), pp. 572-615.

2 Probability theory and Euclidean field theory, Constructive Quantum Field Theory, Springer-Verlag, 1973.

NERETIN, Yu.A.

1 Representations of Virasoro algebra and affine algebras, Itogi nauki i techniki. **22**, pp. 163-244 (Russian).

NESSONOV, N.I.

1 Description of representation of a group of invertible operators on a Hilbert space which contains a unity representation of a unitary subgroup, Funktsional. Anal. i Prilozhen. **17** (1983), No. 1, pp. 79-80.

2 Examples of factor-representations of the group $GL(\infty)$, Mathematical physics and functional analysis, "Naukova dumka", Kiev, 1986 (Russian).

3 A complete classification of representations of $GL(\infty)$ which contain a unity representation of a unitary subgroup, Mat. Sb. **130** (1986), pp. 131-150 (Russian).

VON NEUMANN, J.

1 On infinite direct products, Compos. Math. **6** (1939), pp. 1-77.

2 Mathematische grundlagen der quantenmechanik, Springer, 1981.

NIELSEN, O.A.

1 A note on the product measures and representations of the canonical commutation relations, Communs. Math. Phys. **22** (1971), pp. 23-36.

NIZHNIK, L.P., and M.D. POCHINAĬKO

1 Nonlinear space two-dimensional Schrödinger equation as a Hamilton system, Uspekhi Mat. Nauk **37** (1982), No. 4, pp. 111-112 (Russian).

2 Nonlinear space two-dimensional Schrödinger equation as an integrable Hamilton system, Preprint 85.24 Inst. Mat. Akad. Nauk Ukr. SSR, Kiev, 1985 (Russian).

ODESSKIĬ, A.V.

1 On a certain analog of Sklyanin algebra, Funktsional. Anal. i Prilozhen. **20** (1986), No. 2, pp. 78-79 (Russian).

OL'SHANSKIĬ, G.I.

1 Unitary representations of infinite-dimensional classical groups $U(p,\infty)$, $SO_0(p,\infty)$, $S_p(p,\infty)$, and of the corresponding groups of motion, Dokl. Akad. Nauk SSSR **238** (1978), No. 6, pp. 1295-1298 (Russian).

2 Unitary representations of infinite-dimensional classic groups $U(p,\infty)$, $SO_0(p,\infty)$, $S_p(p,\infty)$, and of the corresponding groups of motion, Funktsional. Anal. i Prilozhen. **12** (1978), No. 3, pp. 32-44 (Russian).

3 A construction of unitary representations of infinite-dimensional classic groups, Dokl. Akad. Nauk SSSR **250** (1980), No. 2, pp. 284-288 (Russian).

4 Description of unitary representations for the groups $U(p,q)$ with a highest weight, Funktsional. Anal. i Prilozhen. **14** (1980), No. 3, pp. 32-44 (Russian).

5 Unitary representations of infinite-dimensional classic groups and R. Howe's formalism, Dokl. Akad. Nauk SSSR **269** (1983), pp. 33-36 (Russian).

6 Infinite-dimensional classic groups of a finite R-rang: a description of representations and the asymptotic theory, Funktsional. Anal. i Prilozhen. **18** (1984), No. 1, pp. 28-42 (Russian).

7 Unitary representations of the group $SO_0(\infty,\infty)$ considered as a limit of unitary representations of the groups $SO_0(n,\infty)$ for $n \to \infty$, Funktsional. Anal. i Prilozhen. **20** (1986), No. 4, pp. 46-57 (Russian).

OMORI, H.

1 Infinite dimensional Lie transformation groups, Lecture Notes in Math. **427**, Springer-Verlag, 1974.

OSTROVSKIĬ, V.L.

1 An analog of Nelson theorem for nuclear nilpotent Lie algebras of currents, Spectral operator theory and infinite-dimensional analysis, Inst. Mat. Akad. Nauk Ukrain. SSR, Kiev, 1984, pp. 120-131 (Russian).

2 Irreducible representations of the group of infinite upper triangular matrices, Ukrain. Mat. Zh. **38** (1986), pp. 255-258 (Russian).

3 A construction of quasi-invariant measures on a certain class of groups which are not locally compact, Ukrain. Mat. Zh. **38** (1986), pp. 524-526 (Russian).

OSTROVSKIĬ, V.L., and Yu.S. SAMOĬLENKO

1 An application of a projection spectral theorem to non-commutative families of operators, Ukrain. Mat. Zh. **40** (1988), pp. 421-433 (Russian).

2 Families of unbounded self-adjoint operators, connected by non-Lie relations, Funktsional. Anal. i Prilozhen. **23** (1989), pp. 67-68 (Russian).

OVS'YANNIKOV, L.V.

1 Group analysis of differential equations, "Nauka", Moscow, 1978 (Russian).

PARRY, W., and K. SCHMIDT

1 A note on cocycles of unitary representations, Proc. Amer. Math. Soc. **55** (1976), pp. 185-190.

PARTHASARATHY, K.R.

1 Infinitely divisible representations and positive definite functions on a compact group, Communs Math. Phys. **16** (1970), pp. 148-156.

2 Probability theory on the closed subspaces of a Hilbert space, Matematika **14** (1970), pp. 102-112.

PARTHASARATHY, K.R., and K. SCHMIDT

1 Factorizible representations of current groups and the Araki-Woods imbedding theorem, Acta Math. **128** (1972), pp. 53-71.

2 A new method for constructing factorizible representations for current groups and current algebras, Commun. Math. Phys. **51** (1976), pp. 167-175.

PEARCY, C.

1 A complete set of unitary invariants for the operators generating finite W^*-algebras of type I, Pacif. J. Math. **12** (1962), pp. 1405-1416.

2 W^*-algebras with a single generator, Proc. Amer. Math. Soc. **13** (1962), pp. 831-832.

3 On certain von Neumann algebras which are generated by partial isometries, Proc. Math. Soc. **15** (1964), pp. 393-395.

PILLIS, J.

1 Noncommutative Markov processes, Trans. Amer. Math. Soc. **125** (1966), pp. 264-279.

PIROGOV, S.A.

1 The states related to the Ising model, Teor. Mat. Phisika **11** (1972), pp. 421-426 (Russian).

PLESNER, A.I.

1 Spectral theory of linear operators, "Nauka", Moscow, 1965 (Russian).

PLESNER, A.I., and V.A. ROKHLIN

1 Spectral theory of linear operators, Uspekhi Mat. Nauk **1** (1946), No. 1 (II), pp. 71-191 (Russian).

PONTRYAGIN, L.S.

1 Topological groups, GITTL, Moscow, 1954 (Russian).

POWERS, R.J.

1 Representations of uniformly hyperfinite algebras and their associated von Neumann rings, Ann. Math. **86** (1967), pp. 138-171.

2 Self-adjoint algebras of unbounded operators, I, II, Communs. Math. Phys. **21** (1971), pp. 85-124.

POWERS, R.J., and E. STÖRMER

1 Free states of the canonical anticommutation relations, Communs. Math. Phys. **16** (1970),

pp. 1-33.

PRESTON, C.J.

1 Gibbs states on countable sets, Cambridge University Press, 1974.

PROKHOROV, Yu.V.

1 Convergence of random processes and limit theorems in probability theory, Teor. Veroyatnost. i Primenen. **1** (1956), pp. 177-238 (Russian).

PUKANSZKY, L.

1 On the theory of exponential groups, Trans. Amer. Math. Soc. **126** (1967), pp. 487-507.

2 On the unitary representations of exponential groups, J. Funct. Anal. **2** (1968), pp. 37-113.

RAMSAY, A.

1 Virtual groups and group actions, Adv. Math. **6** (1971), pp. 253-322.

REED, M.C.

1 A Gårding Domain for quantum fields, Communs. Math. Phys. **14** (1969), pp. 336-346.

2 On self-adjointness in infinite tensor product spaces, J. Funct. Anal. **5** (1970), pp. 49-124.

3 Torus invariance for the Clifford algebras, I, II, Trans. Amer. Math. Soc. **154** (1971), pp. 177-183; J. Funct. Anal. **8** (1971), pp. 450-468.

4 Functional analysis and probability theory, Constructive Quantum Field theory **25**, Springer-Verlag, 1973.

REED, M., and B. SIMON

1 Methods of modern mathematical physics, vol. 1, Academic Press, 1972.

2 Methods of modern mathematical physics, vol. 2, Academic Press, 1975.

3 Methods of modern mathematical physics, vol. 3, Academic Press, 1979.

4 Methods of modern mathematical physics, vol. 4, Academic Press, 1978.

RENAULT, J.

1 A groupoid approach to C^*-algebras, Lecture Notes Math. **793** (1980).

REYMAN, A.G., and M.A. SEMENOV-TYAN-SHANSKY

1 Reduction of Hamiltonian system, affine Lie algebra and Lax equations I., Inv. Math. **54** (1979), pp. 81-100.

2 Current algebra and nonlinear partial differential equations, Dokl. Akad. Nauk SSSR **251** (1980), No. 6, pp. 1310-1314 (Russian).

RICHTER, P.

1 Unitary representations of countable infinite dimensional Lie groups, Karl-Marx-Universitat, Leipzig, 1977.

2 Zur Extremalzerlegung von Zustanden, Wiss. Zeits Karl-Marx-Univers. **3** (1978), pp. 293-297.

RIDEAU, G.

1 On some representations of anticommutation relations, Communs. Math. Phys. **9** (1968), pp. 229-241.

RIEFFEL, M.A.

1 On the uniqueness of the Heisenberg commutation relations, Duke Math. J. **39** (1972), pp. 745-753.

2 Commutation theorems and generalized commutation relations, Bull. Soc. Math. France **104** (1976), pp. 205-224.

RIESZ, F., and B. SZ.-NAGY

1 Lecons d'analyse fonctionelle, 2nd ed., Akad. kiado, Budapest, 1952.

ROCCA, F., P.N.M. SISSON and A. VERBEURE

1 A class of states on the boson-fermion algebra, J. Math. Phys. **17** (1976), pp. 665-667.

ROĬTER, A.V.

1 Boxes with involution, representations and quadratic forms, Kiev, 1979, pp. 124-126 (Russian).

ROKHLIN, V.A.

1 Unitary rings, Dokl. Akad. Nauk SSSR **59** (1948), No. 4, pp. 643-649 (Russian).

ROSS, K.A.

1 Hypergroups and centers of measure algebras, Inst. Naz. Alta Mat. (Symposia Math) **22** (1977), pp. 189-203.

RUELLE, D.

1 Statistical mechanics. Rigorous Results, W.A. Benjamin, Inc., 1969.

SAITO, J.

1 Some remarks on a representation of a group, I, II, Tohoku Math. J. **12** (1960), pp. 383-388; **17** (1965), pp. 206-209.

SAKAI, Sh.

1 C^*-algebras and W^*-algebras, Springer-Verlag, 1981.

SAMOĬLENKO, Yu.S.

1 Matrix-valued kernels of the Wightman functional type, Methods of Functional Analysis in Problems of Mathematical Physics, Inst. Mat. Akad. Nauk Ukrain. SSR, Kiev, 1971, pp. 201-254 (Russian).

2 Locally dependent representations of families of normal operators, Operators of Mathematical Physics and Infinite-dimensional Analysis, Inst. Mat. Akad. Nauk Ukrain. SSR, 1979, pp. 110-114 (Russian).

3 On infinite-dimensional elliptic operators of second order with constant coefficients, Spectral theory of operators, ELM, Baku, 1979, pp. 162-166 (Russian).

4 Self-adjoint representations of graphs and their currents, Mathematical models of statistical physics, Tumen', 1982, pp. 141-142 (Russian).

5 Anticommuting self-adjoint operators, School on the theory of operators in functional spaces, Minsk, 1982, pp. 173-174 (Russian).

6 On spectral theory of operator-functions, inductive limits of locally compact groups, and on their unitary representations, Direct and inverse scattering problems, Inst. Mat. Akad. Nauk Ukrain. SSR, Kiev, 1982, pp. 117-130 (Russian).

7 On countable families of self-adjoint operators, Spectral theory of operators and infinite-dimensional analysis, Inst. Mat. Akad. Nauk Ukrain. SSR, 1984, pp. 132-137 (Russian).

SAMOĬLENKO, Yu.S., and A.M. KHARITONSKIĬ

1 On a representation of relations $AB = Bf(A)$ with bounded self-adjoint operators, Applications of the functional analytic methods to problems of mathematical physics, Inst. Mat. Akad. Nauk Ukrain. SSR, Kiev, 1987, pp. 53-60 (Russian).

SAMOĬLENKO, Yu.S., and G.F. US

1 Differential operators with constant coefficients on functions of infinitely many variables, Spectral analysis of differential operators, Inst. Mat. Akad. Nauk Ukrain. SSR, Kiev, 1970, pp. 156-177 (Russian).

2 On differential operators with constant coeffcients on functions of infinitely many variables, Generalized functions and their applications in Mathematical Physics, VTS Akad. Nauk SSSR, Moscow, 1981, pp. 476-481 (Russian).

3 Differential operators with constant coefficients on functions of a countable number of variables, Uspekhi Mat. Nauk **39** (1984), No. 1, pp. 155-156 (Russian).

SANKARAN, S.

1 The $*$-algebra of unbounded operators, J. London Math. Soc. **34** (1959), pp. 337-344.

SARYMSAKOV, J.A.

1 Introduction to quantum probability theory, "Fan", Tashkent, 1985 (Russian).

SAZONOV, V.V.

1 A remark on characteristic functionals, Teor. Veroyatnost. i Primenen. **3** (1958), No. 2, pp. 201-205 (Russian).

SCHEUNERT, M.

1 The theory of Lie Superalgebras, Lect. Notes in Math. **716**, 1979.

SCHMÜDGEN, K.

1 The ordered structure of topological $*$-algebras, Repts Math. Phys. **7** (1975), pp. 215-227.

2 Uniform topologies and strong operator topologies on polynomial algebras and on the algebra of CCR, Repts Math. Phys. **10** (1976), pp. 369-384.

3 On trace representation of a linear functionals of unbounded operator algebras, Communs. Math. Phys. **63** (1978), pp. 113-130.

4 Two theorems about topologies on countably generated Op^*-algebras, Acta Math. Acad. Sci. Hungar **35** (1980), pp. 139-150.

5 On topologization of unbounded operator algebras, Repts Math. Phys. **17** (1980), pp. 359-371.

6 Positive cones in envoloping algebras, Repts Math. Phys. **14** (1978), pp. 385-404.

SCHRADER, R., and D.A. UHLENBROCK

1 Markov structure in Clifford algebras, J. Funct. Anal. **18** (1975), pp. 369-413.

SEGAL, G.

1 Unitary representations of some infinite dimensional groups, Communs. Math. Phys. **80** (1981), pp. 301-342.

SEGAL, G., and G. WILSON

1 Loop groups and equations of KdV type, Publ. Math. I.H.E.S. **21** (1985), pp. 1-64.

SEGAL, I.E.

1 Irreducible representations of operator algebras, Bull. Amer. Math. Soc. **53** (1947), pp. 73-88.

2 The structure of a class of representations of a unitary group on a Hilbert space, Proc. Amer. Math. Soc. **88** (1958), pp. 197-203.

3 Distributions in Hilbert spaces and canonical systems of operators, Trans. Amer. Math. Soc. **88** (1958), pp. 12-76.

4 Representations of the canonical commutations relations, Cargese Lecture in Phys. **3** (1967), pp. 107-170.

5 Mathematical problems of relativistical Physics, "Mir", Moscow, 1968 (Russian).

6 A non-commutative extension of abstract integration, Ann. Mat. **57** (1953), pp. 401-457.

SEMENOV-TYAN-SHANSKIĬ, M.A.

1 What is a classical R-matrix, Funktsional. Anal. i Prilozhen. **17** (1983), pp. 17-33 (Russian).

SERGEĬCHUK, V.V.

1 Representations of simple involutive quivers, Representations and quadratic forms, Kiev, 1979, pp. 127-148 (Russian).

SHALE, D.

1 Linear symmetries of free Boson fields, Trans. Amer. Math. Soc. **103** (1962), pp. 149-167.

2 Invariant integration over the infinite dimension orthogonal group and related spaces, Trans. Amer. Math. Soc. **124** (1966), pp. 148-157.

SHALE, D., and W.F. STINESPRING

1 States of the Clifford Algebra, Ann. Math. **80** (1964), pp. 365-381.

SHERMAN, T.

1 Positive linear functionals on $*$-algebras of unbounded operators, J. Math. Anal. appl. **22** (1968), pp. 285-318.

SHILOV, G.E.

1 Laplacian for functions on countable dimensional space, Uspekhi Mat. Nauk **24** (1969), No. 4, pp. 158 (Russian).

2 Mathematical Analysis. A second special course, "Nauka", Moscow, 1965 (Russian).

SHILOV, G.E., and TIN' FAN DYK

1 Integral, measure and derivative on linear spaces, "Nauka", Moscow, 1967 (Russian).

SHIMOMURA, H.

1 On the construction of invariant measure over the orthogonal group on the Hilbert space by the method of Cayley transformation, Publ. Res. Inst. Math. Sci. Kyoto Univ. **10** (1975), pp. 413-424.

2 Some new examples of quasi-invariant measure on a Hilbert space, Publ. Res. Inst. Math. Sci. Kyoto Univ. **11** (1976), pp. 635-649.

3 An aspect of quasi-invariant measures on $\mathbb{R}^\infty$, Publ. Res. Inst. Math. Sci. Kyoto Univ. **11** (1976), pp. 749-774.

4 Linear transformations of quasi-invariant measures, Publ. Res. Inst. Math. Sci. Kyoto Univ. **12** (1977), pp. 777-800.

5 Ergodic decomposition of quasi-invariant measures, Publ. Res. Inst. Math. Sci. Kyoto Univ. **44** (1978), pp. 359-382.

SHUBIN, M.A.

1 Pseudodifferential operators and spectral theory, "Nauka", Moscow, 1978.

SIMON, B.

1 The $P(\phi)_2$ Euclidean (Quantum) Field Theory, Princeton University Press, 1974.

SIMON, J.

1 On the integrability of representations of finite dimensional real Lie algebras, Communs. Math. Phys. **28** (1972), pp. 39-46.

2 Gårding domain for representations of some Hilbert-Lie groups, Letters Math. Phys. **1** (1975), pp. 23-29.

3 Analyticité jointe et separée dans les representations des groupes de Lie reductifs, C.R. Acad. Sci. Paris ser A **285** (1977), pp. 199-202.

SINAÏ, Ya.G., and V.V. ANSHELEVICH

1 Some questions of noncommutative ergodic theory, Uspekhi Mat. Nauk **31** (1976), No. 4, pp. 151-167 (Russian).

SIRUGUE-COLLIN, M, and M. SIRUGUE

1 On the Probabilistic Structure of Quasi-free States of a Clifford Algebra, Communs Math. Phys. **48** (1978), pp. 131-136.

SKLYANIN, E.K.

1 On some algebraic structures related to Yang-Baxter equation, Funktsional. Anal. i Prilozhen. **16** (1982), pp. 27-34 (Russian).

2 On some algebraic structures related to Yang-Baxter equation. Representations of a quantum algebra, Funktsional. Anal. i Prilozhen. **17** (1983), pp. 34-48 (Russian).

SKOROKHOD, A.V.

1 Constructional methods to define stochastic processes, Uspekhi Mat. Nauk **20** (1965), pp. 67-87 (Russian).

2 On admissible shifts of measures in Hilbert space, Teor. Veroyatnost. i Primenenia **15** (1970), No. 4, pp. 577-598 (Russian).

3 Integration in Hilbert space, "Nauka", Moscow, 1975 (Russian).

SLAWNY, J.

1 On Factor Representations and the C^*-algebra of Canonical Commutation Relations, Communs. Math. Phys. **24** (1972), pp. 151-170.

SMOLYANOV, O.G.

1 On measurable multilinear and exponential functionals on some linear spaces with a measure, Dokl. Akad. Nauk SSSR **170** (1966), No. 3, pp. 526-529 (Russian).

SMOLYANOV, O.G., and S.V. FOMIN

1 Measures on linear topological spaces, Uspekhi Mat. Nauk **31** (1976), No. 4, pp. 3-56 (Russian).

SOBOLEV, S.L.

1 Introduction into the theory of bulk formulas, "Nauka", Moscow, 1974 (Russian).

STERNHEIMER, D.

1 Proprietes spectrales dans les representations de groupes de Lie, J. Math. Pures et Appl. **47** (1968), pp. 289-319.

STONE, M.H.

1 On one-parameter unitary groups in Hilbert space, Ann. Math. **33** (1932), pp. 643-648.

STÖRMER, E.

1 On the Jordan structure of C^*-algebras, Trans. Amer. Math. Soc. **120** (1965), pp. 438-447.

STRĂTILĂ, S., and D. VOICULESCU

1 Representations of AF-algebras and of the group $U(\infty)$, Lect. Notes Math. **486**, 1975.

STRĂTILĂ, S., and L. ZSIDO

1 Lectures on von Neumann Algebras, Abacus Press, Bucuresti, 1979.

STREATER, R.F.

1 Current commutation relations and continuous tensor product, Nuovo Cimento **53** (1968), pp. 487-495.

2 A continuum analogue of the lattice gas, Communs. Math. Phys. **12** (1969), pp. 226-232.

3 Infinitely divisible representations of Lie algebras, Z Wahrscheinlichkeitstheorie verw. Geb. **19** (1971), pp. 67-80.

STREIT, L.

1 Test function spaces for direct product representations of the canonical commutation relations, Communs. Math. Phys. **4** (1967), pp. 22-31.

SUDAKOV, V.N.

1 Linear sets with a quasi-invariant measure, Dokl. Akad. Nauk SSSR **127** (1969), No. 3, pp. 524-525 (Russian).

2 A characterization of quasi-invariant measures on a Hilbert space, Uspekhi Mat. Nauk **18** (1963), No. 1, pp. 188-190 (Russian).

TAKEDA, Z.

1 On the representation of operator algebras, II, Tohoku Math. J. **6** (1954), pp. 212-219.

2 Inductive limit and infinite direct product of operator algebras, Tohoku Math. J. **7** (1955), pp. 67-86.

TAKESAKI, M.

1 Conditional expectation in von Neumann algebra, J. Funct. Anal. **9** (1972), pp. 306-321.

2 Duality and von Neumann algebras, I, Lecture Notes in Math. **247**, Springer-Verlag, 1972.

3 Theory of operator algebras, I, Springer-Verlag, 1979.

THOMA, E.

1 Die unzerlagbaren, positiv-definiten klasenfunctionen der abzahlbar unendlichen, symmetrischen Gruppe, Math. Z. **85** (1964), pp. 40-61.

TOMITA, M.

1 Spectral theory of operator algebra, I. II., Math. J. Okayama Univ. **9** (1959), pp. 62-98, **10** (1960), pp. 19-60.

TOMIYAMA, J.

1 A remark on representations of CCR algebras, Proc. Amer. Math. Soc. **19** (1968), pp. 1506.

TOPPING, D.M.

1 *UHF* algebras are singly generated, Math. Scand. **22** (1968), pp. 224-226.

TORRESANI, B.

1 Unitary positive energy representations of the Gauge group, Letters in Math. Phys. **13** (1987), pp. 7-15.

UHLMANN, A.

1 The "transition probability" in the state space of $*$-algebras, Repts. Math. Phys. **9** (1976), pp. 273-279.

UMEGAKI, H.

1 Conditional expectations in an operator algebra I, Tohoku Math. J. **6** (1954), pp. 177-181.

UMEMURA, Y.

1 Measures on infinite dimensional vector spaces, Publ. Res. Inst. Math. Sci. Kyoto Univ. **1** (1965), pp. 1-47.

US, G.F.

1 Spectral decomposition of self-adjoint operators which are functions of operators acting with respect to different variables, Dokl. Akad. Nauk SSSR **229** (1976), pp. 812-815 (Russian).

2 On functions of operators acting with respect to different variables, Operators of mathematical physics and infinite dimensional analysis, Inst. Mat. Akad. Nauk Ukrain. SSR, Kiev, 1971, pp. 121-138 (Russian).

3 On the infinite dimensional Laplacian in a space with Gaussian measure, Applications of methods of functional analysis to the problems of mathematical physics, Inst. Mat. Akad. Nauk Ukrain. SSR, Kiev, 1987, pp. 70-85 (Russian).

VAǏNERMAN, L.I.

1 Harmonic analysis on hypercomplex systems with a compact and discrete basis, Spectral operator theory and infinite-dimensional analysis, Inst. Mat. Akad. Nauk Ukrain. SSR, Kiev, pp. 19-32 (Russian).

2 Duality of the algebras with involution and generalized shift operators, Itogi nauki i tekhnici **24**, Mat. Anal., pp. 165-205 (Russian).

VAǏNERMAN, L.I., and G.I. KATS

1 Non-unimodular ring Hopf-von Neumann groups and algebras, Mat. Sb. **94** (1974), No. 2, pp. 194-225 (Russian).

VAKHANIYA, N.N.

1 Probability distributions on linear spaces, "Mentsniereba", Tbilisi, 1971 (Russian).

VAKSMAN, L.L, and Ya.S. SOǏBELMAN

1 An algebra of functions on the quantum group $SU(2)$, Funktsion. Anal. i Prilozhen. **22** (1988), pp. 1-14.

VAN DAELE, A.

1 Quasi-equivalence of quasi-free states on Weyl algebra, Communs. Math. Phys. **21** (1971), pp. 171-191.

VAN DAELE, A., and A. VERBEURE

1 Unitary equivalence of representations of the Weyl algebra, Communs. Math. Phys. **20** (1971), pp. 268-278.

VASILESCU, F.-H.

1 Anticommuting self-adjoint operators, Rev. Roum. Math. Pures Appl. **28** (1983), pp. 77-91.

2 Analytic functional calculus and spectral decompositions, Buc EA Ltd. etc, Reidel, 1982.

VASIL'EV, A.N.

1 On representation theory of a topological (non-Banach) algebra with involution, Teoret. Mat. Phizika **2** (1970), No. 2, pp. 153-168 (Russian).

VASIL'EV, N.B.

1 C^*-algebras with finite irreducible representations, Uspekhi Mat. Nauk **21** (1966), No. 1, pp. 135-154 (Russian).

VERDIER, J.-L.

1 Les representations des algebras de Lie affines: applications a quelques problemes de physique (d'apres E. Date, M. Jimbo, M. Kashiwara, T. Miwa), Seminaire Bourbaki **34** (1981/82), No. 596, pp. 1-13.

VERGNE, M.

1 Groupe symplectique et second quantification, C.R. Acad. Sci. **285** (1977), ser. A, pp. 191-194.

VERSHIK, A.M.

1 Duality in measure theory in linear spaces, Dokl. Akad. Nauk SSSR **170** (1966), No. 3, pp. 497-500 (Russian).

2 Countable groups close to finite, Greenleaf, F: Invariance in mean on topological groups, "Mir", 1973, pp. 112-135 (Russian).

3 Metagonal and Metaplectic infinite groups. General notions and metagonal groups, Differential geometry, Lie groups, and Mechanics, "Nauka", Leningrad, 1983 (Russian).

4 Algebras with quadratic relations, Spectral theory of operators and infinite dimensional analysis, Inst. Mat. Akad. Nauk Ukrain. SSR, Kiev, 1984, pp. 32-56 (Russian).

5 Representation theory of groups and algebras. Skew products, J. von Neumann, Collected works, II vol., "Nauka", Moscow, 1987, pp. 342-348.

VERSHIK, A.M., I.M. GEL'FAND and M.I. GRAEV

1 Representations of the group $SL(2,R)$ where R is a function ring, Uspekhi Mat. Nauk **28** (1973), No. 5, 83-128 (Russian).

2 Irreducible representations of the group G^* and cohomologies, Funktsional. Anal. i Prilozhen. **8** (1974), No. 2, pp. 67-69 (Russian).

3 Remarks on representations on a group of functions, which take the values in a compact Lie group, Preprint 17, Inst. Priklad. Mat. Akad. Nauk SSSR, Moscow, 1975 (Russian).

4 Representations of a group of diffeomorphisms, Uspekhi Mat. Nauk **30** (1975), No. 6, pp. 3-50 (Russian).

5 Remarks on the representations of groups of functions with values in a compact Lie group, Compos. Math. **35** (1977), pp. 299-334.

6 A commutative model of the current group, $SL(2,\mathbb{R})^X$ related to a unipotent subgroup, Funktsional. Anal. i Prilozhen. **17** (1983), No. 2, pp. 70-72 (Russian).

VERSHIK, A.M., and S.V. KEROV

1 An asymptotic of a Plancherel measure of the symmetric group and a limiting form of the Young tables, Dokl. Akad. Nauk SSSR **233** (1977), No. 6, pp. 1037-1040 (Russian).

2 Characters and factor representations of an infinity symmetric group, Dokl. Akad. Nauk SSSR **257** (1981), No. 5, pp. 1037-1040 (Russian).

3 Asymptotic theory of characters of a symmetric group, Funktsional. Anal. i Prilozhen. **15** (1981), No. 4, pp. 16-27 (Russian).

4 Asymptotic of a maximal and typical dimension of irreducible representations of a symmetric group, Funktsional. Anal. i Prilozhen. **19** (1985), No. 1, pp. 25-36 (Russian).

VILENKIN, N.Ya.

1 Theory of characters of topological abelian groups with a given boundness, Izvestia Akad. Nauk SSSR **15** (1951), Ser. Mat., pp. 439-462 (Russian).

2 Special functions and the theory of group representations, "Nauka", Moscow, 1975 (Russian).

VISHIK, M.I.

1 Parametrix of elliptic operators with an infinite number of independent variables, Uspekhi Mat. Nauk **26** (1971), No. 2, pp. 155-174 (Russian).

VLADIMIROV, V.S.

1 Generalized functions in mathematical physics, "Nauka", Moscow, 1976 (Russian).

VREM, R.C.

1 Harmonic analysis on compact hypergroups, Pacif. J. Math. **85** (1979), pp. 239-251.

2 Free states and automorphisms of the Clifford algebra, Communs. Math. Phys. **44** (1975), pp. 53-72.

WEIL, A.

1 L'intégration dans les groupes topologiques et ses applications, Hermann, Paris, 1940.

WEYL, H.

1 Quantenmechanik und Gruppentheorie, Zet. Fysik **46** (1927), pp. 1-46.

2 The Classical groups. Their invariants and representations, Inst. Adv. Study, 1939.

WIDOM, H.

1 Approximately finite algebras, Trans. Amer. Math. Soc. **83** (1956), pp. 170-178.

WILDE, I.F.

1 The free Fermion field as a Markov field, J. Funct. Anal. **15** (1974), pp. 12-21.

WOLFE, J.C.

1 Invariant states and conditional expectation of the anticommutation relation, Communs. Math. Phys. **44** (1975), pp. 53-72.

WOODS, E.J.

1 Continuity Properties of the Representations of the Canonical Commutation Relations, Communs. Math. Phys. **17** (1970), pp. 1-20.

WORONOWICZ, S.L.

1 On a theorem of Mackey, Stone and von Neumann, Studia Math. **24** (1964), pp. 101-105.

2 The Quantum Problem of Moment I, Repst. Math. Phys. **1** (1970). pp. 135-145.

3 The Quantum Problem of Moment II, Repts. Math. Phys. **1** (1971), pp. 175-183.

4 Twisted $SU(2)$ group. An example of a non-commutative differential calculus, Preprint 1.86, Univ. Warszawski, Warszawa, 1986.

XIA DAO-XING

1 Measure and integration theory on infinite-dimensional spaces, Academic Press, 1972.

YAKOVLEV, V.S.

1 On representations of some commutative nuclear $*$-algebras, Ukrain. Mat. Zh. **39** (1987), No. 2, pp. 235-243 (Russian).

2 On representation of some $*$-algebras of smooth functions, Ukrain. Mat. Zh. **39** (1987), No. 3, pp. 400-403 (Russian).

YAMASAKI, Y.

1 Invariant Measure of the Infinite Dimensional Rotation group, Publ. Res. Inst. Math. Sci. Kyoto Univ. **8** (1972), pp. 131-140.

2 Projective limit of Haar measures on $O(n)$, Publ. Res. Inst. Math. Sci. Kyoto Univ. **8** (1972), pp. 141-149.

3 Quasi-Invariance of Measures on an Infinite dimensional Vector Space and the Continuity of the Characteristic Functions, Publ. Res. Inst. Math. Sci. Kyoto Univ. **16** (1980), pp. 767-783.

ZAV'YALOV, O.I., and V.N. SUSHKO

1 Canonical coordinates in the physics of infinite-dimensional systems, Teoret. Mat. Phys. **1** (1969), No. 2, pp. 153-181 (Russian).

ZHELOBENKO, D.P.

1 Compact Lie groups and their representations, "Nauka", Moscow, 1970 (Russian).

ZHOLTKEVICH, G.N.

1 Representations of simple *AF*-algebras, Doklady Akad. Nauk Ukrain. SSR, Ser. A, **4** (1984), pp. 9-11 (Russian).

INDEX